80:20

Development in an Unequal World

Editor: Colm Regan

Authors: Bertrand Borg, Mary Rose Costello, Tony Daly, Frik De Beer, Michael Doorly, John Dornan, Valerie Duffy, Knut Elvatun, Omar Grech, Beatrice Maphosa, Antonella (Toni) Pyke, Ciara Regan, Colm Regan, Scott Sinclair, Roland Tormey

Graphics and Design: Dylan Creane

Cartoons: Brick (John Clark) and Martyn Turner

Photos: Gareth Bentley and Garry Walsh

Published by:
80:20 Educating and Acting for a Better World, Bray, Ireland
UNISA Press, Pretoria, South Africa

ISBN: 978-0-9567185-4-9

www.8020.ie
www.developmenteducation.ie

Our Cover

95:1 the ratio of the average per capita wealth of the world's richest and poorest people

32:1 the average rate at which an individual in the 'Developed World' consumes resources and generates waste when compared with an individual in the 'Developing World'

1/2 for every one country in which inequality has reduced it has increased in more than two since the 1980s

5:50:500 the ratio of aid from voluntary agencies and states going to the 'Developing World' against the transfer of wealth from there to the 'Developed World'

80:20 over 80% of the world's population now lives in the 'Developing World' and less than 20% in the 'Developed World'

28 years ago, this cartoon by Martyn Turner appeared on the cover of the first edition of this book, then entitled 75:25. After its publication, we received many complaints that the cartoon stereotyped people from the Third World and that it reduced the complexity and diversity of that world to a set of simplistic and inaccurate images. Interestingly, we received no complaints that it treated the First World in a similar manner. Since that first edition, the cartoons have continued to elicit considerable comment - not always positive. By its very nature, cartooning trades in caricatures and stereotypes - of politicians, businesspeople, Americans, Africans, Irish, Jews, Arabs, environmentalists etc. In order to get to a core truth or reality, good cartoonists do strip away detail and therefore do reduce complexity. We are pleased to continue this tradition in the 6th edition of this book.

CONTENTS

CONTENTS OF SUPPORT DVD
developed by Dylan Creane, Tony Daly, Ciara Regan, Colm Regan and Scott Sinclair

This edition of 80:20 is dedicated with respect and admiration to educators worldwide who continue to teach not just what is but also what is possible. It is especially dedicated to those teachers with whom we have had the privilege to work.

For Hilda, Aife and Ciara

ACKNOWLEDGEMENTS

80:20 would like to thank our funders for their support – Irish Aid and Concern Worldwide

The views expressed in this publication do not necessarily reflect those of our funders or of 80:20.

Printed in Belfast, Northern Ireland by GPS (FSC and PEFC accredited and Ireland's first carbon neutral printing company).

DVD designed and printed by Genprint, Ireland.

Producing 80:20 Development in an Unequal World is a major task only made possible over the years by a large team of people; we owe them a huge debt:

Gareth Bentley, John Clark (Brick), Anne Cleary, Dylan Creane, Linda Cornwall, Frik De Beer, Jennifer D'Arcy, Vaughan Dodd, Michael Doorly, John Dornan, Gerry Duffy, Gerry Fenlon, Louise Gaskin, Elna Harmse, Anne Kane, Kevin Kelly, Frank Kirwan, Grainne McGettrick, Una McGrath, Tony Meade, Ray O'Sullivan, Ray O'Sullivan jnr., Dermot O'Brien, Myles O'Grady, Adrian O'Neill, Charlie O'Reilly, Hetta Pieterse, Brendan Rogers, Clifton Rooney, Scott Sinclair, Peter Stewart, Roland Tormey, Martyn Turner and Garry Walsh.

Those who gave permission for use of video and poster materials, these are individually listed in the support DVD.

FOREWORD
BY PRESIDENT MICHAEL D. HIGGINS

In South Africa there is a term known as 'Ubuntu' for which direct definition or translation is difficult. It represents more of an ethical concept or philosophy than a mere single word. 'Ubuntu' has its origin in the Bantu languages of Southern Africa and beautifully captures the sense of people's allegiances and relations with each other. Archbishop Desmond Tutu explained it as the essence of being human. Ubuntu speaks particularly about the fact that you can't exist as a human being in isolation. Ubuntu speaks about our interconnectedness with each other, with our community and with the wider world. Nelson Mandela further clarified that although 'Ubuntu' implies generosity, it does not mean that people should not enrich themselves. The key question for Mandela is whether people can do so in a way that also enables the community around them to thrive and prosper.

In other languages in other parts of Africa there are similar words to capture this concept of community solidarity. In Rwanda, there is 'Umuganda' which refers, again conceptually, to the notion of community solidarity and social service. A practical application of this concept in that country results in thousands of Rwandans, rich and poor, leaders and citizens, meeting in their villages and suburbs one Saturday every month to carry out community improvement projects such as collecting litter, planting shrubs, and clearing drains.

Here in Ireland, we also have a beautiful word which captures this spirit. Traditionally, the Irish word 'Meitheal' signified a work team, gang, or party and described the co-operative labour system in rural parts of the country where groups of neighbours help each other in turn with farming work, such as harvesting crops. Neighbours who give their work to others in turn are helped with their own heavy seasonal tasks so the heart of the concept is community unity through cooperative work and mutually reciprocal support. 'Meitheal' is the Irish expression of the ancient and universal notion of cooperation in response to social need, one which can also have a global application, and like 'Ubuntu' and 'Umuganda' can point towards a different way of living our lives and viewing the world around us. I mention these words because they capture the essence of what this excellent book is seeking to achieve. 80:20 provides the facts and figures, the stories and the pictures that provide a compelling case for thinking deeply about the inequalities, the challenges but also the opportunities of the wider world we live in. It provides a useful framework to examine the world around us, the inequalities and injustices, but also the potential that exists to bridge the gap between poverty and plenty. In 80:20, the devastating consequences of persistent and growing inequality for the world's poor - 'the bottom billion' (estimated to be a staggering 1.4 billion people living on less than $1.25 a day) - are presented to the reader in a unique and engaging manner. Using evidence from the 2010 UNDP Human Development Report, the book presents many stark facts. Although the incidence of global poverty has been declining, the truth is that this decline in poverty varies considerably across regions, and the overall welcome improvements in human development indicators mask continued decline and much suffering in many parts of the world. Increasingly more and more people are faced with the effects of the inequality that characterises not only relationships between the Third World and the so-called 'West', but which is also increasingly reflective of societies everywhere.

The very title of the book neatly sums up the reality of the situation in which we find ourselves in 2012: that 80% of the world's people live in the underdeveloped world but that the richest 20% of the world's people share 84% of the world's wealth. The fact that this is the sixth edition of the book speaks volumes. The first edition in 1984 was called 75:25, and reflected the distribution of wealth at that time. It is clear that despite all of the evident progress made in terms of technological advancement and economic and social development, the world has been getting more unequal.

Such harsh facts about the world often appear too large and too stark to internalise. We read statistics about child and maternal mortality, about malnutrition and hunger but in many ways have become immune to the true-life stories behind the numbers, the accounts of real people seeking to lead ordinary but fulfilling lives in the face of great adversity. 80:20 intelligently combines a rigorous and comprehensive presentation of factual information on development with an incisive analysis of the causes of, and potential solutions to, under-development. More importantly, however, the book engages the reader in a narrative in which he or she can be an active participant rather than a passive recipient. By presenting facts and analysis alongside real life stories of hope and survival, 80:20 wants to move the reader beyond a reaction of shock and paralysis. Because the intention of the work is to 'educate' rather than merely to inform, the reader is regarded more as a user of vital information, a potential actor and change-maker, than as a passive beneficiary of fatalistic commentary. The purpose of the book is to inform, but also to equip the reader with the skills necessary to make change at local as well as global levels.

80:20 has become a highly respected essential text for those interested in or studying issues of international development , but the main premise and justification for this book is that education is fundamental to understanding the shape and nature of our unequal world, to interacting with that world, as well as to imagining and shaping a different world. The authors of the book believe that education about the world we live in does three things: It creates choice, it generates capacity and it supplies motivation to take action. In short, it creates 'Ubuntu'.

Irish people have a deep personal interest in the vital subject matter of this publication. With our own history of conflict, famine and poverty, we have first-hand experience of the effects of under-development and the need for international solidarity and a sense of global community. Ireland has a strong voice internationally on development issues. Through our aid programme and through the work of many NGOs and missionary organisations, Irish people have consistently shown solidarity with those who face extreme poverty. Not only is it the right thing to do but it is also in our interests as an outward-looking nation whose future depends on the strength of our political, economic and cultural partnerships worldwide – partnerships between peoples. I am deeply committed to ensuring that we continue to use that strong voice, to advocate for change for the better in lives of poor people, both at home and abroad.

But at its heart, this is a book of hope and courage – a testimony to the enormous efforts that the world has made in tackling problems of inequality and injustice but also setting out clearly the enormous challenges that still face us. Above all, it addresses the central challenges to all of us to be informed and to understand how injustice and inequality continue to persist. Only when we understand can we take the action needed to transform our world and ensure that the central statistic underlying the title of this book becomes a thing of the past.

I hope that you will be as inspired and motivated as I am in reading this publication and I recommend it as an invaluable resource for all those interested in the pursuit of a more just and equitable world.

Michael D. Higgins
Uachtaráin na hÉireann
President of Ireland
20 January 2012

INTRODUCTION

The evidence of progress is impressive – the evidence is there in the realities of poverty reduction, increased literacy, enhanced education opportunities for girls, the expansion and deepening of democracy, growing awareness of environmental challenges and a continuing public interest in *'our place and duty'* in the world. Sadly, the alternate evidence is equally obvious – the life threatening realities of hunger and absolute debilitating poverty; increasing inequality; the resolute refusal to share wealth reasonably; continued and growing inequality in consumption globally, bloated and corrupt arms and finance industries with their inevitable consequences and the range of ongoing public 'resistances' that sustain our 80:20 world.

And, it is a world made all the more obscene by the evidence of what can be done, what is possible, often at little cost, but what appears to be beyond our collective will and reach.

John Berger argues:

'The poverty of our century is unlike that of any other. It is not, as poverty was before, the result of natural scarcity, but of a set of priorities imposed upon the rest of the world by the rich. Consequently, the modern poor are not pitied…but written off as trash. The twentieth-century consumer economy has produced the first culture for which a beggar is a reminder of nothing.'

To take our lead from modern philosophers, readily available analysis shows that the problems of world poverty and inequality are simultaneously *'amazingly small and amazingly large'*. Amazingly small in economic terms – for example, the estimated annual cost of saving the lives of the 529,000 women who die annually and unnecessarily from complications during pregnancy, childbirth or immediately after is US$1.2 billion, less than the cost of just one single Stealth Bomber; the yearly cost of scaling up health investments to reduce child mortality rates by 2/3rds is $US7.5 billion or the cost of supplying adequate clean water and effective sanitation for all amounts to $US7 billion as against the $US12.4 billion spent in 2007 in the US on breast augmentation, liposuction, nose reshaping, eyelid surgery and tummy tuck.

According to the World Bank, the shortfall from the World Bank's *(higher)* poverty line of $US2 per day for the 40% of human beings who now live below this line is barely $US300 billion annually, a tiny fraction of the estimated $US1.29 trillion cost of the wars in Iraq and Afghanistan as of the end of 2011.

On the other hand, the problem of world poverty is amazingly large in terms of its human impact and consequences, accounting for one third of all human deaths and the majority of human deprivation, morbidity, and suffering worldwide.

Such realities beg hard questions – *why does this reality continue, why have we not effectively challenged and ended these obscenities?* The answers are not easy or comfortable.

Clearly, the self-interest of particular groups and sectors generates these realities while significant groups of people benefit directly and indirectly from the dominant current economic, political and social system. But this is an insufficient explanation, these interests could not be sustained if enough of us objected and objected effectively. Our current 80:20 world is built and sustained on another set of realities – misinformation, disinformation, complacency, ignorance, fear and bewilderment. These realities are as important as the dominant economics and politics. We consistently fail to make the connections – between our well-being and the ill-being of others, between conspicuous consumption and deprivation, between *'them'* and *'us'*. We acquiesce in a set of 'resistances' that paralyse and disempower, that tell us change is not possible despite the evidence of change before us.

Education plays a crucial role here. When faced with the realities of 80:20, education cannot remain unengaged or *'neutral'* – from whatever perspective this is not a tenable position. Education needs to not only engage with the realities described but also with the *'resistances'* that sustain them. Education cannot simply be about facts and skills; it must engage equally with values and emotions not just with *'what we know'* but how *'we feel'* about what we know. Education is not just about the right to know but also about the right and the need to dissent. These arguments are at the core of 80:20 Development in an Unequal World and at the heart of development education.

None of us can now realistically or morally claim *'not to know'*; we are daily surrounded by 80:20 realities. The chapters in this book seek to engage with these issues; to explore, debate and review; they are by no means comprehensive or complete – they represent introductions to topics and are designed to stimulate and to challenge. As such 80:20 is an exercise in *'educational activism'* – making use of education not as preparation for engagement in the world but as engagement itself. Our hope is that this resource acts as a catalyst for further exploration and extended engagement.

Colm Regan
February 2012

ON THE STRUCTURE OF 80:20 DEVELOPMENT IN AN UNEQUAL WORLD

Part One introduces and debates a range of fundamental human development issues and values – *How do we define human development; how have we constructed it as both theory and strategy; the growing challenge of sustainability, human rights and women's rights as they relate to development and the issue of justice (and injustice) in the context of development.* Part One seeks to outline the values base around which discussion and debate on human development needs to revolve.

Part Two adopts an issues-based approach to analysing development reviewing, in turn, issues of aid, HIV and AIDS, human security, hunger, resource-based conflict and education. It also explores development issues through the lens of literature, art, film and music in addition to approaching the challenge of change itself.

The DVD accompanying 80:20 provides a broad range of additional materials and activities designed to support the use of this resource in education settings. It includes over 100 activities related directly to the chapter themes, a selection of video materials, an annotated list of useful websites and ideas for debating the issues.

[*'A country's success, or an individual's well-being, cannot be evaluated by money alone; life expectancy, health, education and freedom are vitally important.'*]

Chapter 1

80:20 – 'EXTREME POVERTY, GRACIOUS LUXURY'

John Dornan and Colm Regan

Analysing human development since 1970, this chapter highlights the convergence that has occurred internationally in many key dimensions of development but within a context of ever-growing inequality. The chapter presents core data and analysis on wealth and poverty with a particular emphasis on the poorest worldwide – 'the bottom billion'. Evidence from the UNDP's 2010 Human Development Report is utilised to describe many of the key aspects of human development since 1970, including a 'balance sheet' of progress and failures to date. The chapter also explores different ways of measuring development and sets out a basis for much that follows in subsequent chapters.

KEYWORDS

Convergent yet unequal development, inequality, India, the 'bottom billion', measuring development, human development 1970-2010, 'a balance sheet of human development', Human Development Index, globalisation, consumption

INTRODUCTION

In 1994, Rajni Kothari wrote in The Times of India a commentary on 'development' in India:

'It is emerging that we are a nation of skewed priorities, trying to become a highly industrialised economy and a great power, claiming to have one of the world's largest scientific and technological capabilities, yet one in which a majority of people are living in sub-standard conditions, physically, medically and in respect of basic nutrition, sanitation and cleanliness. We are producing a high-rise, five-star culture which seems to be in perfect harmony with ever more deteriorating conditions of latrines and other waste disposal. More and more of our cities are reeling under heavier and heavier weight of what I would call the refuse of progress. The net result is the outbreak of a whole variety of epidemics – malaria is back, cholera is on the increase, gastro-enteritis, conjunctivitis – the incidence of blindness and other disabilities is growing, we are having more weak and physically debilitated children, a breakdown of a generation of citizens that is supposed to carry the burden of future India.'

Quoted in Seabrook (1995) In the Cities of the South, London.
Verso: 10

Despite frequent references to the significant economic growth experienced by India in the period since Kothari wrote his commentary, according to a 2007 survey published by the UN Department of Economic and Social Affairs, inequality has continued to increase. The report's authors note:

'...though the richer sections of the population benefitted in the post-liberalisation period, there has been a stagnation of incomes for the majority, with the bottom rung of the population severely negatively affected by this process. There is also evidence that, both at the national and state levels, income disparities between the urban and rural sectors increased during this period.'

Parthapratim Pal and Jayati Ghosh (2007) Inequality in India: A Survey of Recent Trends, DESA working paper 45:25

In many fundamental ways, India mirrors patterns and trends across the world – while many key economic indicators document and highlight improvement (significantly in the social and political areas), nonetheless, inequality especially as regards income continues to grow with inevitable and devastating consequences for the world's poor. The 2010 UNDP Human Development Report notes:

'And despite convergent trends in health and education, gaps in human development are huge. A person born in Niger can expect to live 26 fewer years, to have 9 fewer years of education and to consume 53 times fewer goods than a person born in Denmark. While the Danes have elected their parliament in free and open elections since 1849, Niger's president dissolved parliament and the Supreme Court in 2009 – and was then ousted in a military coup. More than 7 of 10 people surveyed in Niger say there were times in the past year when they did not have enough money to buy food for their families. Very few Danes would be in such straits.'

Human Development Report 2010: The Real Wealth of Nations: Pathways to Human Development, Basingstoke and New York, Palgrave Macmillan.

This chapter explores and analyses key dimensions of this reality and lays the foundation blocks for much of what follows it subsequently; it outlines what has been learned about human development in the past three to four decades, and it highlights the impact of inequality on development.

95:1 - POVERTY, INEQUALITY & DEVELOPMENT

A detailed study of world household wealth in 2006 by the World Institute for Development Economics of the UN University provides a graphic overview of the realities of wealth and poverty in the world. The study was the first to measure wealth distribution in each country as opposed to just income and included the most significant elements of household wealth, including financial assets and debts, land, buildings and other property. It highlighted that:

- The richest 1% of adults in the world own 40% of the planet's wealth and that those in financial services and the internet sectors predominate among the super-rich
- Europe, the US and some Asia Pacific nations account for most of the extremely wealthy with more than a third living in the US, 27% in Japan, 6% in the UK and 5% in France
- Half the world's adult population owned barely 1% of global wealth

- Near the bottom of the list were India, with per capita wealth of $1,100, and Indonesia with assets per head of $1,400
- Many African nations as well as North Korea and the poorer Asia Pacific nations were places where the worst-off lived.

Source: World Institute for Development Economics (2006) The World Distribution of Household Wealth, United Nations University)

Table 1:1 below outlines the changing distribution of wealth worldwide between 1995 and 2005 by income group (according to the latest World Bank figures) and highlights the fact that the percentage of world wealth controlled by low income and lower middle income countries increased marginally from 7.18% to 9.13% while that of the upper middle income and high income countries decreased marginally from 90.9% to 89%. However the table also highlights that the ratio of average per capita wealth between the world's poorest and richest people in 2005 stood at 95:1.

Continued on page 15 →

Table 1:1 - Wealth and Per Capita Wealth by Income Group, 1995 and 2005

GROUP	1995			2005		
	TOTAL WEALTH US$ BILLIONS	PER CAPITA WEALTH US$	%OF WORLD TOTAL	TOTAL WEALTH US$ BILLIONS	PER CAPITA WEALTH US$	%OF WORLD TOTAL
LOW INCOME	2447	5290	0.48	3597	6138	0.53
LOWER MIDDLE INCOME	33950	11330	6.7	58023	16903	8.6
UPPER MIDDLE INCOME	36794	73540	7.3	47183	81354	7.0
HIGH INCOME OECD	421641	478445	83.6	551964	588315	82.0
WORLD	504548	103311		673593	120475	

Note: Figures are based on a set of countries for which wealth accounts are available from 1995 to 2005. Data does not include high-income oil exporters.

Source: World Bank (2011) The Changing Wealth of Nations: Measuring Sustainable Development in the New Millennium, Washington.

INEQUALITY & DEVELOPMENT

WHY INEQUALITY MATTERS FOR DEVELOPMENT

In his 2008 analysis From Poverty to Power, Oxfam activist Duncan Green outlines how inequality such as that outlined above is bad for human development and for growth (and, by implication, how equality is good for both); he suggests five reasons:

- Inequality wastes talent: if women are excluded from top jobs, half the talent of any nation is squandered. By one estimate, if all states in India were to perform as well as the best (Karnataka) in eradicating gender discrimination in the workplace, national output would increase by one third. When banks refuse to lend to poor people, good economic opportunities are wasted.

- Inequality undermines society and its institutions: in an unequal society, elites find it easier to 'capture' governments and institutions, and use them to further their own narrow interests, rather than the overall economic good.

- Inequality undermines social cohesion: 'vertical inequality' between individuals is linked to rises in crime, while 'horizontal inequality' (for example, between different ethnic groups)

increases the likelihood of conflicts that can set countries back decades.

- Inequality limits the impact of growth on poverty: a one-percentage point increase in growth will benefit poor people more in an equal society than in an unequal one.

- Inequality transmits poverty from one generation to the next: most cruelly, the poverty of a mother can blight the entire lives of her children. Each year in developing countries around 30 million children are born with impaired growth due to poor nutrition during foetal development. Babies born with a low birth weight are much more likely to die, and should they survive, are more likely to face a lifetime of sickness and poverty.

See Duncan Green (2008) From Poverty to Power: how active citizens and effective states can change the world, Rugby, England, Practical Action Publishing and Oxfam International and for an extended discussion on how inequality undermines development and how equality promotes it, see Richard Wilkinson and Kate Pickett (2009) The Spirit Level, Penguin.

FIGURE 1:1 - *POPULATION AND WEALTH SHARES BY REGION, 2006*

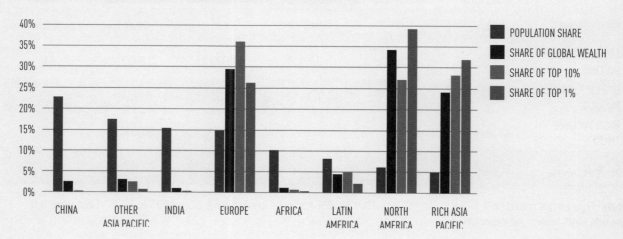

Source: UN University 2006

THE 'BOTTOM BILLION'

Estimates by the World Bank show about 1.4 billion people living below the international poverty line of US$1.25 a day in 2005. This is equivalent to over one quarter of the developing world's population and replaces the previous estimate of poverty at 'a dollar a day'. The incidence of poverty has declined from 52% of global population in 1981 to 42% in 1990 and 26% in 2005 but this latter figure does not include the effects of more recent sharp increases in food and fuel prices. One estimate suggests that rising food prices may have increased total world poverty by 105 million people between 2005 and 2007 alone.

Using the new international poverty line of US$1.25 a day, the decline in poverty varies considerably across regions; led by China, the East Asia and Pacific Region has made dramatic progress, with poverty incidence dropping from 80% to 18% between 1981 and 2005. At the other extreme is Sub-Saharan Africa where the poverty rate remained at 50% in 2005 (no lower than in 1981), although with some recent signs of limited progress. The $1.25 poverty rate fell from 58% in 1996 to 50% in 2005, though this was not sufficient to bring down the number of people living in absolute poverty. Even if Millennium Development Goal 1 (to halve the proportion of people living in absolute

Continued on next page →

Table 1:2 - 'The Bottom 20'

	2009 GNI per capita (US$) and position in bottom 20		2009 PPP and and position in bottom 20		2009 Human Development Rank in the bottom 20 (of 169)
Burundi	150	1	392	2	4
Liberia	160	2	396	3	8
Democratic Republic of Congo	160	3	319	1	2
Malawi	280	4	858	9	17
Afghanistan (2008)	310	5	1,321	23	15
Ethiopia	330	6	934	11	13
Sierra Leone	340	7	808	7	12
Niger	340	8	675	4	3
Mozambique	440	9	885	10	5
Gambia	440	10	1,415	25	19
Nepal	440	11	1,155	17	32
Togo	440	12	850	8	31
Central African Republic	450	13	757	5	11
Uganda	460	14	1,217	20	27
Rwanda	460	15	1,070	14	18
Burkina Faso	510	16	1,187	19	9
Guinea Bissau	510	17	1,070	15	6
Bangladesh	580	18	1,416	26	41
Mali	680	19	1,185	18	10
Benin	750	20	1,508	29	36

Gross National Income (GNI) measures the value of the gross domestic product (the market value of all goods and services produced in a country in a given period) together with income received from other countries (e.g. interest and dividends), less similar payments made to other countries.

PPP = Purchasing Power Parity is a measure that indicates what this income will actually buy locally; it equalises local currencies against the cost of a given basket of goods in terms of US dollars

Human Development Index combines the indicators of life expectancy, educational attainment and income into one composite 'human development' index. The HDI sets a minimum and a maximum for each of 3 dimensions - health, education and living standards - and then shows where each country stands in relation to these.

The education component is measured by mean of years of schooling for adults aged 25 years and expected years of schooling for children of school going age. Life expectancy at birth is calculated using a minimum value of 20 years and maximum value of 83.2 years (the recorded maximum value among countries for the period 1980 to 2010). The decent standard of living dimension is measured by GNI per capita (PPP US$) instead of GDP per capita (PPP US$).

The scores for the three dimensions are then compiled together into a composite index.

Sources: World Bank (2010) World Development Indicators database and United Nations Development Programme, Human Development Report 2010

poverty) is achieved, there would still be 900 million poor people in 2015 or up to 1.13 billion if global economic recovery is weak.

Using the more recent Multi-dimensional Poverty Index (MPI – see below) the Human Development Report for 2010 argues that there are 1.7 billion poor people: 51% in South Asia, 28% in sub-Saharan Africa. By this measure, South Asia has almost twice the number of poor people as Africa and 8 states in India have as many poor people (421 million) as the 26 poorest African countries (410 million).

While the perception may often have been that there are large numbers of poor countries and only a small number of richer 'developed' countries, the reality has also changed radically. The large majority of the world's poorest people (61%) now live in stable Middle-Income Countries, many of which have substantial domestic resources of their own and just 28% in poor and fragile Low-Income Countries. By comparison, in 1988-90, it has been estimated that 93% of the world's poor lived in LICs.

Table 1:3 - The Human Development Index - bottom and top 10 countries compared

COUNTRY	HDI	LIFE EXPECTANCY AT BIRTH (YEARS)	MEAN YEARS OF SCHOOLING (YEARS)	EXPECTED YEARS OF SCHOOLING (YEARS)	GNI PER CAPITA US$, 2008
BOTTOM 10					
Zimbabwe	0.140	47.0	7.2	9.2	176
Democratic Republic of Congo	0.239	48.0	3.8	7.8	291
Niger	0.261	52.5	1.4	4.3	675
Burundi	0.282	51.4	2.7	9.6	402
Mozambique	0.284	48.4	1.2	8.2	854
Guinea Bissau	0.289	48.6	2.3	9.1	538
Chad	0.295	49.2	1.5	6.0	1,067
Liberia	0.300	59.1	3.9	11.0	320
Burkina Faso	0.305	53.7	1.3	5.8	1,215
Mali	0.309	49.2	1.4	8.0	1,171
TOP 10					
Norway	0.938	81.0	12.6	17.3	58,810
Australia	0.937	81.9	12.0	20.5	38,692
New Zealand	0.907	80.6	12.5	19.7	25,438
United States	0.902	79.6	12.4	15.7	47,094
Ireland	0.895	80.3	11.6	17.9	33,078
Liechtenstein	0.891	79.6	10.3	14.8	81,011
Netherlands	0.890	80.3	11.2	16.7	40,658
Canada	0.888	81.0	11.5	16.0	38,668
Sweden	0.885	81.3	11.6	15.6	36,936
Germany	0.885	80.2	12.2	15.6	35,308

Source: United Nations Development Programme, Human Development Report 2010

The poverty line of US$1.25 a day represents the benchmark for poverty in the poorest countries in the world. A less basic standard of US$2 per person per day (the median poverty line for all developing countries) is more appropriate for middle income countries and regions such as Latin America and Eastern Europe. The share of global population living below US$2 a day (at 2005 prices) has fallen from 70% in 1981 to 48% in 2005. However, the number of people living below US$2 a day remained unchanged at around 2.5 billion between 1981 and 2005.

The World Bank estimates that the number of people living between US$1.25 and US$2 doubled from about 600 million to 1.2 billion between 1981 and 2005. Those who live just above the poverty line are particularly vulnerable to the effects of rising food and fuel prices and the impact of climate change.

Commenting on changing statistics such as those above, economist Amartya Sen reminds us:

'Even if the poor were to get just a little richer, this would not necessarily imply that the poor were getting a fair share of the potentially vast benefits of global economic interrelations. It is not adequate to ask whether international inequality is getting marginally larger or smaller. In order to rebel against the appalling poverty and the staggering inequalities that characterise the contemporary world – or to protest against the unfair sharing of benefits of global cooperation – it is not necessary to show that the massive inequality or distributional unfairness is also getting marginally larger. This is a separate issue altogether.'

The American Prospect, January 2002

The implications of international inequality for human development are developed further in the 2010 Human Development Report.

Continued on page 20 →

WHERE THE WEALTH IS

According to a 2006 detailed study of world household wealth carried out by the World Institute for Development Economics Research of the United Nations University:

CANADA

UNITED STATES

MEXICO

CENTRAL AMERICA

BRAZIL

REST OF SOUTH AMERICA

ARGENTINA

› The richest 2% of adults world own more than half of global wealth

› The richest 1% of adults owned 40% of global assets in the year 2000

› The richest 10% of adults accounted for 85% of total world wealth

› The bottom half of the world adult population owned barely 1% of global wealth

› Assets of $2,200 per adult placed a household in the top half of the world wealth distribution; to be among the richest 10% of adults in the world required $61,000 and more than $500,000 was needed to belong to the richest 1% (a group of 37 million worldwide)

› Average wealth amounted to $144,000 per person in the USA in the year 2000 and $181,000 in Japan. Lower down (among countries with wealth data) are India with per capita assets of $1,100 and Indonesia with $1,400 per capita

› Wealth is heavily concentrated in North America, Europe, and high income Asia-Pacific countries. People in these countries collectively hold almost 90% of total world wealth

› While North America has 6% of world population, it accounts for 34% of household wealth. Europe and high income Asia-Pacific countries also own disproportionate amounts of wealth. In contrast, the overall share of wealth owned by people in Africa, China, India, and other lower income countries in Asia is considerably less than their population share, sometimes by a factor of more than ten

DISTRIBUTION OF WORLD WEALTH (2006)
PERCENT OF TOTAL, WITH QUINTILES OF POPULATION RANKED BY HOUSEHOLD WEALTH

RICHEST (1ST) QUINTILE	93.8%
2ND QUINTILE	4.2%
3RD QUINTILE	1.3%
4TH QUINTILE	0.5%
POOREST (5TH) QUINTILE	0.12%

THE DISTRIBUTION OF WORLD GDP (PPP)

SOURCE: WORLD DEVELOPMENT INDICATORS DATABASE, WORLD BANK - 29 SEPTEMBER 2010

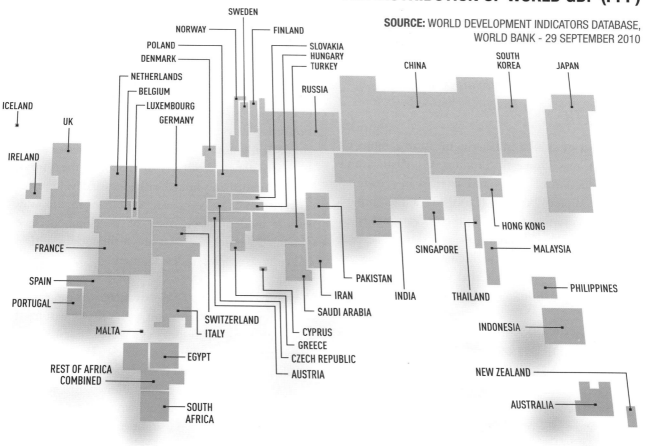

WHO'S CONSUMING WHAT?

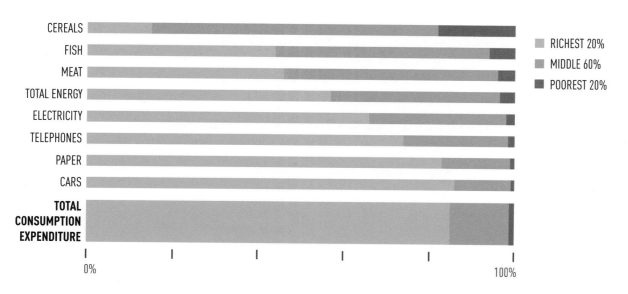

CEREALS
FISH
MEAT
TOTAL ENERGY
ELECTRICITY
TELEPHONES
PAPER
CARS
TOTAL CONSUMPTION EXPENDITURE

RICHEST 20%
MIDDLE 60%
POOREST 20%

0%   100%

Sources: Credit Suisse Research Institute, Global Wealth Report 2010; United Nations University and World Institute for Economics Research The World Distribution of Household Wealth, 2006; World Bank, World Development Indicators Database 2010; Worldwatch Institute Vital Signs 2011.

MEASURING DEVELOPMENT

This section explores three frequently used ways of measuring and mapping development, as well as a range of other development indicators. At the end of the section, data for selected countries illustrate how these indicators compare.

1: WEALTH

A number of similar indicators measure a country's wealth. Gross Domestic Product (GDP) measures the production of goods and services in a country. Gross National Product (GNP) is a measure of income, a country's GDP plus the net earnings from overseas; the World Bank now uses Gross National Income (GNI). All are measured in US$; they are often expressed per capita, that is, proportional to a country's population.

In 2010 the United States was the world's largest economy (GDP $14,657,800 million), while Qatar had the highest GDP per capita ($88,559) and the Democratic Republic of Congo the lowest ($328). The world map on the previous page, drawn proportional to countries' share of world GNP, shows vividly the distribution of the world's wealth, dominated by the USA, Western Europe and Japan. China has grown rapidly and has become the second largest world economy.

Wealth per capita is widely used as a measure of development and is straightforward to understand. For example, it is used by the World Bank as a basis for its categorisation of low, lower middle, upper middle and upper income countries. This is a more sophisticated view of the world than the often simplistically used 'developed/developing' or 'north/south' divisions. Wealth is an economic measure of development, firmly linked with ideas of development as wealth and, by comparing countries in this way, with a modernisation view of development (see Chapter 2). It has less to say about the social, political and cultural aspects of development.

Wealth has a number of disadvantages as a measure of development. In common with other measures it is an average for the country as a whole, so it does not show inequalities within countries. Perhaps more significantly, wealth data do not include those forms of production that are not accounted for, such as subsistence agriculture, unpaid work (for example in the home), or work in the 'informal' economy. These aspects of the economy are likely to be comparatively more significant in Third World countries.

A further weakness is that comparisons in wealth are made between countries with huge differences in living costs; for example you can buy much more for US $1 in India than in the USA. So a variation is to make comparisons using Purchasing Power Parity, which accounts for differences in the cost of living.

Table 1:4 - Similar human development, different income 2010

COUNTRY	HDI	GNI per capita (PPP US$)
Norway	0.938	58,810
Australia	0.937	38,692
Thailand	0.654	8,001
Gabon	0.648	12,747
Haiti	0.404	949
Angola	0.403	4,941

Table 1:5 - Similar income, different human development

Country	GNI per capita (PPP US$)	HDI
Norway	58,810	0.938
United Arab Emirates	58,006	0.815
Dominican Republic	8,273	0.663
South Africa	9,812	0.597
Kenya	1,628	0.470
Cote d'Ivoire	1,625	0.397

Source: UNDP Human Development Report, 2010

2: HUMAN DEVELOPMENT

The Human Development Index (HDI) was devised by the United Nations Development Programme (UNDP) and is now widely used as an indicator of human progress and quality of life. It is based on a score derived from three measures: life expectancy, education (literacy and years of schooling) and income (purchasing power in parity dollars). As the HDI includes both social and economic aspects, it is considered by many to embrace a broader view of development than those based on wealth alone. The HDI is focused on people and their needs, and so is linked with views of development focused on social justice. The UNDP uses the HDI to categorise the world into high, medium and low human development.

The 2010 Human Development Report predictably shows high HDI values in Western Europe, North America and Australasia, and low human development in much of Africa. However, parts of Central and South America, the Caribbean, the Middle East, and East and South-East Asia also have high HDI scores. In 2010 Norway had the highest HDI score (0.938), and Zimbabwe the lowest (0.140).

In 2010, the HDI was amended to take inequality into account with the development of the Inequality Adjusted Human Development Index (IHDI). Using this new measure highlighted the fact that, on average, the losses in human development due to inequality amounted to 22% with the highest losses in Mozambique (45%) and the lowest in the Czech Republic (6%) and with the greatest losses occurring in Sub-Saharan Africa.

Although in general the wealthiest countries have the highest levels of human development, there is not always a straightforward link between wealth and high HDI especially when reviewed over time. The top movers in the HDI since 1970 include several countries in East and South Asia and the Arab States (in North Africa and the oil-rich Gulf region). Oman heads the top 10 list followed by China, Nepal and Indonesia. Reviewing the top 10 in non-income HDI terms highlights some interesting case studies, for example, Ethiopia, Iran and Algeria which score highly in health and education improvements as distinct from those in income. Countries such as Botswana and India score highly on the income improvement dimensions of the HDI. This is because several countries make it into the top 10 listings as a result of their high achievements in health and education despite unexceptional growth performance.

Table 1:6 - *Human Development: high-performing countries 1970-2010*

	HDI	Non-income HDI	Income
1	Oman	Oman	China
2	China	Nepal	Botswana
3	Nepal	Saudi Arabia	South Korea
4	Indonesia	Libya	Hong Kong, China
5	Saudi Arabia	Algeria	Malaysia
6	Lao PDR	Tunisia	Indonesia
7	Tunisia	Iran	Malta
8	South Korea	Ethiopia	Vietnam
9	Algeria	South Korea	Mauritius
10	Morocco	Indonesia	India

Table 1:7 - *Progress in HDI (based on a modified (hybrid) HDI used to monitor changes over time)*

	HDI value 2010	% change 1970-2010
All Developing Countries	.64	57
Arab States	.66	65
East Asia and the Pacific	.71	96
Europe and Central Asia	.75	13
Latin America and the Caribbean	.77	32
South Asia	.57	72
Sub- Saharan Africa	.43	53
Developed Countries	.89	18
OECD	.89	18
Non-OECD	.86	24

Table 1:8 - World population and human development (percentage of world total)

	1990	2010	2030 estimate
Very High Human Development	17.6	15.3	13.6
High Human Development	16.5	15.2	14.1
Medium Human Development	51.8	52.0	51.0
Low Human Development	12.7	16.0	19.6
Least Developed Countries	9.9	12.4	15.3

Source: UNDP Human Development Report, 2010

3: UNDER-FIVE MORTALITY

This figure measures the number of children per 1,000 live births who die before their fifth birthday; it is similar to the Infant Mortality Rate (IMR) which measures the proportion of children who die before their first birthday. UNICEF argues that these measures are important as they indicate the end result of the development process, as it impacts on children. It is also a useful indicator of a population's health and nutritional status, and of social progress through health-care and educational programmes; high infant and under-five mortality rates closely correlate with high adult mortality and low life-expectancy. It is based on what is often described as a basic needs approach to development. In 2007, Iceland and Singapore had the lowest under-five mortality rates (at 3 per 1000) and Sierra Leone and Afghanistan the highest (at 262 and 257 per 1000).

INTERPRETING INFORMATION

The table on the page opposite (Table 8) shows data for 21 countries, chosen to highlight a geographical range, as well as high, medium and low levels of human development and a variety of different paths to development.

This data has many important limitations:

- It only highlights just over 10% of the countries represented in the main pages of the UNDP Human Development Report, and only a tiny fraction of the huge range of statistics available for them
- Statistics such as these are also subject to the weaknesses and inaccuracies inherent in presenting each country as the 'average' of its people
- The statistics routinely include inaccuracies and bias in the collection and presentation of information
- Most of the statistics are a snapshot in time and tell a limited amount about short or long term trends or progress (or dis-improvement) in development, particularly in countries such as China or Zimbabwe that are changing fast.

So, although in some ways this data represents our best guess at presenting some information about development in a single table, it is important to remember what it does not tell us about the world, and to remember that different sets of data – for example those for wealth and health - serve different purposes and tell very different stories about development.

So when exploring information, think about these ideas:

- Use questions to help guide your inquiry; for example, which countries seem to have lower/higher HDI scores?

- Look for patterns and trends, perhaps by ranking countries in one category; for example, which countries seem to have the best record in primary education?

- Look for interesting exceptions to patterns that you might investigate further, for example in the data for child mortality

- Try focusing on one or two data sets, or just compare a few countries; for example, compare the picture for HDI and GNI in the six African countries

- Investigate whether two data sets correlate; for example, does good access to safe water match with high life expectancy?

- Use some of the websites at the end of this chapter to find other sets of data that might tell a different story, or add to the picture; for example, what might data on debt or aid tell us?

Table 1:9 - Human development worldwide 2010

	Quality of Life HDI value 2010	Gross National Income per capita US$ 2008	Life Expectancy (years, 2010)	Maternal Mortality Rate per 100,000 live births 2003-2008	Under 5 Mortality Male/ Female 2005-10 per 1,000 live births	Adult Literacy Rate, % of Adults 15 and over 2005-2008	% of Population with Access to Improved Water Supply	Per capita expenditure on health, 2007, US$	Energy Consumption per capita 2006 (in kilos of oil equivalent in tonnes per capita)
USA	0.902	47,094	79.6	11	7/8	-	99	7,285	7,768
Norway	0.938	58,810	81	7	5/14	-	100	4,763	5,598
Ireland	0.895	33,078	80.3	1	6/6	-	100	3,424	3,628
Japan	0.884	34,692	83.2	6	5/4	-	100	2,696	4,129
UK	0.849	35,087	79.8	8	6/6	-	100	2,992	3,814
Czech Republic	0.841	22,678	76.9	4	5/4	-	100	1,626	4,485
Malta	0.815	21,004	80.0	8	7/7	92.4	100	4,053	2,153
Qatar	0.803	79,426	76.0	12	10/10	93.1	100	3,075	22,057
Argentina	0.775	14,603	75.7	77	17/14	97.7	97	1,322	1,766
Saudi Arabia	0.752	24,726	73.3	18	26/17	85.5	-	768	6,170
Mexico	0.750	13,971	76.7	60	22/18	92.9	94	819	1,702
Iran	0.702	11,764	71.9	140	33/35	82.3	-	689	2,438
Brazil	0.699	10,607	72.9	110	33/25	90.0	97	837	1,184
Botswana	0.633	13,204	55.5	380	60/47	83.3	95	762	1,054
South Africa	0.597	9,812	52.0	400	79/64	89.0	91	819	2,739
India	0.519	3,337	64.4	450	77/86	62.8	88	109	510
Bangladesh	0.469	1,587	66.9	570	58/56	55.0	20	42	161
Nigeria	0.423	2,156	48.4	1,100	190/184	60.1	58	131	726
Zambia	0.395	1,359	47.3	830	169/152	70.7	60	79	625
Ethiopia	0.328	992	56.1	720	138/124	35.9	62	30	-
Liberia	0.300	320	59.1	1,200	144/136	58.1	68	39	289
World	0.624	10,631	69.3	273				869	

Sources: UNDP (2010) Human Development Report 2010: The Real Wealth of Nations: Pathways to Human Development, NY Palgrave MacMillan
UNFPA (2009) State of World Population 2009: Facing a changing world: women, population and climate, NY, UNFPA
World Bank (2010) Development Indicators

THE THIRD WORLD HAS:

> " The fact that development leaves behind, or in some ways even creates, large areas of poverty, stagnation, marginality and actual exclusion from social and economic progress is too obvious and too urgent to be overlooked. "
>
> Gustavo Esteva, 1992

88% OF HIV & AIDS
82% OF THE WORLD'S
80% OF MAJOR CONFLICTS
78% OF THE REFUGEES
75% URBAN DWELLERS
60% OF CEREAL PRODUCTION
45% OF MOBILE PHONES
38% OF CARBON DIOXIDE EMISSIONS
37% OF WORLD WEALTH
28% OF ELECTRICITY CONSUMPTION
9% OF EDUCATION SPENDING
0.5% OF BROADBAND INTERNET CAPACITY

PEOPLE

" Today, for two thirds of the peoples of the world, underdevelopment is a threat that has already been carried out; a life experience of subordination and of being led astray, of discrimination and subjugation. "

Report of 1969 meeting of experts on social policy and planning, United Nations 1971

- 88% OF PEOPLE LIVING WITH HIV AND AIDS IN 2009 (UNAIDS 2010)
- 82% OF THE WORLD'S PEOPLE IN 2008 (POPULATION REFERENCE BUREAU, 2010)
- 80% PLUS OF THE WORLD'S MAJOR CONFLICTS IN 2008-2009 (STOCKHOLM INTERNATIONAL PEACE RESEARCH INSTITUTE, 2010)
- 78% OF THE WORLD'S REFUGEES IN 2009 (OFFICE OF UN HIGH COMMISSIONER FOR REFUGEES, 2010)
- 75% PLUS OF THE WORLD'S URBAN DWELLERS IN 2010 (UN POPULATION DIVISION, 2010)
- 60% OF WORLD CEREAL PRODUCTION IN 2008 (FOOD AND AGRICULTURE ORGANISATION, 2010)
- 45% OF THE WORLD'S MOBILE PHONES IN 2010 (INTERNATIONAL TELECOMMUNICATIONS UNION, 2010)
- 38% OF CARBON DIOXIDE EMISSIONS IN 2008 (US ENERGY INFORMATION ADMINISTRATION, 2010)
- 37% OF WORLD WEALTH IN 2010 (CREDIT SUISSE DATA, 2010)
- 28% OF WORLD ELECTRICITY CONSUMPTION IN 2008 (INTERNATIONAL ATOMIC ENERGY AGENCY, 2010)
- 9% OF WORLD EDUCATION SPENDING IN 2007 (UNESCO, 2009)
- 0.5% OF WORLD BROADBAND INTERNET CAPACITY IN 2010 (INTERNATIONAL TELECOMMUNICATIONS UNION, 2010)

HUMAN DEVELOPMENT EXPLORED

The HDI was developed in 1990 by a team led by Pakistani economist Mahbub ul Haq and the ideas of Nobel prize-winning economist Amartya Sen were also influential in the development of the HDI.

Mahbub ul Haq and his team sought to develop an index which would measure 'human development', going beyond previous ideas of development being synonymous with economic growth. In his 1999 book Development as Freedom, Amartya Sen asked the question: 'What is the relationship between our wealth and our ability to live as we would like?' Clearly, although wealth plays an important part in our general well-being, it is not the only factor.

Expanding the choices people have in terms of life choices is central to the definition of well-being underpinning the concept of human development. In order to expand those choices, however, people must be empowered to do so – most notably by being healthy and by being educated. The HDI therefore takes these three dimensions (economic, health, education) and creates a composite index by which a country's level of human development can be measured and then compared to other countries'. Thus, for the creators of the HDI, human development has three fundamental components:

- Well-being: expanding people's real freedoms so they can flourish
- Empowerment and agency: enabling people and groups to act—to drive valuable outcomes
- Justice: expanding equity, sustaining outcomes over time and respecting human rights and other goals of society.

These dimensions of development are explored further in the various chapters that follow.

A number of criticisms have been levelled at the HDI. It has been attacked for failing to include any ecological dimension, thus ignoring the interdependence between human well-being and ecological conservation. Others believe that by focusing on individual nations and ranking them, it fails to consider development from a global perspective. It has been criticised in the past for not including a gender dimension (i.e. not considering inequalities between the well-being of men and women), while others believe that the data and measurement techniques it is based on are flawed. A final group of critics have no problem with the HDI itself, but believe it to be a reinvention of the wheel, showing nothing that wasn't evident prior to its creation.

The Human Development Index (HDI) is an aggregate measure of progress in three dimensions – health, education and income. In the 2010 report, however, the indicators in education and income have been modified and the collection method has also been changed.

- Mean years of schooling replaced literacy
- Gross enrolment is shown as expected years of schooling which gives the years of schooling that a child can expect to receive given current enrolment rates
- To measure the standard of living, the gross national income (GNI) per capita replaces gross domestic product (GDP). Income earned and remittances received as well as sizeable aid flows lead to large differences between the income of a country's residents and its domestic production
- The collection method in the three dimensions of human development was also changed from an arithmetic mean to a geometric mean. As well as recognising that health, education and income are all important, poor performance in any dimension is now directly reflected in the reporting.

Three other major changes were also introduced in the 2010 Report – the HDI focused increasingly on deprivation, vulnerability and inequality and included the Inequality Adjusted Human Development Index (this measures the losses in human development due to inequality in health, education and income); the Gender Inequality Index (this reveals gender-based disparities in reproductive health, empowerment and labour market participation) and the Multi-Dimensional Poverty Index (which identifies overlapping deprivations suffered by households in health, education and living standards). The changes reflect advances in knowledge and information and allow for innovation in measuring multidimensional inequality and poverty, which can then be applied internationally to enable comparisons to be made and to provide new understandings and insights on human development.

LEARNING ABOUT 'HUMAN DEVELOPMENT' 20 YEARS ON – THE 2010 HUMAN DEVELOPMENT REPORT

Insisting that *people are the real wealth of a nation*, the Human Development Index came into being in 1990 with the publication of the first United Nations Development Programme *Human Development Report* – the new index challenged the dominance of economic measurements of development (which measured 'development' via Gross National Product per capita) and instead argued that development should be measured not simply by national income but also life expectancy and literacy. Despite its limitations (reliance on national averages which concealed skewed distribution, the weightings given to different components of the Index and the absence of 'a quantitative measure of human freedom') and the many criticisms directed at the HDI, the Index significantly transformed understanding and measurement of development.

Through the Index, a concern with equity in human development translates directly into an explicit focus on inequality as a primary concern. On this basis, the 2010 Report argues that while:

'In some basic respects the world is a much better place today than it was in 1990 – or in 1970. Over the past 20 years many people around the world have experienced dramatic improvements in key aspects of their lives. Overall, they are healthier, more educated and wealthier and have more power to appoint and hold their leaders accountable than ever before', nonetheless *'nearly 7 billion people now inhabit the earth. Some live in extreme poverty—others in gracious luxury.'*

Today, the underpinning assumptions of the Index are almost universally accepted - a country's success or an individual's well-being cannot be evaluated by money alone; life expectancy, health, education and freedom are vitally important.

Using the Index to survey human development over the past three decades provides many important insights into key patterns and trends.

- Convergence is occurring and, overall, poor countries are catching up with rich countries in key respects while divergence continues as regards income

- Not all countries have experienced the same progress and the variations are striking (from 20% improvement to 65%) with the slowest progress being experienced in the countries of sub-Saharan Africa affected by the HIV and AIDs pandemic and the countries of the former Soviet Union where mortality rates have risen

- Many countries have displayed remarkable progress in the HDI – China, Indonesia and South Korea, Nepal, Oman, Tunisia, Ethiopia, Botswana etc.

- Of 135 countries surveyed for the period 1970–2010, with 92% of the world's people, only 3 - the Democratic Republic of the Congo, Zambia and Zimbabwe—have a lower HDI today than in 1970

- Advances in health have been very significant but are now slowing down with the overall slowdown largely due to dramatic reversals in 19 countries - in 9 of them - 6 in Sub-Saharan Africa and 3 in the former Soviet Union where life expectancy has fallen below 1970 levels.

- There has been widespread and substantial progress in education reflecting improvement in the quantity of schooling available, getting children into school and also in the equity of access to education for girls and boys

- Progress as regards income has varied very considerably due to the fact that on average rich country incomes have grown faster than poor country incomes and the huge divide between developed and developing countries continues. A small number of countries have remained at the top of the world's income distribution, and only a handful of countries that started out poor have joined that high-income group.

- The share of formal democracies worldwide has increased from fewer than a third of countries in 1970 to half in the mid-1990s and to three-fifths in 2008.

- Since the 1980s, income inequality has risen in many more countries than it has fallen; for every country where inequality has improved in the past 30 years, in more than two it has worsened (especially in former Soviet Union countries). Most East Asian and the Pacific countries have higher income inequality today than a few decades ago. Latin America and the Caribbean are an important exception - long characterised as the region with the widest income and asset disparities, major recent improvements have led to more progressive public spending and targeted social policies.

These realities led to the introduction of three new measures of human development in the 2010 Human Development Report – the Inequality Adjusted Human Development Index (which takes account of inequality in its measurement - IHDI), the Gender Inequality Index (which measures progress or otherwise by gender) and the Multi-dimensional Poverty Index (this measures multiple dimensions of poverty and how they overlap – MPI). These tools seek to capture important key aspects of the diversity of human development in recent decades. For example, their use highlights the following aspects of development:

- People in sub-Saharan Africa suffer the largest losses in human development because of substantial inequality across all 3 dimensions of the HDI whereas in other regions, losses are more directly attributable to inequality in one single dimension e.g. as for health in South Asia.

- Gender inequality varies hugely across countries and the losses in achievement due to gender inequality range from 17% to 85%; the Netherlands tops the list of the most gender-equal countries, followed by Denmark, Sweden and Switzerland while those doing badly include the Central African Republic, Haiti and Mozambique.

- Sub-Saharan Africa has the highest incidence of multidimensional poverty; the level ranges from a low of 3% in South Africa to a massive 93% in Niger; the average share of deprivations ranges from about 45% (in Gabon, Lesotho and Swaziland) to 69% (in Niger). Yet half the world's multi-dimensionally poor live in South Asia (844 million people), and more than a quarter live in Africa (458 million).

SELECTED STATISTICAL SOURCES
(MOST ARE AVAILABLE ANNUALLY AND ONLINE)

Food and Agriculture Organisation State of World Food Insecurity

Instituto del Tercer Mundo Social Watch Report (a Third World perspective)

Instituto del Tercer Mundo The World Guide (bi-annual)

UNDP Human Development Report

World Bank World Development Report

UNICEF State of the World's Children Report

UNESCO Statistical Yearbook

READING

Human Development Report 2010: The Real Wealth of Nations: Pathways to Human Development, Basingstoke and New York, Palgrave Macmillan

World Institute for Development Economics (2006) The World Distribution of Household Wealth, United Nations University

Duncan Green (2008) From Poverty to Power: how active citizens and effective states can change the world, Warwickshire, Practical Action Publishing

Richard Wilkinson and Kate Pickett (2009) The Spirit Level: Why More Equal Societies Almost Always Do Better, London, Penguin.

Amartya Sen (1999) Development as Freedom, Oxford University Press

Michel Chossudovsky (2003) The Globalization of Poverty and the New World Order, Ontario, Global Outlook

Maggie Black (2007, 2nd edition) The No-Nonsense Guide to International Development, London, New Internationalist and Verso

Jeremy Seabrook (2007) The No-Nonsense Guide to World Poverty, London, New Internationalist and Verso

HUMAN DEVELOPMENT
A BALANCE SHEET

PROGRESS = ■
DEPRIVATION = ■

LIFE EXPECTANCY

Average life expectancy has increased by 21% to 68 years in developing countries overall since 1970

The increase has been highest in the Arab World and in South Asia

Life expectancy declined to below 1970 levels in 2010 in 9 countries – 6 in Sub-Saharan Africa and 3 in former Soviet Union countries

LITERACY

Literacy has increased by 61% between 1970 and 2010 for all developing countries with the greatest increases being recorded in Sub-Saharan Africa (183%) and the Arab States (149%)

Literacy rates remain drastically low in some countries – in Bangladesh 20%, Ethiopia 35% and Zambia 60%.

EDUCATION

Educational enrolment has increased by 28% for all developing countries in the period 1970 – 2010 with increases of 109% in Sub-Saharan Africa and 89% in the Arab States.

Since 1990 average years of schooling have risen by two years internationally and gross enrolment ratios by 12% while literacy rates have risen from 73% to 84% with no country worldwide experiencing declines in literacy or years of schooling since 1970

Spending displays enormous disparity – the annual average for 2005-2009 was nearly $4,611 per pupil worldwide but only $184 in Sub-Saharan Africa, roughly 1/8th that in Latin America and less than 1/40th that of developed countries. And the gap in spending per pupil is widening.

HUMAN DEVELOPMENT

All but 3 of the 135 developing countries have a higher level of human development today than in 1970

The fastest progress has been in East Asia and the Pacific, followed by South Asia and the Arab States

The citizens of 3 countries - the Democratic Republic of the Congo, Zambia and Zimbabwe now have lower levels of human development than in 1970

INCOME

Average income increased for all developing countries 1970-2010 by 184%, most spectacularly in East Asia and the Pacific (1,183%) followed by South Asia (162%)

Since 1970, 155 countries (95% of the world's people) have experienced increases in real per capita income; the annual average today is $10,760, almost 1.5 times its level 20 years ago and twice its level 40 years ago

While average income for all developed countries increased by 126% in the same period, that for Sub-Saharan Africa increased by just 20%

From 1970 to 2010 per capita income in developed countries increased 2.3% per year on average and 1.5% for developing countries. In 1970 the average income of a country in the top quarter of the world income distribution was 23 times that of a country in the bottom quarter but by 2010 it Increased to 29 times

The real average income of people in 13 countries in the bottom 25% of today's world income distribution is lower than in 1970.

Sources:

World Bank (2010) Poverty Reduction and Equity: Statistics and Indicators
UNDP (2010) Human Development Report, 2010
Arab Human Development Report 2009, Challenges to Human Security in the Arab Countries, NY, UNDP
UNICEF (2008) State of the World's Children Report 2009, New York

POVERTY

The world is on track to meet the first Millennium Development Goal (MDG) target of halving the 1990 poverty rate by 2015,

Based on World Bank projections, the number of people living on less than $1.25 a day is projected to be 883 million in 2015, lower than previous estimates

The poverty rate declined to 25% of global population in 2005, compared to 42% in 1990 and 52% in 1981. By 2015, that number is expected to drop to less than 15%

The decline in poverty between 1981 and 2005 varied considerably – while China, East Asia and the Pacific Region made huge progress, Sub-Saharan Africa, had only a modest decline in the poverty rate ($1.25 a day) between 1981 and 2005. Due to population growth, the number of poor people living in Sub-Saharan Africa almost doubled and its share of the world's poor increased from 11% to 27%.

With a poverty level of $2 per person per day in Latin America, the Caribbean, Eastern Europe and Central Asia, the share of global population living below that level fell from 70% 1981 to 47% in 2005 but population growth caused the numbers of poor people to remain at about 2.5 billion.

The share of democratic countries worldwide increased from less than a third in the early 1970s to more than half in 1996 and to three-fifths in 2008

The largest advances were in Europe and Central Asia (in 1988 the only democratic country was Turkey but since then 13 of the 23 became democracies); in Latin America and the Caribbean (in 1970 most countries were not democratic but by 2008 87% were.

6% of governments in the Pacific and Sub-Saharan were democratic in 1970 but by 2008 the share had risen to 44% in East Asia and the Pacific and 38% in Sub-Saharan Africa

CIVIL & POLITICAL RIGHTS

In its report for 2008, the Arab Organisation for Human Rights (AOHR) cites examples of the violation of the right to life in eight Arab states. In addition to Iraq and the Occupied Palestinian Territory, these states were Egypt, Jordan, Kuwait, Morocco, Saudi Arabia and Syria. The UN High Commission for Human Rights reported the continued use of torture in Algeria, Bahrain, Morocco and Tunisia

HEALTH

Infant mortality rates (per 1,000 live births) fell from 89 to 45 between 1970 and 2006 while adult female mortality rates fell from 257 to 164 (per 1,000 adults) and that for male adults from 308 to 237.

There are 8 times more infant deaths per 1,000 live births in developing countries than in developed countries

Health progress has declined dramatically in 19 countries

FOOD & NUTRITION

The share of undernourished people in developing countries fell from 25% in 1980 to 16% in 2005

The absolute number of malnourished people hardly changed from 850 million since 1980 with over 1 billion now hungry- 63% are in Asia and the Pacific, 26% in Sub-Saharan Africa and 1% in developed countries

WOMEN

School enrolment for girls has increased faster than for boys over the past few decades, and from 1991 to 2007 the ratio of female to male primary enrolment rose in all regions. Both primary and secondary school completion rates have improved more rapidly for girls

However, inequality continues - of the 156 countries with data, only 87 have primary school enrolment ratios for girls close to or above those for boys

For the bottom 20 countries on the HDI, average maternal mortality ratio is about 915 deaths (per 100,000 live births) well above the global average of 273

For the same bottom 20 countries, there is only one woman for every eight men in parliament

Photo: Garry Walsh

[*'What tends to 'inflame the minds' of suffering humanity cannot but be of immediate interest both to policy-making and to the diagnosis of injustice.'*]

Amartya Sen

Chapter 2

DEVELOPMENT: THE STORY OF AN IDEA

Tony Daly and Colm Regan

This chapter explores the current thinking and debates about the nature and scope of 'development' having initially sketched out its evolution and parameters from early world views via colonialism, the decolonisation process and the emergence of the 'age of development' following the Second World War. The emergence of modernisation theory, dependency theory and broader 'Third World' critiques of the theory and practice of development are also explored and debated. The issues surrounding the divergent definitions of development along with key phrases and debate such as that on the term 'Third World' are addressed.

KEYWORDS

Nature and history of development, development definitions and debates, gender, environment, human rights and development, modernisation theory, dependency theory, Third World perspectives

INTRODUCTION

The half century between 1950 and 2000 has been characterised by many analysts as the 'age of development', one, it is now argued, that has been superseded by the 'age of globalisation'. In the aftermath of the ending of World War Two, the re-building of Europe, the rise of the United States and the beginning of the process of decolonisation, international 'development' became a dominant focus in international relations. This vision and agenda is perhaps best captured in the 1949 inaugural address of US President Harry S Truman who argued:

'We must embark on a bold new program for making the benefits of our scientific advances and industrial progress available for the improvement and growth of underdeveloped areas. The old imperialism – exploitation for foreign profit – has no place in our plans. What we envisage is a program of development based on the concept of democratic fair dealing ... More than half the people of the world are living in conditions approaching misery. Their food is inadequate. They are victims of disease. Their economic life is primitive and stagnant. Their poverty is a handicap and a threat both to them and to more prosperous areas. For the first time in history, humanity possesses the knowledge and the skill to relieve the suffering of these people.'

The decades that followed were characterised by the Cold War (with contrasting views of, and strategies for 'underdeveloped' countries); the establishment of the institutions and structures of the United Nations (especially those directly addressing development); widespread decolonisation; the eventual collapse of the Soviet Union and the opening up of Eastern Europe. While, in development terms, much was achieved in many developing countries (in health, literacy, basic needs and education), the fundamental divide and inequalities that characterise the world continued to grow and deepen. So much so that commentators such as Gustavo Esteva have claimed:

'Development has evaporated. The metaphor opened up a field of knowledge and for a while gave scientists something to believe in. After some decades, it is clear that this field of knowledge is a mined, unexplorable land. Neither in nature nor in society does there exist an evolution that imposes transformation towards 'ever more perfect forms' as a law. Reality is open to surprise. Modern man has failed in his effort to be god.'

Esteva in W. Sachs, ed., 2010, The Development Dictionary, London, Zed Press

In compiling the second edition of the Development Dictionary in 2010, editor Wolfgang Sachs insisted:

'The idea of development stands like a ruin in the intellectual landscape. Delusion and disappointment, failures and crimes, have been the steady companions of development and they tell a common story: it did not work. Moreover, the historical conditions which catapulted the idea into prominence have vanished; development has become outdated. But, above all, the hopes and desires which made the idea fly are now exhausted; development has grown obsolete. Nevertheless, the ruin stands there and still dominates the scenery like a landmark..."

Despite such dismissals, development remains a key frame of reference in international relations as well as in the popular imagination; it also forms a core element in the architecture of development cooperation at both governmental and non-governmental levels. Despite the fact that development is regularly deemed to have failed (especially when viewed from the perspective of the poor) it continues to be financed, debated, measured, monitored and evaluated – development remains big business!

THREE KEY PERSPECTIVES

In the course of the past six decades, the idea of development has been expanded to encompass a great variety of perspectives and emphases, many of which are explored below but at this stage, three particular perspectives are worthy of note – the 'feminist'; the 'ecological' and the 'human rights'.

The feminist perspective has been outlined by many commentators, few more authoritatively than Gita Sen and Caren Grown in 1988:

'The perspective of poor and oppressed women provides a unique and powerful vantage point from which we can examine the effects of development programmes and strategies … if the goals of development include improved standards of living, removal of poverty, access to dignified employment, and reduction in societal inequality, then it is quite natural to start with women. They constitute the majority of the poor, the underemployed and the economically and socially disadvantaged in most societies. Furthermore, women suffer from additional burdens imposed by gender-based hierarchies and subordination.'

Sen, Gita and Caren Grown (1985) DAWN, Development, Crisis and Alternative Visions; Third World Women's Perspective. Stavanger: Verbum

The environmental or ecological perspective has always paralleled the 'development story' but, in more recent decades it has become intrinsically central to it and vice versa. This view was elegantly outlined by Indian activist Vandana Shiva in 1989:

'Among the hidden costs (of development) are the new burdens created by ecological destruction, costs that are invariably heavier for women, both in the North and the South. It is hardly surprising, therefore, that as GNP rises, it does not necessarily mean that either wealth or welfare increase proportionately … In actual fact, there is less water, less fertile soil, less genetic wealth as a result of the development process.'

Vandana Shiva (1989) Staying Alive: Development, Ecology and Women in Micheline R. Ishay, (ed., 1997) The Human Rights Reader, Routledge, London.

In more recent years, the interface between human rights and human development has come more to the fore with significant implications for both. This perspective was coherently outlined by the UNDP Human Development Report for 2000:

'If human development focuses on the enhancement of the capabilities and freedoms that the members of a community enjoy, human rights represent the claims that individuals have on the conduct of individual and collective agents and on the design of social arrangements to facilitate or secure these capabilities and freedoms.'

One area in which there is almost unanimous agreement is that the definition of development (and the practices and structures it has generated) is both controversial and contested – there is little agreement as to its precise definition and meaning with different groups emphasising different dimensions at different times. For those who argue that we now live in a 'post development age', it is time to abandon the very concept of development while for those who challenge 'post-developmentalism'; it is far too soon to write its obituary.

The remainder of this chapter explores the debate.

EARLY IDEAS ABOUT DEVELOPMENT

Today, development is strongly associated with the idea of change but this has not always been so. The Ancient Greeks (from whom much of the West's thinking springs, especially from the philosophy of Aristotle) argued that all things have an essential and inherent nature and that life is a matter of cycles; all things that are born and grow will also fade and die, in a perpetual series of new beginnings followed by decline and decay. Arabic philosophers argued similarly. Gilbert Rist, in his influential History of Development, argues that there are three important links or continuities between modern development thinking and historical philosophy:

- one, 'development' is seen as natural and necessary
- two, linking development with nature and the natural
- three, the links implied between science and myth (development as scientific progress but also as a religion)

However, with the rise of the Judaeo-Christian tradition, the idea of a God intervening in human history became crucial. This implied a couple of key ideas which are still with us, and which are still hotly debated.

- First, there is a universal history for all humankind (is there?)
- Second, history is a historical progression from creation to the end of time (is it?)
- Third, this is God's plan, unalterable by humans (is this true?)

This increasingly linear view of history (and of development) was strengthened further during the 17th and 18th centuries, particularly by the social evolutionism of the Enlightenment. In general, social evolutionism argued that the progress of civilisation is a 'one way street' that all societies follow, from hunter gatherers to mid-nineteenth century England and Western Europe. Comparing societies with each other, scientists concluded that the march of civilisation could be divided into a series of sequential steps. The assumption that all human cultures develop along a single or unilinear path is illustrated, for example, by the ideas of American Anthropologist Lewis Henry Morgan who identified a series of steps from 'savagery and barbarism' to 'civilisation' with each society moving from step to step over time.

Rist comments:

'Belief in the natural and inevitable 'development' of societies prevented them from being considered for themselves, with their own specificities (... deprived both of their history and of their culture ...). Instead, they were simply judged in accordance with the western reference.'

He also argues that:

'... the practices which are today claimed as new have a long history behind them, and that control over the lands of the South has long dressed itself up as high minded internationalism.'

COLONIALISM AND DEVELOPMENT

The high point of colonisation occurred at the end of the 19th century spearheaded primarily by Britain, France and Germany with many different views and policies as regards the role and development of their colonies. While some liberal economists questioned the costs and the supposed benefits of having colonies, the armed forces, the merchants and the missionaries supported colonisation in their own interests.

From the beginning, colonisation was characterised by ambiguity - on the one hand, colonies were there

'... the practices which are today claimed as new have a long history behind them, and that control over the lands of the South has long dressed itself up as high minded internationalism.'

to be exploited in the interests of Europe but, against this, colonisation was also seen to be a moral responsibility of Europe – the famous 'White Man's Burden'. The colonisers offered a vision of worldwide 'civilisation' for all and argued that there was common cause between coloniser and colonised. However, the reality of colonialism pivoted around resource, economic, political and military agendas while intellectual and popular commentators argued that the 'development' of a nation (especially 'undeveloped' nations) paralleled that of human life and was, thus both inevitable as well as a moral duty.

This remained the situation until the end of World War I in 1918 when Germany lost its colonies to Britain and France. In this context, the League of Nations was founded (the precursor to the United Nations) with security and the settlement of disputes by peaceful means as its primary goals (it also had a number of other 'development' related goals). While the United States did not join the League, it was very influential in 'redefining and reshaping' colonialism especially in the creation of a 'mandate system' where some countries were deemed ready for independence; where others could achieve it in the future and where a third group were seen to have little prospect of independence. Article 22 of the founding articles of the League stated that in territories which ceased to be under the sovereignty of states, inhabited by peoples:

'...not yet able to stand by themselves under the strenuous conditions of the modern world', the tutelage of such

peoples 'should be entrusted to advanced nations' and that the character of the mandate should differ 'according to the stage of the development of the people...'

Thus, the concept of 'stages of development' was introduced, justifying a classification system with 'developed' nations at the top of the ladder. The language used also exhibited a strong humanitarian and religious tone with words such as 'civilisation', 'material and moral well-being', and 'social progress', 'sacred trust' legitimising western intervention in other regions and countries.

Colonisation meant not only taking control of people and places but also of peoples' minds – in the West as much as elsewhere. It contributed significantly to the emergence of ways of thinking and a language about development which has lasted to the present time. It created the illusion of a world united and it also created the basis for the emergence of the Third World, although that term only emerged later (see page 46).

THE 'AGE OF DEVELOPMENT' BEGINS

In development terms, the period following World War II heralded the modern age of development – one that had much in common with the past while also rejecting some of its key characteristics (e.g. its racial overtones and much of its social evolutionism) and creating much of the architecture of international development today. The politics and ideology of the world were fundamentally reshaped following 1945. The Allied Powers sought to ensure that many of the conditions which gave rise to WWII – mass unemployment, protectionist policies, competitive currency devaluations, collapse of commodity prices – would not recur.

Pre-eminent among the creations of the Allies were the International Bank for Reconstruction and

Development (better known as the World Bank) and the International Monetary Fund (IMF), both located, tellingly, in New York, representing growing US hegemony internationally. The Bank and the Fund were to be specialised agencies within the United Nations (established in 1945) much like the Food and Agricultural Organisation (FAO) or the International Labour Organisation (ILO), located in Rome and Geneva respectively. The UN was to be the focal point for international economic management and the World Bank was set up to promote economic growth and development. At that stage, no distinction was made between these two closely related ideas.

By now, the idea of development was inextricably linked with economics, an economics that assumed economic growth was the fundamental necessity. Over time, the World Bank and the IMF became the dominant power brokers as regards development

and, unlike the UN (which is run on a one-country, one-vote system) they were run on a one-dollar, one vote system. In this way, the Bank and the Fund came to represent powerful financial interests and the development role of the UN was steadily transferred to them.

It was against this background that the idea of 'development' (and 'underdevelopment') came to be defined, most notably in the inaugural speech of President Truman in 1949 (referred to earlier). Truman's speech was important for a number of reasons - firstly, for the first time the adjective 'underdeveloped' appears (meaning economically 'backward') in an official document and subsequently the idea of 'underdevelopment' was introduced. These new words actually altered the way the world was seen. Development was now something one agent could do to another, whereas underdevelopment was apparently a 'natural' cause-less state. This obscured

VIEWPOINT: URUGUAYAN WRITER AND COMMENTATOR EDUARDO GALEANO (2001)

The equator did not cross the middle of the world map that we studied at school. More than half a century ago, German researcher Arno Peters understood what everyone had looked at but no one had seen: the Emperor of Geography had no clothes. The map they taught us gives two-thirds of the world to the north and one-third to the south. Europe is shown as larger than Latin America, even though Latin America is actually twice the size of Europe. India appears smaller than Scandinavia, even though it's three times as big. The United States and Canada fill more space on the map than Africa, when in reality they cover barely two-thirds as much territory.

The map lies. Traditional geography steals space just as the imperial economy steals wealth, official history steals memory, and formal culture steals the world.

historical processes which created the conditions of development and underdevelopment, a view that was contested by many critics (for example, Guyanese historian and activist Walter Rodney in his influential book How Europe Underdeveloped Africa). The old dichotomy between coloniser and colonised was now replaced by that between 'developed and underdeveloped'. The latter group of countries could achieve the development status of the former through adopting a set of strategies and institutions characteristic of 'developed' states. The key to prosperity and happiness for all would be achieved through increased production, measured by the new standard of Gross National Product (GNP). Later, broader and more 'sensitive' indices of development were created (see Chapter 1). The consequences for newly independent states were they now had to follow the path of development mapped out for them by others – that of western capitalism or, alternatively, the communist Soviet Union. The 'age of development' created a situation where countries were no longer, in the words of Gilbert Rist '... *African, Latin American or Asian (not to speak of Bambera, Shona, Berber, Quechua, Aymara, Balinese or Mongol), they were now simply 'underdeveloped'.*

Predictably, given the nature and politics of the anti-colonial struggle in many parts of the world, this view of development and of recent world history was rejected as well as its constructions of 'development' and 'underdevelopment'.

DEVELOPMENT AS 'MODERNISATION'

With the rise of these new economic and political institutions and world order (post 1945 and especially in the context of the emergent 'Cold War' precisely at the time when the field of 'development' was taking shape), one of the most influential theories claiming to provide a universal model of development applicable to all countries

was that of economist Walt Rostow in his 1960 Stages of Economic Growth, tellingly sub-titled A Non-Communist Manifesto. Rostow argued that all societies developed through a series of stages beginning with stage one (a traditional society), via stages two (preconditions for take-off – based on a reading of earlier European history), three (when societies integrated 'growth' as an integral part of the habits and structures of that society), four ('the drive to maturity' when a society gears itself to 'efficient' modern production) to a final stage five (that of high mass consumption) when that society became 'developed'.

A central criticism of Rostow's theory was that it was based on the particular case of a limited number of Western economies and was not only inappropriate to other societies but was fundamentally inapplicable. It was also heavily criticised for failing to adequately integrate the realities of colonialism and imperialism, historical realities that not many states could emulate. A third criticism was that not all societies value the accumulation of material goods equally or indeed seek to establish an age of high mass consumption and yet these are deemed to be fundamental to development. Another problem was that presenting development and 'modernisation' as a way of increasing people's choices obscured what had been lost, or choices that no longer existed.

Nevertheless, Rostow was a product of his time. Ten years prior to his 'take-off' thesis the report prepared by the United Nations group of experts tasked with designing concrete policies and measures "for the economic development of underdeveloped countries" stated:

'There is a sense in which rapid economic progress is impossible without painful adjustments. Ancient philosophies have to be scrapped; old social institutions have to disintegrate; bonds of cast, creed and race have to burst; and large numbers of persons who cannot keep

up with progress have to have their expectations of a comfortable life frustrated. Very few communities are willing to pay the full price of economic progress.'

United Nations, Department of Social & Economic Affairs 1951: 15

Rostow's theory was but one in a school of thought (and planning) that viewed development as 'modernisation' in the mode and manner of western 'capitalist' societies. For example, US political scientist David Apter offered a political analysis linking democracy and good government with modernisation while US psychologist David McClelland argued that modernisation is impossible in a society that does not value free enterprise and success while Israeli sociologist Shmuel Eisenstadt argued that:

'Historically, modernisation is the process of change towards those types of social, economic, and political systems that have developed in Western Europe and North America from the seventeenth century to the nineteenth and have then spread to other European countries and in the nineteenth and twentieth centuries to the South American, Asian, and African continents.'

S. Eisenstadt 1966, Modernisation, Protest and Change

Stages theory and modernisation theory were ultimately criticised for being theories about the 'westernisation' of the world rather than about development per se; for emphasising capitalism and western values over others and for the maintenance of western dominance worldwide. Logically and inevitably a range of counter-theories emerged, the most important of them being 'Dependency Theory'.

DEVELOPMENT AND DEPENDENCY – THE THIRD WORLD CRITIQUE

The anti-colonialist and anti-imperialist movements that emerged worldwide post 1945 gave rise to many detailed critiques of western theories of growth, development and modernisation as being essentially about maintaining and strengthening western capitalist dominance. Forged in the furnace of political and armed struggle, they highlighted the violence and inequities that western development generated and imposed on others. They challenged the core idea that Third World countries could (or should) follow free market models as 'benign' or level playing fields in which developing countries could compete. They also highlighted the need for strong state intervention in support of development and were, in many fundamental ways the intellectual and practical opposites of western modernisation theory. The intellectuals and activists who promoted these theories gradually came to focus on US policy and were routinely referred to a 'Third Worldists'.

Dependency Theory emerged from North American Marxists and Latin American intellectuals during the 1950s as a critique of modernisation – it represented a general school of thought followed by different intellectuals from a variety of disciplines and countries and was never set out as a coherent theory. They studied real history in concrete circumstances with countries being treated as part of an international structure and not as individual entities. They viewed colonialism and imperialism as processes in which 'things fall apart' (echoing Nigerian writer Chinua Achebe in his 1958 novel Things Fall Apart) and where development is not contrasted with tradition but rather 'underdevelopment' and contrary to commentators such as Rostow, the expansion of western values and institutions does not produce development but rather its opposite - underdevelopment. In the words of German-American economist Andre Gunder Frank:

'...historical research demonstrates that contemporary underdevelopment is in large part the historical product of past and continuing economic and other relations between the satellite underdeveloped and the now developed metropolitan countries. Furthermore, these relations are an essential part of the capitalist system on a world scale as a whole.'

Andre Gunder Frank (1966) The Development of Underdevelopment, New York, Monthly Review Press.

...it became a strong component of a growing Third Worldist critique of western ideas and practices around the 'development idea' and it strengthened the view that underdevelopment was not a 'natural' state but could actually be created and, in the world of the 1960s and 70s was being created by the West

Dependency theorists also argued that the development of young or emerging economies required their withdrawal from the structure of exploitation that existed worldwide and in many cases the adoption of socialism rather than capitalism. Since capitalism was inextricably intertwined with colonialism and imperialism, then anti-colonialism needed to adopt socialism – this analysis was also fuelled significantly by the realities and ideologies of the Cold War. Dependistas also argued that industrial capitalism was inextricably linked to finance capital and created a concentration of wealth and power in giant transnational corporations that controlled markets; they hoped that a world revolution would change this situation, but the steps to be taken to achieve this were not clear.

As with modernisation theory, there were significant problems associated with dependency theory. The wealth of the developed countries was not solely due to colonial exploitation, but was also

due to state regulation and the growth of domestic markets. Also, it did not pay sufficient attention to the cultural aspects of development or to the ecological consequences of treating industrialisation as necessary for collective well-being. In this way it did not offer an idea of development based on assumptions fundamentally different to that of modernisation.

But it became a strong component of a growing Third Worldist critique of western ideas and practices around the 'development idea' and it strengthened the view that underdevelopment was not a 'natural' state but could actually be created and, in the world of the 1960s and 70s was being created by the West (for an excellent introduction to this literature and analysis, see Majid Rahnema and Victoria Bawtree eds. 1997, The Post Development Reader, London, Zed Press).

In the African context, Guyanese historian and activist Walter Rodney offered a scorching critique of colonialism and its impact on development in his How Europe Underdeveloped Africa (1972) in which he argued that European powers deliberately and systematically exploited and underdeveloped Africa and offered a new African insight into the issue. He wrote:

'African development is possible only on the basis of a radical break with the international capitalist system, which has been the principal agency of underdevelopment of Africa over the last five centuries.'

Walter Rodney (1983) How Europe Underdeveloped Africa preface

As a trained psychiatrist, the French philosopher and revolutionary Franz Fanon had taken a different view based on his own Algerian experience. The colonial state was an implant that imposed itself on the colonised society by restructuring social arrangements and reimaging the past to fit its own imperial agendas. Natives were actively alienated from their historical experiences and new social divisions were put in place. Moreover, Fanon believed that Africans should no longer be tempted by development by European design or as inspiration; his was a radical confrontation with the Western model of progress.

'The basic confrontation which seemed to be colonialism versus anti-colonialism, indeed capitalism versus socialism, is already losing its importance. What matters today, the issue which blocks the horizon, is the need for a redistribution of wealth. Humanity will have to address this question, no matter how devastating the consequences may be.'

Franz Fanon (1961) The Wretched of the Earth

Summarising much of the above, Wolfgang Sachs captures the core of the debate when he states:

'Since the Euro-Atlantic model of wealth emerged under exceptional conditions, it cannot be generalised to the world at large. In other words, the model requires social exclusion by its very structure; it is unfit to underpin equity on a global scale. Therefore, development-as-growth cannot continue to be a guiding concept of international politics unless global apartheid is taken for granted.'

Wolfgang Sachs (2010, 2nd edition) The Development Dictionary, London, Zed Press

'DEVELOPMENT' REDEFINED

The 1980s and 1990s witnessed the emergence of a strong challenge to the dominant economic models and analyses of development through what later became defined as the human development approach (known more popularly for its measurement index – the Human Development Index (HDI) and the annual Human Development Report). This approach is associated primarily with the United Nations Development Programme (UNDP) and the work of economists Mahbub Ul Haq and Amartya Sen. The first Human Development Report published in 1990 made explicit its approach and values base *'to shift the focus of development economics from national income accounting to people centred policies'* and its opening lines stated: *'People are the real wealth of a nation'* (United Nations Development Programme 1990); people were not simply the *'beneficiaries'* of economic and social progress in a society, but were active agents of development and change.

Within the, then dominant economic model, characteristic of the vast majority of 'official' documents and policies whether from the World Bank, the IMF, the World Trade Organisation or many bi-lateral governmental aid organisations, the purpose of 'development' was to stimulate growth and from that 'growth' benefits would 'trickle down' to society. The primary purpose of the state should be to provide support and an *'enabling environment'* for that growth; other areas of focus - social, cultural or environmental issues should be of only secondary importance. The economy, rather than human beings occupied centre stage and while much (if not all) alternative academic and NGO analysis vehemently disagreed with this model, orthodox development theory and practice continued to promote it.

In contrast, the approach advocated by the human development perspective fundamentally challenged this view and offered a much broader agenda - a comprehensive approach to all aspects of development, a set of policy priorities, tools of analysis and measurement and a conceptual framework. A report published by the UNDP prior to the 1992 Earth Summit in Rio de Janeiro highlighted massive inequalities in human development, showing, for example, that the richest 20% of people received 150 times the income of the poorest 20%. The report demonstrated that global economic growth rarely *trickles down*; it illustrated that 10 countries with more than 70% of the world's poor people received only 25% of global aid and showed that only 6.5% of Overseas Development Assistance was focused on human development concerns such as basic education, primary health care, safe drinking water, family planning and nutrition.

The foundations of the human development approach were rooted in the approach of Indian economist Amartya Sen who, in 1989, defined human development as a process of enlarging people's *functionings and capabilities to function, the range of things that a person could do and be in her life* later expressed in terms of expanding 'choices'. Sen went on to argue (in 1999) that key *freedoms are not only the primary ends of development, they are also among its principal means*. Development should be seen as a process of expanding such freedoms. Development, he argued requires the removal of poverty, tyranny, lack of economic opportunities, social deprivation, neglect of public services, and the machinery of repression and, additionally, *'the formation of values and the emergence and evolution of social ethics are also part of the process of development'*. His approach and that of many others concerned with the human development approach is, perhaps best illustrated in his emphasis on the role of women in development. Nothing, he argued, is *'as important today in the political economy of*

development as an adequate recognition of political, economic and social participation and leadership of women. This is indeed a crucial aspect of 'development as freedom'.

The concept of human development became far more complex, extensive and political/social than previous dominant models; development became about people being able to live in freedom and dignity; being able to exercise choice, pursue an engaged and creative life. Priorities in development thus came to focus on removing limitations such as illiteracy, ill-health, and lack of access to and control over resources as well as increasing participation in the community, in decision-making and in strengthening the ability to challenge social, cultural and political oppression. In 1995, the UNDP declared that *'one of the defining movements of the 20th century has been the relentless struggle for gender equality, led mostly by women, but supported by growing numbers of men. . . . Moving toward gender equality is not a technocratic goal – it is a political process'.*

The era of the predominance of economic models of development had come to an end.

Additional views and debates on development are included elsewhere in other chapters, particularly Chapter 5 on women and development, Chapter 3 on sustainable development and Chapter 4 on human rights and development.

ON THE DIFFICULTY OF 'DEFINITIONS' OF DEVELOPMENT

Attempting to provide a clear and comprehensive definition of development that might be agreeable to the majority of people worldwide is a well-nigh impossible task as different contexts; timeframes and circumstances as well as perceptions and aspirations inevitably vary so hugely. For example, we could usefully consider how development might be conceived by a poor farmer in Zambia, a mother living in a slum in Sao Paulo, an environmentalist living in San Francisco or a chief executive of a transnational company based in London.

As Gilbert Rist points out, different understandings of 'development' depend on how each individual (or group of individuals) picture the ideal conditions for social existence (we should also add environmental existence). Rist argues that conventional thinking on development swings between two extremes:

- The expression of a general wish among all peoples to live and experience a better life (however defined) – something which ignores the fact that the very different ways of achieving this general objective or vision would immediately encounter very different political, economic and environmental ways of achieving it.

- The vast range of (often conflicting) actions which are theoretically designed to achieve the greatest happiness for the largest number of people.

The central issue is not to simply point out that one set of countries (labelled 'developed') have more of some things (schools, roads, average calorie consumption, computers, democracy, industrial employment, supermarkets etc.) and less of others (illiteracy, poverty, high infant and maternal deaths during pregnancy etc.) while others (labelled 'developing') have the reverse. What is crucially important to note is the processes that underpin and reproduce such contrast. Rist also makes the obvious comment

that 'development' is not simply a priority for poorer countries but equally for richer countries.

Rist then proceeds to outline a number of key components of a definition of development which is broad enough to capture history and different systems and practices and to pose the question whether 'development' might be viewed as a part of modern religion:

'Development' consists of a set of practices, sometimes appearing to conflict with one another, which require – for the reproduction of society – the general transformation and destruction of the natural environment and of social relations. Its aim is to increase the production of commodities (goods and services) geared, by way of exchange, to effective demand.'

He concludes:

'Development' thus appears to be a belief and a series of practices which form a single whole in spite of contradictions between them.'

This discussion takes us some distance away from mechanical or simplistic definitions of development into the fields of not just economics and politics but also philosophy and history and even into the area of myth.

For Swedish economist Bjorn Hettne:

'Development is a contested concept, which implies that it has meant different things from one historical situation to another and from one actor to another.'

In his 2009 analysis Thinking About Development (London, Zed Press), he insists:

'A critical approach is also necessary because much harm has been done to people in the name of development. Development practice in the so-called developing countries is ultimately rooted in colonialism, and has therefore sometimes contained a good measure of

paternalism, not to speak of arrogance and racism.'

Echoing much of the approach of Rist, Hettne extends the debate on development significantly beyond the search for a definition:

'Development thinking in fact constitutes an exceptionally rich tradition in social science, encompassing important theoretical debates on the dynamics of social change, as well as an ambition to represent a global experience of empirical conditions in different local corners of the world.'

He concluded in 1995:

'There can be no fixed and final definition of development; only suggestions of what it should imply in particular contexts.'

B. Hettne (1995) Development Theory and the Three Worlds, Essex, Longman

VIEWPOINT

'I wish you wouldn't squeeze so,' said the Dormouse, who was sitting next to her, 'I can hardly breathe'

'I can't help it,' said Alice very meekly: 'I'm growing'

'You've no right to grow here,' said the Dormouse

'Don't talk nonsense,' said Alice more boldly: 'you know you're growing too.'

'Yes, but I grow at a reasonable pace,' said the Dormouse: 'not in that ridiculous fashion.'

Alice's Adventures in Wonderland
Lewis Carroll, 1865

ON LANGUAGE AND LABELS

'Developing countries' is the name that experts use to designate countries trampled by someone else's development...'

This is the telling comment of development critic and 'third worldist' Eduardo Galeano in his challenging book *Upside Down: A Primer for the Looking Glass War*. In the book, he relentlessly challenges many 'western' constructions and understandings of the world and questions how we 'label' that world. In this he enters the contested agenda of how the world is divided and how such divisions are 'named' and 'labelled'. As 2005 Nobel Prize winner for Literature, Harold Pinter has commented *Language ... is a highly ambiguous business. So often, below the word spoken, is the thing known and unspoken'* and no-where is this more apparent than in the world of 'international development'.

It is a minefield of language in trying to describe both the diversity and the uniformity of those countries most frequently described as the 'Developing World' – even the use of the word 'developing' is not without difficulty. 'Developing' towards what and from what? If we can have a 'developing country' can we have an 'underdeveloping country' – a country going backwards in terms of human development? From the perspective and experience of the world's very poorest people (all too often, the majority in a given country), the term 'developing' could well be interpreted as a deliberate insult to their experience of life and a deliberate refusal to accurately describe their condition, famously captured in the words of Franz Fanon as the 'Wretched of the Earth'.

In a world where the language of 'communication' has become pre-eminent and where words conceal as much as they reveal and where 'spinning' is a highly professionalised 'business', it is important that we do not lose the capacity to see behind words and beyond the dominant consensus, especially in the context of our 80:20 world.

Throughout the history of the interaction between cultures worldwide, how one culture meets, 'understands' and describes 'The Other' has always been problematic and fraught with, on the one hand, fascination, awe, wonder and inspiration and, on the other hand, misunderstanding, arrogance, bewilderment and fear. Nowhere is this clearer than in the relationship between the 'the West' and 'the Rest' in the field of development (and underdevelopment). The politics involved in this interaction are highlighted in the debates around the use of language and labels.

For decades there has been a vigorous debate on how to describe the divisions that characterise the world today. Is it accurate to describe the 'West' as 'the Developed World' (given the many characteristics of both underdevelopment and overdevelopment present) and the 'Rest' as the 'Developing World' (given the historical sophistication and complexity of many of its 'societies' and economies)? At another level altogether, it is possible to find a phrase that adequately encompasses the massive diversity that exists within both these worlds. Whatever phrases are chosen, it is clear that political choices and agendas cannot be avoided.

In recent years, one of most vigorous debates, at least in the West, has been over the use of the phrase 'Third World' to describe the countries and agendas of Africa, Latin America and Asia collectively. Critics opposed to the use of the phrase describe it as pejorative (it places these countries 'third' in an international pecking order), negative (it fuels negative stereotypes and attitudes), outdated (in that the Second World – the Communist Bloc no longer exists) and inaccurate (it ignores the real achievements of such countries). As an alternative, such critics offer phrases such as Majority World and/or Global South. Clearly, given what has been said above, there is obviously no 'right' or 'wrong' phraseology but the debate is telling as regards the politics of the 'development/underdevelopment' story today.

The phrase 'Third World' was introduced into the story of development by the French anti-colonialist writer and activist Albert Sauvy echoing the thinking and writing of many other earlier commentators such as Aime Cesaire (Martinique born poet and teacher), Ho Chi Minh (President and 'father' of modern Vietnam) and Franz Fanon who sought to highlight the common 'anti-colonialist' agenda of that world. Sauvy chose his phrase to echo the political struggles of French history where the monarchy had divided its servants into the First Estate (clergy), the Second Estate (aristocracy) and the Third Estate (the bourgeoisie). Sauvy emphasised the political nature of his categorisation by arguing that the:

'...ignored, exploited, scorned Third World, like the Third Estate, demands to become something as well'.

The key point in Sauvy's analysis was that just as the Third Estate had been excluded from having a voice in the political life of France, so too had the Third World been similarly excluded. Sauvy was placing primary emphasis on the political agenda internationally. Jawaharlal Nehru (Prime Minister of India) echoed Sauvy's analysis in 1958 when he emphasised the need for countries outside the two dominant blocs of the capitalist and US-led West and the Communist, Soviet-led alliance to 'collect together' to oppose their plans for possession of the Third World. He argued that '*...it is right that countries of a like way of thinking should come together, should confer together, should jointly function in the United Nations or elsewhere'.*

The political and economic agendas of the Third World acting as a bloc of countries in opposition to the dominant blocs and interests was carried forward well into the 1990's by the Group of 77 or,

the G-77. Established in 1964 by 77 developing countries the G77 (now made up of 130 countries - the original name has been kept because of its historic significance) is the largest intergovernmental organisation of developing states in the United Nations. It provides a platform for the countries of the Developing World to state and promote their collective economic interests and enhance their joint negotiating power in international meetings and agendas.

Proponents of the term 'Third World' continue to emphasise that political agenda today in that the countries of that world remain largely excluded from international political and economic decision-making and that it is only in the last two decades that India and China (along with Brazil and post-Communist Russia) have become important and recognised political voices in international fora and strategies for international development.

Those who prefer to use the phrases Majority World or Global South argue that the terms are non-judgemental and 'neutral', arguments rejected by others. Critics of these categories argue that they:

- Are weak analytically and have little substance in describing the realities, histories and agendas of the countries included

- Have the result of de-politicising a complex and rich agenda and reducing it to a dualism of simple Majority and Minority or to a geographical term Global South (but which must, of necessity, eliminate countries actually in the Global South)

- Ignore and even deny the history of the (unequal) relations between 'developed and developing countries'

- Ignore the ongoing reality of the 'exclusion' of the poor from the key debates about the structure and functioning of the world today.

VIEWPOINT: EDUARDO GALEANO

'It was the promise of the politicians, the justification of the technocrats, the illusion of the outcast. The Third World will become like the First World – rich, cultivated and happy if it behaves and does what it is told, without saying anything or complaining. A prosperous future will compensate for the good behaviour of those who died of hunger during the last chapter of the televised serial of history. WE CAN BE LIKE THEM, proclaimed a gigantic illuminated board along the highway to development of the underdeveloped and the modernisation of the latecomers.

'But 'what can't be, can't be, and more than that is impossible', as Pedro el Gallo, the bullfighter so rightly said. If the poor countries reached the levels of production and waste of the rich countries, our planet would die. Already it is in a coma, seriously contaminated by the industrial civilisation and emptied of its last drop of substance by the consumer society.

'The best of the world lies in the many worlds the world contains, the different melodies of life, their pains and strains: the thousand and one ways of living and speaking, thinking and creating, eating, working, dancing, playing, loving, suffering, and celebrating that we have discovered over so many thousands of years.

'Whoever doesn't have isn't. He who has no car or doesn't wear designer shoes or imported perfume is only pretending to exist. Importer economy, imposter culture: we are all obliged to take the consumer's cruise across the swirling waters of the market. Most of the passengers are swept overboard, but thanks to foreign debt the fares of those who make it are billed to us all. Loans allow the consuming minority to load themselves up with useless new things, and before everyone's eyes the media transform into genuine needs the artificial demands the North of the world ceaselessly invents and successfully projects onto the South.

'…There is no country in the world as unequal as Brazil. Some analysts even speak of the 'Brazilianisation' of the planet in sketching a portrait of the world to come. By Brazilianisation they certainly don't mean the spread of irrepressible soccer, spectacular carnivals, or music that awakens the dead, marvels that make Brazil shine brightest; rather they're describing the imposition of a model of progress based on social injustice and racial discrimination, where economic growth only increases poverty and exclusion. 'Belindia' is another name for Brazil, coined by economist Edmar Bacha: a country where a minority lives like the rich in Belgium while the majority lives like the poor of India…'

Eduardo Hughes Galeano is a Uruguayan journalist (born in 1940) who is famous throughout Latin America for his biting political commentary as well as for his fiction, his journalism and his political analysis. Of himself, Galeano comments: 'I'm a writer obsessed with remembering, with remembering the past, remembering the past of America above all, and above all that of Latin America, intimate land condemned to amnesia.'

READING

Maggie Black (2007) No-Nonsense Guide to International Development, Oxford, New Internationalist

Wayne Ellwood (2006) No-Nonsense Guide to Globalisation, Oxford, New Internationalist

Eduardo Galeano (2001) Upside Down: A Primer for the Looking Glass War, New York, Picador

Gilbert Rist (2008) The History of Development: from western origins to global faith, London, Zed Press

Jeffrey Sachs (2008) Common Wealth: Economics for a Crowded Planet, London, Penguin

Wolfgang Sachs (2010) The Development Dictionary, London, Zed Press

Amartya Sen (1999) Development as Freedom, Oxford University Press

Vandana Shiva (2006) Earth Democracy, South End Press, Cambridge

MORE INFORMATION & DEBATE

http://www.ids.ac.uk - Institute of Development Studies, Sussex, UK

http://www.socialwatch.org Uruguay-based international network

http://hdr.undp.org/en home of the Human Development Reports

http://www.worldwatch.org Washington-based institute with focus on environment and development

http://www.guardian.co.uk/global-development Guardian Newspaper UK section on development

[*'The early warning signs are already visible. Today, we are witnessing at first-hand what could be the onset of major human development reversal in our lifetime.'*]

Chapter 3

SUSTAINABILITY: THE DEFINING DEVELOPMENT ISSUE

Roland Tormey

With the concept of sustainable development at its core, this chapter highlights the fundamental interconnections between environment and development and the unsustainable nature of current development models and strategies. The chapter explores different definitions and understandings of 'sustainable development'; the evidence and argument on climate change and its impact; the evidence of ecological foot-printing and the challenge of maintaining bio-diversity. The chapter presents a case study of the current threats to the Amazon Forest as not simply a local, but rather a global issue.

KEYWORDS

Global warming, **climate change**, greenhouse gases, **sustainable development**, 'tipping points', **ecological footprints**, biodiversity, **the Amazon**, the 'global commons'

INTRODUCTION

THE CHANGING CLIMATE ON THE MOUNTAINS OF THE MOON

The Rwenzori Mountains lie on the border between Uganda and the Democratic Republic of Congo. The mountains soar to a height of over 5,000 metres – higher than the highest mountain in the European Alps. Despite being close to the equator, the highest peaks are high enough to have a permanent covering of snow and ice. Almost 2,000 years ago the snow-covered peaks were famous as far away as Europe, where the geographer Ptolemy wrote of these mountains in the heart of Africa which he called the Mountains of the Moon.

Now, the ice caps on the Rwenzori Mountains are melting and, according to Richard Taylor of University College London, may be vanishing completely. By 2005 the glaciers were half the size they were twenty years previously, and one-sixth the size they were a century ago. At present rates of melting, the ice on the Mountains of the Moon will be completely gone by 2025.

The Intergovernmental Panel on Climate Change (IPCC) is a group of internationally renowned scientists, appointed by governments to determine independently what the evidence says about climate change. They have concluded that warming 'of the climate system is unequivocal, as is now evident from observations of increases in global average air and ocean temperatures, widespread melting of snow and ice and rising global average sea level'. The earth's temperature has changed many times over its history for purely natural reasons; however, the IPCC conclude that something different is now happening, 'most of the observed increases in global average temperatures since the mid-20th century

are 'very likely', they say, to be due to greenhouse gas emissions from human sources'.

The region of the Rwenzori Mountains, like Uganda and sub-Saharan Africa more generally, has not contributed significantly to these greenhouse gas emissions. In fact, the biggest contributors to greenhouse gas emissions are industrialised western countries and oil-producing countries (see map, page 58). As a result of a cruel trick of our global climate system, however, these areas are some of the first to be affected. The IPCC have noted that Africa is one of the global regions most likely to be affected by climate change. Climatic change is important everywhere but is a matter of life and death in much of Africa. In a country like Uganda, for example, over 80% of the population live off agriculture, and agriculture is completely dependent upon the climate. The comparatively low levels of technology used in much subsistence farming means that it will be much harder to adapt to the effects of climate change.

According to the IPCC the likely effects of climate change on African countries include:

- A loss of fertile land due to water shortages and a reduction in the number of times crops can be harvested each year due to changes in weather patterns. This could lead to a drop in crop production in some areas by as much as 50% by 2020. Small-scale, poorer farmers are those most likely to be affected.

- Increases in extreme weather events like floods or droughts. These can damage crops and homes, can kill and maim and can destroy livelihoods. Possible increases in sea water levels will also make costal areas more susceptible to flooding. By 2020 it is estimated that 50 million people will live in a string of costal cities stretching from Accra to the Niger delta in West Africa. Areas like these will be susceptible to flooding with Alexandria in Egypt and Lagos in Nigeria likely to be significantly affected. This will probably have a greater impact upon poorer people as they are most likely to be in vulnerable areas.

- The spread of diseases into regions that were not previously badly affected as changing weather patterns affect the spread of insects and other carriers. Malaria may become more prevalent in areas not previously affected, like southern Africa and in highlands in Kenya, Rwanda, Burundi. Other diseases like dengue fever, meningitis and cholera may also spread.

- A reduction in the availability of water and an increase in the number of people who have difficulty in accessing safe water, particularly in the north and in the south of Africa. By 2050 it is estimated that some 350 million to 600 million people on the continent will be at increased risk of water problems.

- A loss of forests and changes in grasslands and in marine environments. These, in turn, are likely to lead to the extinction of a large number of species. It is estimated that by 2080 25% to 40% of mammals in national parks in sub-Saharan Africa will be endangered. This in turn will make Africa far less attractive as a tourist destination and will damage the tourism industry.

This chapter explores and debates some of the key issues as regards sustainable development; what the term means and how it is understood; the debate on 'limits to growth'; the challenge of climate change and how it relates to development; ecological foot printing and what it reveals; deforestation; the debate on the 'global commons' and the issue of biodiversity.

SUSTAINABLE DEVELOPMENT – WHAT DOES IT MEAN?

'Among the hidden costs (of development) are the new burdens created by ecological destruction, costs that are invariably heavier for women, both in the North and the South. It is hardly surprising, therefore, that as GNP rises, it does not necessarily mean that either wealth or welfare increase proportionately ... In actual fact, there is less water, less fertile soil, less genetic wealth as a result of the development process.'

Vandana Shiva in Micheline R. Ishay, (ed., 1997) The Human Rights Reader, Routledge, London

There is a cruel irony at the heart of development: most efforts to improve the living standards of people around the world in the last century have gone hand in hand with increases in industrialisation, in energy usage and in the output of certain gasses ('greenhouse gasses') which are produced by some types of agriculture and when fossil fuels like coal or oil are burned. These greenhouse gasses contribute significantly to climate change and a global temperature increase. This, in turn, increases the risks experienced by the poorest people on the planet. In this way, 'development' in the West in the present will have to be paid for by increased poverty, ill-health and danger for the poorest in the next generation.

In the 1980s this realisation led world leaders to argue that 'development' (i.e., improvements in living standards that was paid for by the poorest of the poor and by future generations) was not enough; what was needed was a model of development that would bring a decent living standard to all in the present and enable future generations to also have a decent living standard. This idea became referred to as 'sustainable development'. It was defined in 1987 by the World Commission on Environment and Development (often called the Brundtland Commission because it was chaired by a former Prime Minister of Norway, Gro Harlem Brundtland) as follows:

'…development that meets the needs of the present without compromising the ability of future generations to meet their needs'.

Professor Charles Hopkins of York University in Toronto often uses a simpler definition: *'Enough, for all, forever'.*

There are a number of important components in this definition:

- Development must meet the needs of people in the present. This means addressing the huge inequalities in wealth and poverty that exist in the present.

- Development must meet the needs, but not necessarily the wants, of people in the present. Given the limited resources available to people on planet earth (and the limits of current technology to make the most of these resources) a fairer distribution of resources may mean that people in western countries cannot continue to consume energy and resources and to pollute in the way we do at present.

- Development requires equality between generations, in that we cannot meet the needs of the present by utilising so much of the earth's resources that we hinder the chances of future generations to meet their needs. This is sometimes called 'trans-generational equality' and it means paying attention to the capacity of the planet to regenerate resources or to absorb the pollutants caused by human actions.

Not everyone agrees that the Brundtland Commission's definition of sustainable development is the best one available and many would argue about some of the core concepts that form part of the definition. In order to better understand these arguments it is necessary to look at the different models that exist for making sense of the appropriate relationship between humans and the planet. One critic, Timothy O'Riordan suggests that there are four different theories as to how humans should relate to the earth, two of which, he says, places ecological laws at the centre of their approach and identifies that humans are subject to these laws (he classifies these approaches as 'eco-centric') and two of which place humans and their capacity to adapt the world to their needs at the centre of the approach (he classifies these as 'techno-centric').

As you will see from the table opposite, the Brundtland Commission's report would tend to be characterised as based on the 'accommodation' approach and would be described by O'Riordan as 'techno-centric'. When reviewing the table, it is useful to reflect on which of the approaches described best captures your own understanding of the relationship between humans and the environment.

TABLE 3:1 - What sort of environmentalist are you?

ECO-CENTRIC

	Gaianism	Communalism
Basic tenets	The earth is a system with many different parts The body of the earth is called Gaia, and we are all part of her Gaia and her parts (animals, plants, rocks, soil) have rights that need to be protected	Global capitalism and international organisations, such as the WTO and the World Bank, are destroying the earth Big technology, such as nuclear power, is out of control We need to move towards small-scale, self-reliant communities We need to use renewable resources and 'appropriate technology'
Who holds this view?	Scientist and environmentalist James Lovelock and often radical greens	Economist E.F. Schumacher, many radical socialists and environmentalists
How widespread is it?	0.1-3% of population in various opinion surveys	5-10% in various opinion surveys
What are they likely to say?	The *"entire range of living matter on Earth…could be regarded as a single living entity, capable of manipulating the Earth's atmosphere to suit its overall needs. This organism, of which human society is a part, but only one part, regulates her activities in a very complex and subtle way"* (J Lovelock, 1979).	*"Ever bigger machines, entailing ever bigger concentrations of economic power and exerting ever greater violence against the environment, do not represent progress: they are a denial of wisdom. Wisdom demands a new orientation of science and technology towards the organic, the gentle, the non-violent, the elegant and beautiful."* (E F Schumacher, 1973).

TECHNO-CENTRIC

	Accomodation	Intervention
Basic tenets	Humans have the right to exploit the earth's resources for their well-being. The earth's resources are finite and must be properly managed. We must carefully assess the potential dangers to the environment and their knock-on effects of humans. We must manage human development to accommodate ourselves to environmental limits.	The earth is robust and can survive human interventions. When resources become scarce, supply and demand will make them expensive and will thereby reduce demand. There will then be a market for businesses to develop new products that will fill any gaps. Business will fund research to find new ways of coping with scarcity and for developing new technologies. There is, therefore no need for major social change, our current way of working will solve any problems.
Who holds this view?	Environmental scientists and liberal-socialist politicians, the Brundtland Commission report of 1987 and many UN Human Development Reports.	This view is often associated with business leaders, right-wing politicians and with some scientists and engineers.
How widespread is it?	55 – 70% in various opinion surveys.	10 – 35% in various opinion surveys.
What are they likely to say?	*"Human Beings are at the centre of concerns for sustainable development. They are entitled to a healthy and productive life in harmony with nature"* (Principle 1 of the 'Rio Declaration of Environment and Development', 1992).	*"It is time for the technical community to abandon its attempts to accommodate irreconcilable opponents [greens] and instead aim to re-establish the idea of scientific authority…"* (M. Grimston, 1990 in the UK Atomic Energy Authority's Magazine Atom).

Source: Adapted and developed from T. O'Riordan (1989) 'The Challenge for Environmentalism' and from D. Pepper (1996) Modern Environmentalism: An Introduction.

ARE THERE LIMITS TO GROWTH?

For much of the period since the beginning of industrialisation, the environment has largely been taken for granted by economists and other writers on development. One exception was the British economist, Thomas Malthus, who wrote around the end of the eighteenth century that, because population had a tendency to increase at a quicker rate than food supply, there would inevitably be 'positive checks' such as starvation, famine or disease that would increase the death rate, or 'preventive checks', like late marriage, which would reduce the birth rate. Both, he felt, were characterised by 'misery and vice'. Although events like the Irish Famine of the 1840s seemed to some people to confirm Malthus' theory, Malthus was broadly inaccurate and the population of the earth has grown massively since his writings, as has food supply. The error in Malthus' theory was that he did not pay sufficient attention to the capacity for new technology to increase food supply to match the increase in population. Today some people argue that we now need to look again at Malthus in the context of world population now reaching 7 billion.

In 1972 a report called The Limits to Growth was published, once more resurrecting the Malthusian hypothesis. Using more modern modelling techniques and factoring in a range of different types of scenario the team argued that the use of resources, industrial output, population and pollution are growing at an ever increasing rate, while the capacity of the earth to accommodate such growth is limited. Even allowing for new technologies, continued economic growth would lead to breaching the earth's limits and a consequent collapse. Although the Limits to Growth report was widely criticised for its methodology and its assumptions, it remained influential (as did the sequel reports, published 20 and 30 years after the original). The authors argued that the key to survival was having accurate feedback mechanisms that would enable impending trouble to be spotted and evasive action taken.

With world population topping 7 billion with suggestions that it will reach something approximating 9 billion by 2050 (although this 'high growth' projection is contested by many), the debate on the relationship between population growth and environment is significantly back on the agenda. Issues of particular concern include climate change, water scarcity, biodiversity loss, rising energy prices, and food security. In a joint statement in 1993, representatives of 58 national scientific academies stressed the complexities of the population-environment relationship and concluded that:

As human numbers increase, the potential for irreversible changes of far-reaching magnitude also increases. ... In our judgment, humanity's ability to deal successfully with its social, economic, and environmental problems will require the achievement of zero population growth within the lifetime of our children.'

While some continue to argue for population controls (especially in the Developing World), the 1994 UN Conference on Population (held in Cairo) abandoned 'control' strategies in favour of those focusing on the health, rights, and well-being of women, especially as regards access to education, family planning services (where international investment was more than halved between 1995 and 2007) and greater equality. The current debate on population and environment is seen to be rooted in our understandings of both human development and human rights especially the rights of women.

✎ MORE INFO

UN Department of Economic and Social Affairs Population Division (2004) World Population to 2300, New York; Robert Engelman (2011) 'An End to Population Growth: Why Family Planning Is Key to a Sustainable Future', Solutions 2:3, April and, for a useful review of many of the main points in the debate on world population see http://www.economist.com/debate/days/view/364

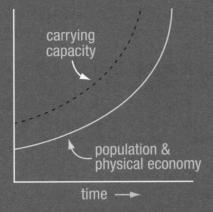

carrying capacity

population & physical economy

time →

What are the possible scenarios for growth and its limits?

Where technology can enable the earth's carrying capacity to be increased along with resource use, industrial output, population and pollution, then the limits will never be reached and growth can continue indefinitely.

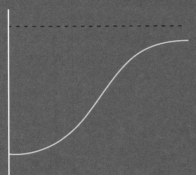

Where there is a limit to the extra carrying capacity that can be developed by technology, and where the feedback mechanisms that would alert people to such limits are quick, accurate and responded to by political leaders, then the limit will be approached but never breached.

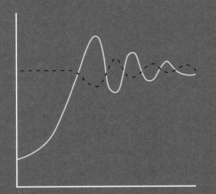

Where the feedback mechanism is delayed, inaccurate or ignored and where the earth can quickly recover from damage, then there will be a period of successive waves of growth and collapse, with economic activity eventually settling near the limits of growth.

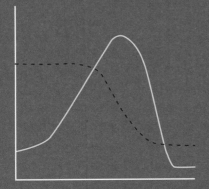

Where the feedback mechanism is delayed, inaccurate or ignored and where the earth cannot recover quickly from damage there will be a significant collapse in people's well being after growth is breached.

BREAKING THE LINK BETWEEN DEVELOPMENT AND CLIMATE CHANGE

Below are two maps: the first maps how countries measure on the Human Development Indicator – one commonly used measure of development. The second shows the size of countries relative to its emissions of carbon dioxide – the bigger the country the more its output of greenhouse gasses. What do the two maps tell us?

Human Development Indicator (HDI – 2010)

Source: http://hdr.undp.org/en/data/map/

Global variation in CO$_2$ emissions

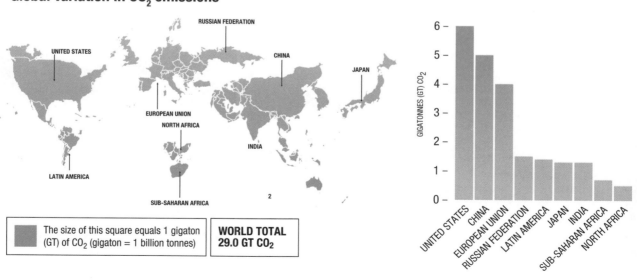

The size of this square equals 1 gigaton (GT) of CO$_2$ (gigaton = 1 billion tonnes)

**WORLD TOTAL
29.0 GT CO$_2$**

Source: Based on Map 1:1, UNDP Human Development Report 2007

DEVELOPMENT AND CLIMATE CHANGE – AN OVERVIEW

The United Nations Development Programme provided a succinct overview of the intimate relationship between development and climate change in its Human Development Report for 2007. For the UNDP, climate change is characterised as the 'defining human development issue of our generation' with the capacity to erode human freedoms and to limit choice. Their assessment is stark:

'The early warning signs are already visible. Today, we are witnessing at first-hand what could be the onset of major human development reversal in our lifetime. Across developing countries, millions of the world's poorest people are already being forced to cope with the impacts of climate change. These impacts do not register as apocalyptic events in the full glare of world media attention. They go unnoticed in financial markets and in the measurement of world gross domestic product (GDP). But increased exposure to drought, to more intense storms, to floods and environmental stress is holding back the efforts of the world's poor to build a better life for themselves and their children.'

The need to address climate change in the context of development is based ultimately on core threats to two particular groups, both with little or no political influence today - the world's poor and future generations. For the UNDP, climate change raises profoundly important questions about social justice, equity and human rights across countries and generations.

'Dangerous climate change is the avoidable catastrophe of the 21st Century and beyond. Future generations will pass a harsh judgement on a generation that looked at the evidence on climate change, understood the consequences and then continued on a path that consigned millions of the world's most vulnerable people to poverty and exposed future generations to the risk of ecological disaster.'

Having argued that the evidence of overloading the carrying capacity of the planet as regards greenhouse gases (exceeding the natural range of the past 650,000 years) and citing the dangers associated with an average global temperature threshold of 2°, beyond which irreversible damage is very difficult to avoid, the Report states 'we are recklessly mismanaging our ecological interdependence'. Rich countries account for the bulk of emissions of greenhouse gases but yet it is poor countries that pay the highest price; if all of the world's people generated such gases at the same rate as some developed countries, we would need nine planets to absorb it.

'When people in an American city turn on the air-conditioning or people in Europe drive their cars, their actions have consequences. Those consequences link them to rural communities in Bangladesh, farmers in Ethiopia and slum dwellers in Haiti. With these human connections come moral responsibilities, including a responsibility to reflect upon - and change - energy policies that inflict harm on other people or future generations.'

The current climate challenge has three distinctive features:

- The combined force of inertia and cumulative outcomes of climate change - once emitted, carbon dioxide (CO_2) and other greenhouse gases stay in the atmosphere for a long time
- Urgency is the second feature of the climate change challenge
- The third important dimension is its global scale

'One tonne of greenhouse gases from China carries the same weight as one tonne of greenhouse gases from the United States - and one country's emissions are another country's climate change problem. It follows that no one country can win the battle against climate change acting alone. Collective action is not an option but an imperative.'

There are grounds for optimism however – whereas before there was widespread debate as to whether or not climate change was taking place and whether or not it was human-induced; now that debate is almost over; there is overwhelming scientific consensus that climate change is both real and man-made and almost all governments are part of that consensus. So, change is on the agenda and the debate now is on the degree of change and whether it is enough or not to make a real and sustained difference. 2012 marks the end of the Kyoto Agreement, and developed countries now need to take the lead with the major *'emerging'* developing countries playing an increased role.

If we continue with the *'business as usual'* model of development the likelihood is that we will trigger large-scale reversals in human development, undermining livelihoods and causing mass displacement; by the end of the 21st Century, *'catastrophic'* ecological impacts could have moved from the bounds of the possible to the probable. Evidence of accelerated collapse of ice sheets in the Antarctic and Greenland, acidification of the oceans, the retreat of rainforest systems and melting of Arctic permafrost all have the potential - separately or in interaction - to lead to *'tipping points'*. With 15% of world population, rich countries account for almost half of emissions of CO_2 and high growth rates in India and China mean rapid *'catching up'*. However, the carbon footprint of the US remains 5 times that of China and over 15 times that of India while in Ethiopia, the average per capita carbon footprint is 0.1 tonnes of CO_2 compared with 20 tonnes in Canada, while overall emissions continue to rise globally. Deep adjustments are needed. And yet, current estimates suggest a carbon intensive energy infrastructure with coal playing the dominant role will be the focus of investment for the immediate decades ahead – the implications of this are highly significant.

'We estimate that avoiding dangerous climate change will require rich nations to cut emissions by at least 80% with cuts of 30% by 2020. Emissions from developing countries would peak around 2020, with cuts of 20% by 2050.'

The cost of achieving this would amount to an average of 1.6% of GDP per year.

Like so much else internationally, vulnerability to climate shocks is unequally distributed: climate change is 'ratcheting up' the risks and vulnerabilities facing the poor placing further stress on already over-stretched coping mechanisms and trapping people in deprivation and strategies for coping with climate change can deepen deprivation.

The Report identifies 5 key mechanisms through which climate change impacts on development negatively:

- Agricultural production and food security - climate change will affect rainfall, temperature and water availability for agriculture in vulnerable areas.

- Water stress and water insecurity – changed run-off patterns and glacial melt will add to ecological stress, compromising flows of water for irrigation and human settlements in the process.

- Rising sea levels and exposure to climate disasters - sea levels could rise rapidly with accelerated ice sheet disintegration.

- Ecosystems and biodiversity - climate change is already transforming ecological systems; around one-half of the world's coral reef systems have suffered 'bleaching' as a result of warming seas.

- Human health - rich countries are already preparing public health systems to deal with future climate shocks, such as the 2003 European heat wave and more extreme summer and winter conditions.

ENVIRONMENTAL TIPPING POINTS

- THE MELTING OF SUMMERTIME SEA ICE IN THE ARCTIC, WHICH WILL AMPLIFY GLOBAL WARMING

- A DECLINE IN SIZE OF GREENLAND'S ICE SHEET.

- THE WEST ANTARCTIC ICE SHEET SLIPPING INTO THE OCEAN

- THE DRYING UP AND WITHERING OF BOREAL FORESTS

- A DECREASE IN THE AMAZON RAINFOREST'S RAINFALL

- AN INCREASE IN THE EL NIÑO EFFECT, CAUSING DROUGHT IN SOUTHEAST ASIA

- A STRENGETHNING IN INDIA'S MONSOONS, AS WARMER AIR CARRIES MORE WATER

- ANY SHIFT IN THE WEST AFRICAN MONSOON WILL IRREVERSIBLY DAMAGE THE SAHARA DESERT

- A DISRUPTION OF THE GULF STREAM WILL LEAD TO A COOLER AND DRIER EUROPEAN CLIMATE

UNDP Human Development Report, 2008

These mechanisms will interact with wider social, economic and ecological processes that shape opportunities for human development.

The Report argues that setting credible targets, backed up by clear policies linked to global mitigation goals is the starting point for the transition to a sustainable emissions pathway but, in this regard the record to date is not encouraging. Areas where significant positive change would be both effective and feasible include pricing carbon emissions, carbon trading, renewable energy, personal transportation, international trade, low carbon technologies, increased energy efficiency and deforestation. In this regard, public information is crucial as is the development of infrastructure to manage risk, especially in developing countries.

The Report concludes that we can avoid 21st Century reversals in human development and catastrophic risks for future generations, but only by choosing to act with a sense of urgency – that sense is currently missing. *'Governments may use the rhetoric of a 'global security crisis' when describing the climate change problem, but their actions – and inactions - on energy policy reform tell a different story'.* Political leadership is vital when we are confronted by *'what may be the gravest threat ever to have faced humanity'.*

See: UNDP (2007) Human Development Report 2007/2008: Fighting climate change: Human solidarity in a divided world, Basingstoke, Palgrave Macmillan.

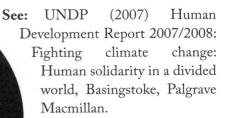

ECOLOGICAL FOOTPRINTS

One way to attempt to measure the impact that human development is having on the planet and on future generations is what is known as an 'ecological footprint'. This statistic seeks to measure the amount of the planet that each person would need to sustainably support:

- the food, timber and other resources they use
- the energy they consume
- the area necessary to absorb their waste
- and the space they use for built-up land

Added together this estimates how much space a person takes up on the planet – their footprint.

The Worldwide Fund for Nature (WWF) estimates that the earth has a productive capacity of about 1.8 global hectares per person. In 2007, they estimated that each person on the planet was using an estimated 2.7 global hectares, an overshoot of about 50%. In fact they estimate than every year since the 1970s the earth's population has been using more of the earth's resources than can be done sustainably.

So, where do the extra resources come from?

The ecological footprint measures how much we can sustainably use. We can use more, though in doing so we damage the capacity of the ecological and natural systems of the planet to restore themselves. This means, in effect, taking from future generations.

1:32 Comparative Carbon Footprint

I SAID, "YOUR TURN"...

SOUTH

NORTH

Figure 3.1 - Ecological footprint by political group

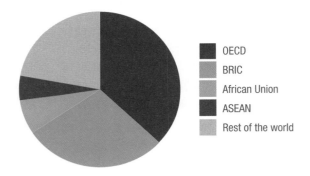

OECD
BRIC
African Union
ASEAN
Rest of the world

Source: Global Footprint Network, 2010

The estimated 2.7 global hectares is an average figure across all people on the planet. It hides a considerable degree of variation. Oil producing countries like the United Arab Emirates and Qatar have ecological footprints of more than 10 hectares per person, about four times the global average and more than five times the sustainable level. Western industrialised countries like Denmark, Belgium, the United States of America have ecological footprints of about 8 hectares per person; Ireland has a footprint of about 6 hectares per person while countries like the United Kingdom and Germany use about 5. At the other end of the scale, countries like Zambia, Malawi, Eritrea, Haiti and India all use far less than the global average.

In total, 37 countries account for about two-thirds of the ecological footprint. In 2007, the 31 OECD countries — which include the world's richest economies — accounted for 37% of humanity's Ecological Footprint. In contrast, the 10 ASEAN countries and 53 African Union countries — which include some of the world's poorest and least developed countries — together accounted for only 12% of the global Footprint (see Figure 3.1).

As well as reflecting the amount of goods and services consumed and CO_2 waste generated by the average resident, Ecological Footprint is also a function of population. As shown in Figure 3.2, the average per-person Ecological Footprint is much smaller in BRIC countries than in OECD countries; however, as there are over twice as many people living in BRIC countries as in OECD countries, their total Ecological Footprint approaches that of OECD countries. The current higher rate of growth in the per person Footprint of BRIC countries means these four countries have the potential to overtake the 31 OECD countries in their total consumption.

Figure 3.2 - Ecological foorprint per person by political group

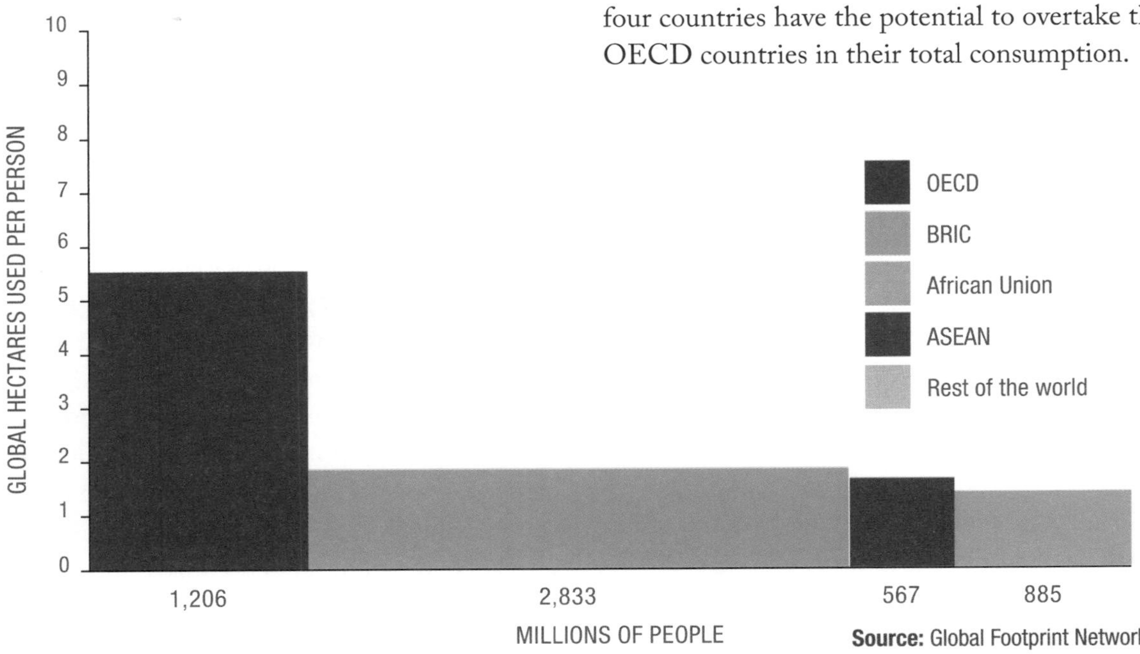

GLOBAL HECTARES USED PER PERSON

10
9
8
7
6
5
4
3
2
1
0

1,206   2,833   567   885

MILLIONS OF PEOPLE

OECD
BRIC
African Union
ASEAN
Rest of the world

Source: Global Footprint Network, 2010

BIODIVERSITY

'The library of life is burning and we don't even know the titles of the books'

Gro Harlem Brundtland

There are an estimated 5 to 30 million different species on planet earth. Only a minority of these have been identified and named.

Different plants and animals are often tangled together in an intricate and delicate balance. If a species is taken from this web, or a new species introduced, the impact upon this balance can be huge. For example, the Nile Perch was introduced into Lake Victoria in east Africa around 1970 as it was thought to be a species that would provide an excellent source of protein for fishing. Instead it turned out to be a voracious predator that destroyed the populations of existing fish. Of the 110 different species found in fishing nets between 1978 and 1982, only 10 were found in 1987. While there have been efforts to restrict the impact of the Nile Perch in more recent years, it is estimated that dozens of species of fish have been wiped out, with a huge impact upon those who seek to make their living on and around the lake.

The WWF compile a 'Living Planet Index' in order to assess the survival of diverse species on the planet. In the same way as a stock market index tracks the health of various share prices, the Living Planet Index seeks to provide a single measure of the health of biodiversity by calculating the changes in population across a range of nearly 8,000 species.

The Living Planet Index shows a decline of about 30% between 1970 and 2007. This hides two very different pictures:

- The index for temperate and polar regions of the earth has actually increased by 29% since 1970.
- The index for tropical regions shows a 60% decline since 1970

What makes this particularly worrying is that tropical regions support two-thirds of all plant species and tropical rainforests alone are thought to be home to between 50 and 90 percent of all species.

There are five major threats to biodiversity:

1. Habitat loss or alteration. This happens through conversion of lands for agriculture or for industrial or urban use, through damming and other changes to rivers, such as for irrigation or hydroelectric power.
2. Over-exploitation of wild species. This happens when species are harvested at a rate quicker than they are replaced, for example the almost total eradication of cod stocks from the North Atlantic due to over-fishing.
3. Pollution. The use of pesticides and excessive fertiliser in agriculture, human waste and industrial pollutants all lead to loss of species.
4. Climate change. The changing of weather patterns leads to the loss of habitat and to loss of species.
5. Invasive species: New species introduced deliberately (like the Nile Perch in Lake Victoria) or accidentally can lead to significant loss of biodiversity.

DEFORESTATION: A BRAZILIAN CASE STUDY

All major tropical forests - including those in the Americas, Africa, Southeast Asia, and Indonesia - are significantly disappearing, largely to make way for human food production, including livestock and crops. Although meeting some of the current needs of development, such deforestation also has long-term, profound and sometimes devastating consequences including conflict, human rights abuses, habitat destruction (leading to the extinction of plants and animals) and climate change. These challenges do not simply affect the immediate region but also have global implications. The current debate on deforestation in the Amazon region of Brazil illustrates many of these points and the debates and challenges they highlight.

The Amazon region makes up 61% of Brazil's land area (5.3 million square kilometres) and has an estimated population of 20 million; the region has the largest continuous tropical forest in the world and is home to about 20% of the world's plant and animal species. Deforestation in the Brazilian Amazon is the result of a series of activities including mining, logging, subsidies for cattle ranching, infrastructure development, land tenure conflicts, poor law enforcement and the high price of grains and meat. In recent years, the massive expansion of cattle ranching, spurred initially by the domestic market but more recently by international demand has been a major issue.

In March 2004, Brazil's President Luis Inácio (Lula) da Silva announced a major new Action Plan to Prevent and Control Deforestation in the Legal Amazon, the plan committed the government to spending US$ 135 million dollars to reduce deforestation. It included action on land use planning, greater enforcement of laws concerning deforestation and the illegal occupation of government lands, deforestation monitoring, reviews of public infrastructure investments, support for indigenous territories and community forestry, support for sustainable agriculture and greater control over credit for ranchers. The Plan was in response to a rise in the deforested land area in the Brazilian Amazon from 41.5 million hectares in 1990 to 58.7 million hectares in 2000 (in just 10 years the country lost an area of forest twice the size of Portugal). Despite optimistic assessments, the rate of deforestation again skyrocketed in 2002.

While there is concern about the expansion of soybean cultivation into the Amazon, it explains only a small percentage of total deforestation; the total area of soybeans in the Legal Amazon in 2002 was only 4.9 million hectares (concentrated in previously deforested areas), while the area in pasture was almost ten times more that amount. According to the Centre for International Forest Research (CIFOR, based in Indonesia) logging rarely leads directly to deforestation as most loggers only remove a small number of trees per hectare; while this often damages the forest, it does not destroy it. Indirectly, logging does contribute to deforestation by making it easier for forests to catch fire and for farmers to move into forested areas. However, for CIFOR, logging is much less damaging than the growth of cattle ranching and they highlight the fact that in 2000 for every one hectare of cropland there were almost six hectares of cattle ranching suggesting a direct correlation between deforestation and ranching.

Cattle expansion in the Amazon in the last two decades has been phenomenal; in the 1990's alone the number of cattle more than doubled, from 26 million in 1990 to 57 million in 2002. The vast majority of the new cattle were concentrated in Brazil's Amazon states of Mato Grosso, Pará, and Rondônia, which also have the highest rates of deforestation. Brazil has become one of the world's largest meat and beef exporters and this has had profound implications for the Amazon region.

The increase in beef production is very recent; in fact the Amazon region did not even produce enough beef to feed its own population until 1991. Subsequently, the region began to produce a surplus for national consumption and exports abroad played only a minor role; as recently as 1995, Brazil exported less than $500 million dollars of beef but by 2003, Brazil was exporting three times as much. Between 1997 and 2003, the volume of exports increased more than fivefold to 1.2 million metric tons and is expected to top 1.8 million in 2011.

According to Tasso Azevedo (Director of the

Brazilian Forest Service until 2009), agribusiness has been one of the key drivers for the 'development' of the region including that of a new infrastructure network especially roads. Current government plans include road paving, new hydropower projects and construction of waterways and ports; a development plan that has the potential to alter drastically the social, economic and environmental landscape of the region. For example, recent studies have shown that more than 70% of deforestation occurs within 50 km of paved roads, while at most 7% occurs along unpaved roads. The deforestation facilitated by road pavement and low law enforcement could also dramatically increase the annual net carbon emissions from the Amazon. The trees in the Amazon forests contain 60 to 80 billion tonnes of carbon, more than the global emissions generated by humans in a decade. Deforestation in the Brazilian Amazon alone releases about 200 million tonnes of carbon annually, accounting for 3% of global net carbon emissions and 70% of national emissions. Recent research suggests that if the 'business as usual' model continues 32 billion tonnes of carbon would be emitted by 2050 (equivalent to four years of current global annual emissions) contrasted with 15 billion tonnes of carbon under a more restrictive government strategy, which is experiencing strong opposition from vested interest groups.

According to Azevedo:

'Forest exploitation and conversion have not brought true development, employment opportunities, better income distribution for local populations or environmental benefits to the region. Currently, about 45% of the population of the Brazilian Amazon has income below the poverty line.'

Since 2004, combined government action and international market trends and pricing have reduced deforestation by 52% but debates about the future of the region are stark. At one extreme is a 'business-as-usual' scenario which includes the pavement of 14,000 km of roads up to 2027, continuing low levels of law enforcement, ongoing agricultural expansion and population growth with in-migration. According to the scenario, 40% of Amazon forests would be lost by 2050. At the other extreme of the debate is a strong government directed scenario which includes pavement of 11, 500 km of roads along with increased law enforcement, agro-ecological zoning (preventing agricultural expansion onto inappropriate areas or 'protected' areas – now some 48% of the Amazon region) and expansion and conservation of protected areas.

The Amazon remains one of the world's most important bulwarks against global warming with estimated 300m hectares of jungle remaining, or 75% of the original forest cover. Current government plans to build a US$16 billion dam, a new Forestry Bill that would give amnesty to those clearing land illegally prior to 2008 and the continued expansion of cattle ranching pose a major threat. According to Marina Silva, a former Green party presidential candidate and environmental minister – *'This is a global question not just a Brazilian one.'*

✎ MORE INFO

David Kaimowitz, Benoit Mertens, Sven Wunder and Pablo Pacheco (2004) Hamburger Connection Fuels Amazon Destruction: Cattle ranching and deforestation in Brazil's Amazon, Center for International Forestry Research, Jakarta, Indonesia - www.cifor.org

C. Azevedo-Ramos (2007) 'Sustainable development and challenging deforestation in the Brazilian Amazon: the good, the bad and the ugly', Vancouver, Canada.

READING

D. Chivers (2011) No Nonsense Guide to Climate Change, Oxford, New Internationalist

Peter Singer (2002) One World: the Ethics of Globalisation, New Haven and London, Yale University Press

E. F. Schumacher (1973) Small is Beautiful: Economics as if People Mattered, New York, Harper Perenniel (1989 edition)

Vandana Shiva (2005) Earth Democracy; Justice, Sustainability, and Peace, New York, South End Press

UNDP (2007) Human Development Report 2007/2008: Fighting climate change: Human solidarity in a divided world, Basingstoke, Palgrave Macmillan

Worldwatch Institute State of the World Report 2011 (annual), New York and London, Norton

World Wildlife Foundation, in association with the Zoological Society of London and the Global Footprint Network (2010) Living Planet Report 2010 Biodiversity, biocapacity and development.

MORE INFORMATION & DEBATE

http://www.worldwatch.org - Worldwatch Institute in Washington

http://www.greenpeace.org/international/en/ - Greenpeace International site

http://www.ipcc.ch - home of the Intergovernmental Panel on Climate Change

http://www.guardian.co.uk/global-development - Guardian Newspaper UK section on development (includes a considerable amount on sustainable development)

http://www.unep.org - site of the United Nations Environment Programme

[*'Peace, in the sense of the absence of war, is of little value to someone who is dying of hunger or cold. It will not remove the pain of torture inflicted on a prisoner of conscience. It does not comfort those who have lost their loved ones in floods caused by senseless deforestation in a neighbouring country. Peace can only last where human rights are respected, where the people are fed, and where individuals and nations are free'*]

The Dalai Lama

Chapter 4

HUMAN RIGHTS AND DEVELOPMENT 'A RIGHT, NOT AN ACT OF CHARITY'

Omar Grech

Having sketched the evolution of the concept of human rights and its emergence to the foreground of international politics post 1945, this chapter focuses primarily on the intrinsic relationship between development and human rights. Inter-alia, it reviews the Universal Declaration of Human Rights and its associated Covenants; regional human rights institutions in Europe and Africa (and the debates surrounding each); the International Criminal Court and the 1986 Declaration on the Right to Development. The chapter also explores what human rights conversations and actions have brought to the human development process and vice-versa.

KEYWORDS

Human rights, **principles**, frameworks and instruments, **universal**, indivisible and inalienable, **civil and political rights**, social, **economic and cultural rights**, the right to development, **European Convention on Human Rights**, African Charter on Human and Peoples Rights, **people's rights and individual rights**

INTRODUCTION

At the very outset, it is necessary to remind ourselves that human rights are not granted by the State nor acquired by agreement; instead they are innate in every human person. The language used in human rights treaties and declarations is that such rights are inherent in every individual by virtue of birth. Commenting on the United States Bill of Rights, US Supreme Court Judge, William J. Brennan noted in 1990:

'The Framers of the Bill of Rights did not purport to 'create' rights. Rather, they designed the Bill of Rights to prohibit our Government from infringing rights and liberties presumed to be pre-existing.'

Such a perspective on rights has profound implications not just for rights per se but for a range of other fundamental issues – including development. In recent decades, the general debate on development has become increasingly influenced by the human rights 'conversation' as well as by human rights' principles, frameworks and instruments. This chapter focuses on human rights and on the ways in which human rights and development impact on each other. The major argument is that development issues need to be examined and analysed through a human rights' lens – implying that real human development is the platform upon which all may enjoy human rights and in this sense human rights and development are one and the same thing.

An understanding of human rights is crucial to a full 'reading' of development – something that is increasingly highlighted, especially in Third World commentary and analysis, with fundamental repercussions for international relations, aid and trade as well as humanitarian interventions. This chapter explores some of these key issues, in particular:

- how we understand and describe human rights
- the evolution of human rights and the role of international law
- the right to development
- some debates and controversies surrounding human rights and the right to development

DESCRIBING HUMAN RIGHTS

Although deceptively simple, defining human rights as 'the rights that every individual possesses simply by virtue of being a human being' is revolutionary in its consequences especially in circumstances such as those of the developing world, where states like to present themselves as the source and origin of all rights and duties, including those related to development. Human rights challenge such assumptions by asserting that the human person is the source of rights independent of, and often in opposition to, state authority. American political scientist, Noam Chomsky comments:

'States are not moral agents, people are, and can impose moral standards on powerful institutions.'

Source: The United States and the "Challenge of Relativity" in Tony Evans (ed.1998) Human Rights Fifty Years On: A Reappraisal, Manchester University Press

If we examine the ways in which rights are acquired in our domestic legal system, it is indeed the State through the laws which it enacts that creates rights or allows rights to be created. The State creates rights through legislation, for instance the State may enact a law which grants children the right to inherit their parents' property. On the other hand the State allows rights to be created when its laws enable individuals to sell or exchange rights. For example contract law allows individuals to acquire rights over property through the consent of two or more individuals.

In recent decades, the UN has begun to emphasise three adjectives in describing human rights; human rights are said to be universal, indivisible and inalienable. The notion of universality of human rights was to the fore in the minds of the drafters of the Universal Declaration of Human Rights of 1948 as the title itself implies - the first paragraph of the Preamble makes an immediate reference to the *equal and inalienable rights of all members of the human family*. In the 1993 World Conference on Human Rights held in Vienna the ideas of indivisibility and interdependence of human rights were highlighted.

'States are not moral agents, people are, and can impose moral standards on powerful institutions.'

Continued on next page →

UNIVERSAL, INDIVISIBLE AND INALIENABLE

UNIVERSAL – if human rights are rights human beings possess by virtue of being human, it follows that all humans possess them - there can be no distinction between human beings based on any criterion such as nationality, sex or sexual orientation, religion, culture etc. This is an important notion in the context of development and the right to development. It links all members of the human race in a chain of rights and responsibilities that have implications for law, justice and morality.

Linked to universality is the notion of equality as emphasised in the Universal Declaration which in Article 1 states, 'All human beings are born free and equal in dignity and rights.' Such ideas - universality, equality and a common humanity - lie at the heart of human rights. Concretely, this means that in the context of development the people of South Africa or India are entitled to the same enjoyment of human rights as people in Ireland or Malta.

INDIVISIBLE – the term human rights refers to a plurality of different rights - economic, civil, social, political and cultural. When we describe human rights as indivisible we mean that all of these various parts form one integral whole which may not be divided. Human rights can therefore be understood as a package which may not be divided into its various components for the purpose of picking and choosing some rights at the expense of others or for the purpose of giving importance to some rights while ignoring the rest. Historically, and especially during the Cold War, the West tended to emphasise civil and political rights at the expense of economic and social rights whilst Communist states tended to reverse the emphasis – with very considerable implications for development.

Indivisibility is linked to the concept of interdependence – meaning that each right forming part of human rights is only fully enjoyed when the other rights are also being enjoyed. That is why we say that human rights are indivisible and interdependent; every right forming part of human rights is equally important, deserves equal protection and promotion and can only be truly enjoyed if all other rights are concurrently implemented.

INALIENABLE – is a legalistic word which simply means 'that which may not be given away'. To alienate means to sell or part with. If something is inalienable it means that it may not be sold or parted with in any other way. Hence when we describe human rights as inalienable we are saying that human rights may not be sold or renounced. This has practical implications for the protection of human rights – for example, if an individual is arrested by the police and is then forced to sign a document in which s/he accepts to be kept under arrest without trial indefinitely, that document would be invalid since no one may alienate or renounce her/his right not to be subject to arbitrary arrest.

THE STORY OF HUMAN RIGHTS

The history of human rights may be viewed from two perspectives. Firstly, there is the history of human rights themselves - lawyers and politicians sometimes like to think that the story of human rights started in 1945 with the establishment of the United Nations or in 1948 with the adoption of the Universal Declaration of Human Rights. However, according to the description of rights above, the history of human rights is essentially as long as that of humanity. The history of human rights can thus be traced through the history of the values which underpin human rights such as freedom, justice and dignity.

Early philosophers and commentators on human rights values included:

- Mo-zi, Chinese, 470BC-390BC, key idea - an individual owes respect to all other individuals and not only to family or clan but universally throughout the world
- Plato, Greek, 427BC-347BC, key idea - the most important thing for a government to enforce is justice. Justice means doing what is right and fair
- Aristotle, Greek, 384BC-322BC, key idea - if liberty and equality, as is thought by some, are chiefly to be found in democracy, they will be best attained when all persons alike share in the government to the utmost
- Cicero, Roman, 106BC-43BC, key idea - justice consists in doing no injury to others and there is decency in giving them no offence
- Al-Farabi, Persian, 870-950, key idea - the perfect city is a society in which all individuals are endowed with rights and live in love and charity with their neighbours.

In addition, all major religions have had important and positive things to say about human rights. The following quote from Micheline Ishay highlights how religion has influenced the development of human rights, especially through a common emphasis on respect for fellow human beings.

'Despite many controversies regarding the origins of human rights, few would dispute that religious humanism, Stoicism, and natural rights theorists of antiquity influenced our secular modern understanding of rights...[M]ost religious texts, like the Bible, Buddhist texts, the New Testament and the Koran incorporate moral and humanistic principles, often phrased in terms of duties...The religious origins of universal ethics are greatly indebted to the Bible, whose teachings are shared by Jews, Christians and Muslims alike...The Ten Commandments represented a code of morality and mutual respect that had far-reaching influence on the Western world...Similar moral and humanistic principles can be found in Buddhism'

Micheline R. Ishay, (ed., 1997) The Human Rights Reader, Routledge, London

In this context then, when we talk of respecting human rights we are essentially thinking about *justice*. A lack of human rights would mean an absence of justice. The genesis of human rights reflected this preoccupation with fairness and equality. The notion of natural law – people are born with rights, and their very existence presupposes the existence of such rights – was originally based upon the concept of natural justice.

Philosophers such as Rousseau and Hobbes, not satisfied with natural law, developed the idea of human rights as part of a social contract – people accept limitations on natural freedoms in exchange for social order and peace. Human rights were part of the *'state of humankind'*, created to ensure a peaceful existence – as well as a just one – that a *'state of nature'* could not. A human rights based approach to development, therefore, is one that is preoccupied with ensuring social justice and equality first and foremost.

Recognising that human rights ideas did not suddenly emerge in 1945, the major developments in the adoption of legal measures, national or international, intended to promote and protect human rights, resulted from political crises, devastating wars or similarly challenging situations. This history of human rights therefore adheres to one of the laws of history outlined by the historian Arnold Toynbee who referred to the historical law of *progress through crisis*.

WORLD WAR II AND ITS AFTERMATH

The well-documented atrocities of the Second World War period forced the international community to reflect on necessary interventions to avoid such atrocities in the future with action being agreed on two fronts - human rights and human responsibilities. The values that were to lay the foundations for the human rights edifice constructed post-1945 had already been developed practically by the Allied Powers during World War II. In particular, President Franklin Roosevelt's 1941 enunciation of the 'Four Freedoms' that ought to prevail globally was an important moment in shaping the post-war agenda. The freedoms outlined by President Roosevelt were:

- freedom of speech and expression
- freedom of worship
- freedom from fear
- freedom from want

The inclusion of the latter two freedoms was especially significant as they departed from the traditional American notion of civil and political freedoms to include conceptions of social, economic and human security, so fundamental to human development.

At the Nuremberg Tribunals of 1945/6, set up to try Nazi war criminals, a key principle was established - individuals have responsibilities and obligations under international law. Therefore, individuals who commit grave breaches of human rights are responsible under international law for such breaches and are liable for them.

The human rights side of the coin was immediately evident in the Preamble to the Charter of the United Nations (1945) which points out that one of the aims of the UN is that of *encouraging respect for human rights and for fundamental freedoms*. The Charter as a whole contains numerous references to the term human rights but there is no definition or description of the term – the task of definition fell to the Commission on Human Rights established by the General Assembly of the UN and led to the formulation of the Universal Declaration on Human Rights, adopted on the 10th December 1948.

The Declaration is of particular interest in a number of respects:

- the Preamble describes the context, the values and the key ideas that underpin the document, in particular it refers to the *'inherent dignity'* of every individual and to *'inalienable rights'*
- the Preamble also introduces the notion of gender equality by referring to the *'equal rights of men and women'*
- it emphasises the equal importance of political and economic rights with phrases such as *'freedom of speech and belief and freedom from fear and want'*

Unlike later human rights documents and instruments, the Declaration brings together both civil and political rights (such as freedom of assembly, of expression and the right to life) as well as economic and social rights (such as the right to work and the right to education). Echoing the Rooseveltian Four Freedoms, the Declaration in its Preamble envisages:

'the advent of a world in which human beings shall enjoy freedom of speech and belief and freedom from fear and want has been proclaimed the highest aspiration of the common people'

Following the adoption of the Declaration, the UN Commission on Human Rights started work on an International Covenant on Human Rights. This Covenant was intended to render the rights listed in the Declaration legally binding on states. The Covenant, unlike the Declaration, would impose direct legal obligation on states.

Originally, the Covenant was intended to be a single text, including both civil and political rights as well as economic and social rights - this would have continued the example set in the Declaration with a single document including all human rights. However, the work of the Commission fell victim to the political agendas of the Cold War. The Western bloc increasingly wanted to give preference to civil and political rights while the Communist bloc insisted on the primacy of economic, social and cultural rights. The notion of the indivisibility of human rights was challenged seriously and persistently.

These political rivalries and Cold War dynamics delayed the drafting of the Covenant and eventually a decision was taken to split the single text into two Covenants:

- the International Covenant on Civil and Political Rights (ICCPR)
- the International Covenant on Economic, Social and Cultural Rights (ICESCR)

These two Covenants were adopted in 1966, 18 years after the adoption of the Universal Declaration and only came into force in 1976, a further 10 years after they were adopted. The United Nations has, over the past 50 years promoted the adoption of a number of treaties and conventions dealing with specific human rights issues. This has been important because it has both kept human rights on the agenda of the UN as well as provided legal protection for individuals suffering racial and gender discrimination, torture, as well as for specific categories, such as children.

A criticism of these Conventions is that the reporting mechanisms they utilise are weak and fail to ensure adequately the implementation of the rights contained in the various Conventions. For instance, none of the Conventions allow for an automatic right of individuals, whose convention rights have been breached, to bring a claim against the state before an international tribunal.

Continued on page 77 →

THREE GENERATIONS OF RIGHTS?

Some commentators have classified human rights into 'three generations of rights' reflecting the different historical periods in which these rights emerged:

- **first generation rights** - associated with civil and political rights (e.g. the right to freedom from arbitrary arrest, to freedom of assembly or freedom of conscience and expression)
- **second generation rights** - those rights which guarantee the economic and social rights of individuals (e.g. the right to health, education, employment and housing)
- **third generation rights** - these rights were not directly included in the Universal Declaration and are usually termed as Group or People's Rights (e.g. the right to self-determination, the right to development etc.,) and are rights which can only be fully achieved within the context of a community. Some commentators deny the status of human rights to this category, claiming that human rights are by definition only those rights pertaining to an individual.

There is now also talk of a Fourth Generation of Rights linked mostly to issues of Intergenerational Justice or the Rights of Future Generations.

CASE STUDY: THE INTERNATIONAL CRIMINAL COURT

Paragraph 9 of the 1986 Declaration on the Right to Development states that the achievement of development is dependent on the *elimination of massive and flagrant violations of the human rights of peoples and individuals*. Article 5 of the Declaration reiterates this by stating that States should take 'resolute steps' to eliminate such massive and flagrant violations. In the same Article we find reference to aggression and war as situations which are likely to result in massive violations of human rights. The necessity of maintaining peace and security in order to achieve true and effective development is emphasised further in Article 7 of the same Declaration.

There is no doubting that wars, international as well as civil, have contributed very significantly to human rights' abuses throughout history. Wars have also been intimately connected to underdevelopment in a number of countries. If we look at the former Yugoslavia and Rwanda in the 1990s we see how both of these armed conflicts have led to death, material and moral destruction, as well as unimaginable human rights abuses.

The establishment of the International Criminal Court in 1998 was a major step in the protection of human rights because for the first time it established a permanent international criminal court where individuals guilty of the worst human rights abuses – genocide, crimes against humanity and war crimes - can be punished.

Genocide is defined as any one of a number of acts (such as killing or preventing births) committed with the intent to destroy a national, ethnic, racial or religious group. The essence of this crime is that it aims to destroy in some way a group of people by reason of their race, ethnicity, nationality or religion. The Holocaust, the genocide of Tutsis in Rwanda and the extermination of Kurds by Saddam Hussein are all examples of genocide.

Crimes against humanity are attacks of a widespread or systematic nature committed against civilians by individuals in pursuance of state or organisational policies. Examples of such acts are extermination, forcible transfer of populations, sexual violence such as rape, and apartheid.

War crimes are, in a nutshell, the most serious breaches of international humanitarian law, the law that regulates armed conflict and provides detailed rules on the means and methods of warfare including the principle that non-combatants (people who are not parties to the conflict or who have ceased to be parties to the conflict due to injury, sickness or surrender) should be protected.

Any person who is suspected of having committed any one of the crimes under the jurisdiction of the court may be tried including heads of state, government ministers, members of parliament and civil servants - no one is immune from prosecution. As of the end of 2011, 119 States are parties to the Statute of the ICC - 26 from Latin America, 33 from Africa, 17 from Asia, 18 from Eastern Europe and 25 from Western Europe and other states.

The ICC may also be viewed as an exercise which, among other things, attempts to increase *'human security'* (see Chapter 11). Human security is a concept which developed since the end of the Cold War and it has, to a considerable extent, become the major security paradigm. It essentially replaces the old state-centric approach to security (which measured security by how safe the state was and felt) with a human-centric approach to security which measures security by how safe individuals are and feel. Human security is linked to both human rights and development by having the human being as its fundamental referent.

THE CRIME OF GENOCIDE

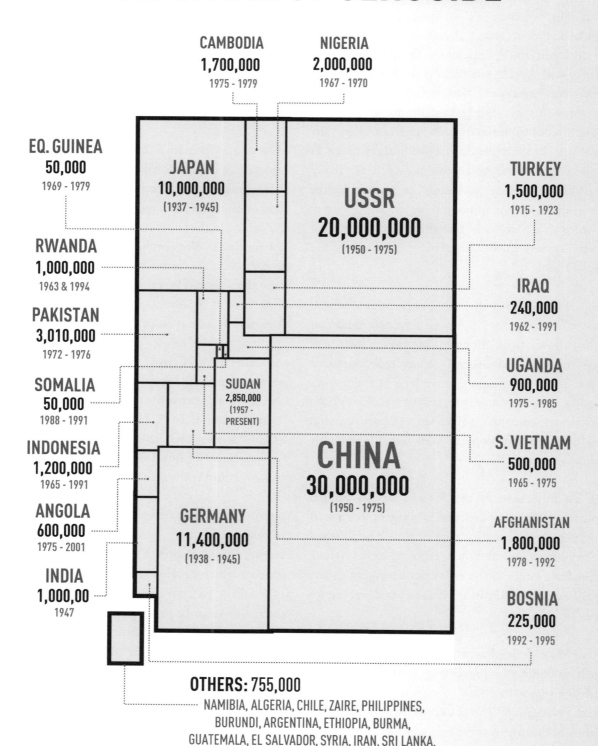

CAMBODIA
1,700,000
1975 - 1979

NIGERIA
2,000,000
1967 - 1970

EQ. GUINEA
50,000
1969 - 1979

JAPAN
10,000,000
(1937 - 1945)

USSR
20,000,000
(1950 - 1975)

TURKEY
1,500,000
1915 - 1923

RWANDA
1,000,000
1963 & 1994

IRAQ
240,000
1962 - 1991

PAKISTAN
3,010,000
1972 - 1976

SUDAN
2,850,000
(1957 - PRESENT)

UGANDA
900,000
1975 - 1985

SOMALIA
50,000
1988 - 1991

CHINA
30,000,000
(1950 - 1975)

S. VIETNAM
500,000
1965 - 1975

INDONESIA
1,200,000
1965 - 1991

ANGOLA
600,000
1975 - 2001

GERMANY
11,400,000
(1938 - 1945)

AFGHANISTAN
1,800,000
1978 - 1992

INDIA
1,000,00
1947

BOSNIA
225,000
1992 - 1995

OTHERS: 755,000
NAMIBIA, ALGERIA, CHILE, ZAIRE, PHILIPPINES,
BURUNDI, ARGENTINA, ETHIOPIA, BURMA,
GUATEMALA, EL SALVADOR, SYRIA, IRAN, SRI LANKA,
LANKA, YUGOSLAVIA

Source: 5:50:500 (2009), 80:20 Educating & Acting for a Better World

REGIONAL HUMAN RIGHTS MECHANISMS

Action to protect human rights can be taken at different levels - domestic (or national), regional and international – for example, regional mechanisms exist in Europe (the European Convention on Human Rights) and in Africa with the African Charter on Human and People's Rights. At the international level, it is normally the UN that takes the initiative to draft and adopt treaties such as the ICCPR or the Covenant for the Elimination of Discrimination Against Women.

The critics of such regional arrangements note:

- *If human rights are universal should there not be also universal protection?*
- *Don't regional mechanisms reduce the relevance of the international mechanisms supported by the UN?*
- *Isn't there a risk of increasing different interpretations of human rights by the different regional mechanisms?*

Supporters of such mechanisms emphasise the positive aspects:

- *It is easier to find agreement amongst a reduced number of states with common cultural/historical traditions*
- *Decisions by regional bodies are more likely to be respected for the same reason*
- *Regional bodies may have, as in the case of the European Court of Human Rights, powers which UN Treaty bodies simply do not have*

The UN was initially suspicious of regional mechanisms fearing they would reduce its own credibility but eventually the UN itself began to advocate the establishment of regional human rights instrument where none existed.

The European Convention on Human Rights of 1950 was adopted under the auspices of the Council of Europe in the context of a war-ravaged Europe. Initially a part-time institution, the European Court developed into a full-time Court of Human Rights capable of entertaining claims from individuals against States. Its main features include the fact that it includes only civil and political rights, it has a fully fledged Court which hears cases and delivers judgments (enforced by the Council of Europe), any person within the territory of a State Party may file a case (and does not need to be a national of that state). The Court (with 3 full-time judges) is considered to be one of the most effective mechanisms for the protection of human rights available

However, critics of the Convention/Court highlight the fact that some of its judgments are only executed after extreme delay, the success of the Court has meant that its work load has increased to an impossible level, social and economic rights are totally excluded (thereby seeming to support the view that some rights are more important than others) while some argue that most social and economic rights are not capable of being enforced in a court of law in any case.

The European Court of Human Rights has been very productive in the number of cases it has dealt with and delivered judgments declaring, for example, that a particular law or part of a law of a State Party conflicts with the Convention and that the State Party should amend its legislation to comply with the decision of the Court. The fact that States are willing to comply with the decisions of a supranational Court is a significant matter. Another productive area has been with reference to criminal law relating to homosexuality in Cyprus, Ireland and the United Kingdom where the Court found that certain provisions in the domestic legal systems of these three countries concerning homosexual behaviour violated Article 8 (the right to privacy and family life). As a result the U.K., Ireland and Cyprus eventually changed their laws.

A further interesting development in the case-law of the Court refers to the notion of the 'extra-territorial' application of the Convention. The Convention, in principle, covers individuals who are on the territory of a State Party. However, over recent years there have been attempts at applying the Convention to individuals who have their rights violated by State Parties outside the same State Parties' territory. For example, in the Loizidou case, a Cypriot citizen who had lost use and possession of her property as a result of the Turkish invasion/intervention in Northern Cyprus requested the Court to rule that Turkey was responsible for her loss of use. The Court ruled that although the territory of Northern Cyprus was not within the normally accepted definition of Turkish jurisdiction, Turkey was responsible as it was in 'effective control' of that territory.

This extension of the jurisdiction of the Court to beyond the territory of the State Parties to situations where State Parties have 'effective control' has created the possibility for State Parties who commit human rights violations outside their territory to be also held responsible.

This extension of the jurisdiction of the Court to beyond the territory of the State Parties to situations where State Parties have 'effective control' has created the possibility for State Parties who commit human rights violations outside their territory to be also held responsible. This is a currently developing area of human rights law in Europe and can have profound implications for human rights in conflict zones where State Parties are involved.

The African Charter on Human and Peoples Rights is the youngest and most challenging of human rights regional mechanisms adopted in 1981 in Banjul (sometimes referred to as the Banjul Charter), it came into force in 1986. The title of the Charter immediately highlights its own particularities:

- It includes people's rights, such as the right to peace and security, as well as the people's right to development
- it includes a list of the duties of individuals including the duty to preserve the harmonious development of the family, the duty to preserve and strengthen positive African cultural values as well as to pay taxes imposed by law in the interest of the society
- the Preamble makes specific reference to African history - in particular colonialism - and emphasises the importance of the right to development
- unlike other regional human rights mechanisms, it includes economic and social rights including the right to health and to education

These characteristics are considered as positive by some in that they produce a text that includes human rights from all generations of rights while acknowledging an African 'flavour' making the Charter 'culturally sensitive'. Criticisms of the Charter have focused on the following:

- the references to certain specific duties owed by the individual to the State weakens human rights and are at any rate inappropriate in a human rights instrument
- the rights guaranteed are subject to wider limitations than the limitations attached to the same or similar rights in the European or Inter-American regional instruments
- the Charter provides a weak monitoring mechanism (the Commission), which has none of the judicial powers or the effectiveness of the

European Court – however, since January 2004, there has been an African Court on Human and People's Rights but it is too early to assess its impact and, unlike the European Court, the African Court does not have jurisdiction to hear individual complaints against a State Party without the consent of the same State

Unlike the European Convention, the African Charter's implementation and enforcement procedures are not well developed and the African Commission on Human and People's Rights has limited enforcement powers. Nevertheless the Commission has delivered important decisions, especially relating to the military government of Nigeria in the mid-1990s. In the Civil Liberties Organisation v Nigeria, the Commission found that Decrees issued by the military government suspending the Constitution and prohibiting Nigerian Courts from examining the validity of the Decrees were in breach of Article 7 of the African Charter (right to a fair trial). This decision was reiterated in Constitutional Rights Project v Nigeria where the Commission recommended that 7 men sentenced to death under one of the Decrees should be released and held that the Decree itself violated the African Charter. The development of the African Court of Human and People's Rights will be of special interest. Drawing on the experience of the European Court, the development of the African Court may have a direct bearing on the development on human rights more generally throughout the African continent. The Court has so far only delivered one judgment (on 15th December 2009) in which it found the complaint lodged inadmissible. This serves to emphasise the point that the degree to which the Court will be successful will depend to some extent on whether the State Parties allow individual petitions by aggrieved individuals. A campaign to lobby for the acceptance of the right to individual petition would thus be a useful starting point.

THE DECLARATION ON THE RIGHT TO DEVELOPMENT

Agreed on the 4th December 1986, this Declaration essentially recognised development as a human right. It proclaims that every individual has a right to develop politically, economically and socially or, to put this in human rights terms, every individual has the right to all rights mentioned in the Universal Declaration and not simply to a selection of those rights depending on the whim of her/his state. The Declaration brings together, in concise form, the two Covenants. Its importance lies in the explicit statement that development of all human beings is a matter of right: not merely of right but of human rights.

A SUMMARY OF THE DECLARATION

The Declaration contains a Preamble and 10 articles which make explicit that there are rights and duties that operate at the individual, state and interstate level.

Article 1 states that, the right to development is one which individuals and peoples should both enjoy and contribute to. This Article also makes clear that development has economic, social, cultural and political dimensions.

Article 2 (2) makes an interesting point concerning the responsibility for implementing the right to development. *All human beings have a responsibility for development, individually and collectively…'*

The subsequent Articles also focus on the primary responsibility of states to:

- formulate appropriate national and international development policies
- create national and international conditions favourable to development
- cooperate with each other in ensuring development and eliminating obstacles to development
- take all steps to eliminate massive violations of human rights
- cooperate with each other to promote all human rights for everyone without distinction
- promote international peace and security and achieve disarmament
- take all necessary measures to ensure the right to development and ensure equality of opportunity in access to education, health, food, housing etc.
- encourage popular participation in every sphere of activity

The Declaration attempts to bring together the individual, the community in which s/he lives and the international system. All these actors have a role to play in ensuring the implementation of the right to development – in this context the debate on the right to development has highlighted a key aspect of modern thinking and practice on development – the fundamental importance and role of civil society. Development is no longer the preserve of the state but also includes all types of civil society organisations.

HUMAN RIGHTS AND DEVELOPMENT

In 2000, the UNDP's Human Development Report focused upon human rights and development. Its conclusions were resoundingly clear: human rights are not some form of 'reward' of development – they are critical to achieving it. The report draws a parallel between human rights and development: they both seek to ensure the freedom, well-being and dignity of all people everywhere.

When human rights and development progress together, they reinforce one another. As the report concludes, *'human development is essential for realising human rights, and human rights are essential for full human development'.*

Human rights add value to the development agenda, by:

- Drawing attention to the accountability of providing all people with rights
- Providing a moral legitimacy to development
- Shifting priority onto the most deprived and excluded societal groups
- Directing attention towards the need for information and the creation of a political voice for all people.

In turn, human development is an asset to human rights through:

- Creating a dynamic, long-term perspective on the fulfilment of rights
- Focusing attention on the socio-economic context required for human rights to be realised, as well as the contexts in which human rights are threatened
- Providing a systematic assessment of the economic and institutional constraints to the realisation of rights.

The report has a number of suggestions in the quest to achieve global human development and ensure that the human rights of all individuals are respected

- **Legislation is not a sufficient safeguard** - every country must strengthen its social arrangements for securing human freedoms. Laws alone do not guarantee human rights: they require adequate institutions to implement legislation, as well as a backing culture of norms and ethics that reinforces, rather than threatens, the legal structure
- The fulfilment of all human rights requires **democracy** that is inclusive, with a separation

of powers and public accountability. Elections are not enough

- The **eradication of poverty** is not just a challenge to development: it is also a central goal of human rights fulfilment
- **Accountability should not be limited to states,** but also extend to non-state actors. Furthermore, states' accountability does not end at their borders. In an integrated world, human rights must be accompanied by global justice
- **Information and statistics** are powerful tools for creating a culture of accountability and realising human rights. Spreading accurate information can prevent myth-making and mobilise changes in policy and behaviour
- Achieving universal human rights will require **action and commitment from all societal groups** – from government, to NGOs, media, the business communities and other opinion leaders
- **Stronger international action is needed:** development cooperation must be rights-based and free of conditionality and existing human rights legislation and mechanisms must be strengthened.

DEVELOPMENT - BETWEEN PEOPLE'S RIGHTS AND INDIVIDUAL RIGHTS

The right to development within the African Charter, to which we referred earlier, is considered to be a 'people's right' rather than an individual right. One may argue that development (together with the right of self-determination) has been a catalyst in introducing the notion of people's rights into human rights language – such rights might be described as those rights persons enjoy as members of their community rather than on a purely individual level.

Development thus provides new perspectives on how human rights work in, and relate to, the state, the community and the individual. Development operates at all of these levels and therefore is a link between them all. The human rights of an individual 'live' in a community that, in its turn, is rooted in a state.

This debate is also relevant within the context of the right to development itself. Some lawyers have argued that the 'beneficiary' of the right to development is the community and not individuals. It may be suggested that the question of who is the beneficiary of the right is wrongly framed. Such a contraposition between the individual and the community in respect of the enjoyment of the right to development presupposes that development is a finite quantity. A more appropriate proposition could revolve around how we are to ensure development of the individual, the community and the state together. The development of each of these is interdependent and interconnected.

Finally, development brings to human rights a more balanced perspective of how human rights function i.e. that rights are not enjoyed by individuals in a vacuum, but within the fabric of society.

A RIGHTS-BASED APPROACH TO DEVELOPMENT

'The satisfaction of the needs of a people should be perceived as a right and not as an act of charity. It is a right which should be made effective by norms and institutions'

The quotation above from Algerian international lawyer and formerly President of the International Court of Justice, Mohammed Bedjaoui (International Law: Achievements and Prospects, 1991) stresses what might be considered the most crucial contribution of human rights to development i.e. the move from considering development as a morally desirable objective to considering development as a right belonging to every person.

To those brought up in the West, contributing money to charity is a widespread and mostly unquestioned activity, images of the starving African child present impulses of moral or religious responsibility. We should do 'good' and thus give of our plenty to those who, happen to be 'less fortunate than ourselves'. There are few suggestions that the starving child has a right to eat or drink, s/he is merely unlucky. The underlying assumption is that we are the active ones, the ones who are central or of most importance – there is little suggestion of an issue of law or rights.

Human rights challenge such assumptions and ask us to look at the image in a different frame, a legal one - the child is having her/his rights violated and, legally, when a right is being violated the corollary is that someone is responsible for that breach and the image now becomes much more demanding and potentially dangerous. The viewer of the photo changes from a potential saviour (by virtue of their good deeds) to a potential rights violator.

A rights framework approaches development in a more rigorous fashion and asks a set of difficult questions:

- what is the legal basis of the right to development?
- who is/ are the beneficiary/ies of the right?
- who is responsible for ensuring that the persons owning the right actually enjoy it?
- how is the right implemented?
- what happens if the right is not implemented?

'For too long the development debate has ignored the fact that poverty tends to be characterised not only by material insufficiency but also by denial of rights. What is needed is a rights-based approach to development. Ensuring essential political, economic and social entitlements and human dignity for all people provides the rationale for policy. These are not a luxury affordable only to the rich and powerful but an indispensable component of national development efforts.'

UN Commission for Social Development, 36th session, 1998

The practical consequences of the existence of the right to development are of utmost importance as they present the greatest challenge to the right itself. Critics of the right to development argue that no such right exists because its content and consequences have not been clarified. As Mohammed Bedjaoui states:

'It is clear, however, that a right which is not opposable by the possessor of the right against the person from whom the right is due is not a right in the full legal sense. This constitutes the challenge which the right to development throws down to contemporary international law…'

The future of the right to development relies, to a considerable extent, on whether and how the challenge presented by Bedjaoui is met. What international norms should be in place stating the exact content of the right to development? What international institutions should be established monitoring the implementation of the right? What sanction should be imposed on those responsible for breaches of the right?

These questions apply equally to other human rights that are, as yet, not adequately implemented. Issues of implementation present core difficulties for international law in general and human rights in particular.

SOME DEBATES: THE RIGHT TO DEVELOPMENT - A CONTROVERSIAL ISSUE

The right to development is one of the so-called new or third generation rights which have come to the fore in the latter part of the 20th century. However, not everyone agrees that this new right is in fact a new right or a right at all. This is due to the difficulties in defining exactly the content of the right and the delicate questions relating to implementation to which we have already referred. In particular, criticism of the right to development centres on the notion of development as a group right rather than a human right.

Traditionally, development was seen as a right pertaining to states and this was of itself contradictory to the idea of human rights that by definition belong to the individual. This was in part due to the tendency of measuring development through a country's GDP per capita. This measurement emphasised the state-centric and economic approach. More modern approaches to measuring development focus on rates of illiteracy, access to education, food etc…These approaches are more human-centric and move away from the purely economic but they highlight how much more needs to be done in clarifying the right to development.

There are two contradictory views – one is that development is the most important of all rights while the other view implies that the right to development can be nothing except the implementation of 'traditional' human rights. According to Mohammed Bedjaoui:

'The right to development is a fundamental right, the precondition of liberty, progress, justice and creativity. It is the alpha and omega of human rights, the first and last human right, the beginning and the end, the means and the goal of human rights, in short it is the core right from which all others stem.'

Or as another commentator, Jack Donnelly notes:

'Suppose that Article 28 [of the Universal Declaration on Human Rights] were to be taken to imply a human right to development. What would such a right look like? It would be an individual right, and only an individual right, a right of persons not peoples, and certainly not States. It would be a right to the enjoyment of traditional human rights, not a substantively new right.'

SOME DEBATES: CULTURAL RELATIVISM

The importance of the universality of human rights is reflected in the title of the Universal Declaration adopted at a time when most of Africa and large tracts of Asia were under colonial rule. In the post-colonial period the emerging states of Africa and Asia tended to view with suspicion the international legal order constructed by European states before their independence.

Former colonial territories considered that the international legal order established by their colonisers was set up for the sole purpose of accommodating the interests of Europe and America. Therefore, anything associated with this Eurocentric order was automatically suspect and human rights (having been formally included in the UN Charter and elaborated in the 1948 Declaration) fell into this category of *'suspect'* concepts.

From the 1970s onwards the stated *'universality of human rights'* began to be challenged from a number of perspectives in Asia and Africa with two schools of thought emerging. The first holds that human rights are a purely Western concept and therefore have no application outside Western culture and society while the second holds that these rights have to be interpreted and implemented in a culturally sensitive way.

Critics of the universality of human rights present the Universal Declaration as an example of how human rights concepts were shaped by an essentially Euro-American group of men (and one woman – although this is not entirely true). In response, the proponents of universality point out that key contributors included such diverse authors as René Cassin (France), Charles Malik (Lebanon), Peng Chun Chang (China), Eleanor Roosevelt (USA), Hernan Santa Cruz (Chile), Alexandre Bogomolov/Alexei Pavlov (Soviet Union), Lord Dukeston/Geoffrey Wilson (United Kingdom), William Hodgson (Australia) and John Humphrey (Canada).

Another response to the critics of universality is that it is irrelevant who drafted the Universal Declaration; the question is rather does the Declaration express the desires, aspirations and principles common to humanity? Proponents of universality maintain that the Declaration does present a set of rights to which all of humanity aspires and wishes to claim.

The controversy around cultural relativism as opposed to universality took centre stage at the 1993 World Conference where a number of Asian and African leaders made emphatic calls in favour of a culturally sensitive approach to human rights. The UN Preparatory Meeting for Asia had already indicated that a number of Asian countries would be challenging the absolute universal nature of human rights. The Final Declaration while nodding to universality immediately challenges it:

'... while human rights are universal in nature they must be considered in the context of a dynamic and evolving process of international norm-setting, bearing in mind the significance of national and regional particularities and various historical, cultural and religious backgrounds'

Final Declaration of the Regional Meeting for Asia of the World Conference on Human Rights, 1993

The Declaration adopted at the Conference tried to bridge the gap between these two opposing views:

'All human rights are universal, indivisible and interdependent and interrelated. The international community must treat human rights globally in a fair and equal manner, on the same footing, and with the same emphasis. While the significance of national and regional particularities and various historical, cultural and religious backgrounds must be borne in mind, it is the duty of States, regardless of their political, economic and cultural systems, to promote and protect all human rights and fundamental freedoms.'

Vienna Declaration and Programme of Action, adopted by the World Conference on Human Rights 1993

The Declaration in Vienna may be regarded as an attempt to paper over the cracks and smooth the tensions between the *'universalists'* and the *'cultural relativists'*, the tensions, however, remain evident.

As it happens, the view that Asian values are quintessentially authoritarian has tended to come, in Asia, almost exclusively from spokesmen of those in power…But foreign ministers, or government officials, or religious leaders, do not have a monopoly in interpreting local culture and values. It is important to listen to the voices of dissent in each society. Aung San Suu Kyi has no less legitimacy – indeed clearly has rather more - in interpreting what Burmese want than have the military rulers of Myanmar, whose candidates she defeated in open elections before being put in jail by the defeated military junta.

DEVELOPMENT, FREEDOM AND RIGHTS

The Universal Declaration of Human Rights, as already noted, is rooted in certain core values including the value of freedom. The linkage between human rights, freedom and justice is highlighted in the first paragraph of the Declaration's Preamble. The recognition of inalienable human rights, the Preamble insists, 'is the foundation of freedom, justice and peace in the world'. This linkage between freedom, rights and justice is worth reiterating because it bears on the debate around what development is and how it should be achieved. Indeed, some human rights treaties speak about 'human rights and fundamental freedoms' (example: The European Convention for the Protection of Human Rights and Fundamental Freedoms). Furthermore, the language in which particular rights are formulated in law also draws frequently on the language of freedom. For example, Article 18 of the Universal Declaration states: *'Everyone has the right to freedom of thought, conscience and religion; this right includes freedom to change his religion or belief, and freedom, either alone or in community with others and in public or private, to manifest his religion or belief in teaching, practice, worship and observance'.*

In Chapter 7, Colm Regan elaborates on the relationship between ideas of justice and development. Thus, the focus in this section will be on the relationship between Freedom, Human Rights and Development.

The crucial role of the value of freedom in shaping the human rights agenda post-World War II has already been alluded to. The relationship between rights, freedom and development in the contemporary world has been articulated coherently and in depth by Amartya Sen. In Development As Freedom Sen presents development *'as a process of expanding the real freedoms that people enjoy'.*

A summary of a few of the main points which Sen makes:

- It is important to clarify what the ends of development are and not to focus only on the means
- Freedom is not only the most important means of attaining development but also its most important end
- Political freedoms, which are sometimes taken to be subsidiary to development, are fundamental to development both in terms of process and outcome
- Freedom is the ability to lead the kind of life we have reason to value.

Sen's approach to development reinforces a conception of development which focuses on the individual as the key actor in its processes and its outcomes. If development means allowing individuals to lead lives they have reason to value, then the life of the human person becomes the benchmark against which development is measured.

By emphasising freedom as both a means and an end to development Sen also makes a contribution to a proper understanding of the relationship between the various rights and freedoms. Political freedoms for example are sometimes 'shelved' in the 'cause' of economic development. Sen argues that, setting aside the dubious link between authoritarianism and economic growth, such an approach is inherently flawed. Political freedoms are 'constitutive' of development in the same manner as other freedoms (such as social freedom or economic freedom). For example political freedoms and participatory freedoms (such as the freedom of expression and freedom to take part in political life) play a role in increasing the capabilities of individuals. Such rights enable individuals to participate in public life and contribute to shape public opinion and societal values. This enhances the ability of people to help themselves and also to influence the world around them (see Chapter 6).

For additional information on human rights today, see pages 208/209.

READING

Mary Ann Glendon (2001) A World Made New: Eleanor Roosevelt and the Universal Declaration of Human Rights, New York, Random House

Audrey Osler and Hugh Starkey (2010) Teachers and Human Rights Education, Stoke on Trent, Trentham Books

Thomas Pogge (2008) World Poverty and Human Rights, New York, Polity Press

Omar Grech (2007) A Human Rights Perspective on Development, Bray, Ireland, 80:20 Educating and Acting for a Better World

Olivia Ball and Paul Gready (2006) The No-Nonsense Guide to Human Rights, London, New Internationalist and Verso

Micheline R. Ishay, (ed.,1997) The Human Rights Reader, London, Routledge

Samantha Power (2002) A Problem from Hell: America and the Age of Genocide, New York, Basic Books

Jeffrey Robertson (1999) Crimes Against Humanity: The Struggle for Global Justice, London, Allen Lane

MORE INFORMATION & DEBATE

http://www.hrw.org - NGO Human Rights Watch

http://www.amnesty.org - Amnesty International site

http://www.african-court.org/en/ - African Court on Human and Peoples' Rights

http://www.ohchr.org – Office of the Un High Commissioner for Human Rights

http://www.un.org/womenwatch/ - UN Inter-Agency Network on Women and Gender Equality

['...a problem of extreme magnitude...']

Chapter 5

30 YEARS OF WOMEN'S RIGHTS AND WRONGS

by Ciara Regan

Against the backdrop of 30 years of CEDAW, this chapter explores the current situation of women worldwide in the areas of health, education, poverty, environment and politics. Utilising a case study of the United States, it analyses the issue of 'reservations' to CEDAW and the debates that continue worldwide today on the issue of the rights of women. The chapter particularly focuses on the challenge of violence against women, discussing issues such as 'honour killings' and female genital mutilation. How women have recently used the legal system in Kenya and the internet in Iran to assert and uphold their rights are also explored.

KEYWORDS

Convention on the Elimination of Discrimination Against Women (CEDAW); Gender Equality; Women's Rights Worldwide; **Reservations;** Gender-based Violence; **Honour Killings;** Female Genital Mutilation, United States; **South Africa; Iran; Kenya.**

VIEWPOINTS

'There is much in our art and literature that romanticizes girls and women and the role they play in our culture. But sadly, in our world today, being female often means being sentenced to a life of poverty, abuse, exploitation, and deprivation. Compared to her male counterpart, a girl growing up in the developing world is more likely to die before her fifth birthday and less likely to go to school. She is less likely to receive adequate food or health care, less likely to receive economic opportunities, more likely to be forced to marry before the age of 16, and more likely to be the victim of sexual and domestic abuse. Girls are forced to stay home from school to work. In fact, two thirds of the nearly 800 million illiterate people in the world are women. Only one in 10 women in Niger can read. Five hundred thousand women die every year from childbirth complications— that's one woman every minute. Girl babies are even killed in countries where males are considered more valuable.'

World Vision Magazine 2007

'This is why, I believe, these days global awareness and sense of feminist responsibility give special credence to CEDAW as a very sound tool to move the women's equality agenda forward. As a global mechanism pooling the experiences and demands of women, monitoring national policies and actions, keeping abreast of developments worldwide, and holding governments accountable to universal standards of gender equality, CEDAW is crucial in promoting women's human rights. And it has every promise to be ever more so, to the extent that it is utilized and supported by the women's movement nationally and internationally.'

Feride Acar, Turkey, member of the CEDAW Committee

INTRODUCTION

December 18th, 2009, marked the 30th anniversary of what UN General Secretary Ban Ki-moon argued is *'one of the most successful human rights treaties ever'* – the Convention on the Elimination of All Forms of Discrimination Against Women. Despite this positive assessment, the Convention continues to face *'great challenges, including a new conservative onslaught that seeks to reverse much of the progress made in favour of gender equality'* (Silvia Pimental, a CEDAW Committee member) and the view (highlighted by the Sri Lankan Permanent representative to the UN, Palitha Kohona) that the Convention is in danger of being politically undermined by the reservations expressed by 22 countries seeking exemptions from some of the Convention's legal obligations.

'A reservation must not defeat the object and purpose of a treaty ... If a state has intrinsic difficulties with a treaty, it has the right not to become a party ... To become a party and then defeat the object and purpose of the treaty is unacceptable'.

So what is the reality of women's rights in the world today? Are they embedded (and observed) in one of the most successful human rights treaties of all times as the UN Secretary General would argue or are they in danger of being undermined and reversed? Are women's rights secure worldwide or do they remain insecure? This chapter briefly reviews and debates many of these issues and explores a number of case studies to illustrate some fundamental realities. Through a series of themes and case studies, the chapter then explores the urgent issue of gender-based violence and its impact on development.

CONVENTION ON THE ELIMINATION OF ALL FORMS OF DISCRIMINATION AGAINST WOMEN (CEDAW)

CEDAW marked the culmination of over thirty years of work by the United Nations Commission on the Status of Women (CSW), which was created in 1946 in order to promote and monitor women's rights. CSW has been instrumental in highlighting areas in both public and private life where women are effectively denied equality with men. The Commission's work for the progress of women's rights has resulted in a number of different declarations (such as the Convention on the Political Rights of Women in 1952 and the Convention on Consent to Marriage, Minimum Age for Marriage and Registration of Marriages in 1962), in which CEDAW is the most central and the most comprehensive instrument as it is significantly broader in scope than any previous treaty. It was the first international human rights instrument which explicitly defined all forms of discrimination against women as fundamental human rights violations. The first global conference on women took place in 1975 in Mexico City.

> *CSW has been instrumental in highlighting areas in both public and private life where women are effectively denied equality with men.*

It is often argued that, historically, one of the main obstacles to the realisation of women's rights has been the presumption that the state should not interfere in the private sphere or in family relations despite the fact that unequal power relations within these areas contribute considerably to gender inequalities. The Convention recognises that these inequalities have a direct impact on every aspect of women's lives and directs states to take certain measures in order to correct this power imbalance. It prescribes, not only an international bill of rights for women, but also a strategic framework, or agenda for action, for nation states to ensure the enjoyment of fundamental rights and freedoms as set out by the Convention. The ethos of the Convention is very much rooted in the goals and values of the United Nations - to reaffirm faith in fundamental human rights, the dignity, value and worth of each human person and the equal rights of both men and women.

The Convention itself consists of a Preamble and 30 articles which initially define discrimination against women and which set out an agenda for action to not simply challenge but effectively end such discrimination. The Preamble argues that *'extensive discrimination against women continues to exist'*, emphasising that this *'violates the principles of equality of rights and respect for human dignity'* before addressing the three key dimensions of the situation of women directly addressed by the Convention:

› The civil rights and legal status of women
› Human reproduction
› The impact of cultural factors on gender relations

The Convention requires state parties to take *'all appropriate measures, including legislation, to ensure the full development and advancement of women, for the purpose of guaranteeing them the exercise and enjoyment of human rights and fundamental freedoms on a basis of equality with men'*. For these principles to become genuinely effective, action is deemed necessary at a number of levels - establishing gender-sensitive laws and policies, as well as abolishing previous discriminatory laws and challenging attitudes and stereotyping as well as practices (by government officials or by individual citizens) that underpin discrimination and inequality.

State parties to the Convention have a three-fold obligation as regards the rights of women:

» **Respect:** The state must abstain from any conduct or activity of it's own that violates human rights.
» **Protect:** The state must prevent violations by non-state actors including individuals, groups, institutions and corporations.
» **Fulfil:** The state must take whatever measures are necessary to move towards the realisation of women's human rights.

DEBATING CEDAW – THE ISSUE OF 'RESERVATIONS'

Upon signing and ratifying CEDAW, states have the right to enter reservations outlining a number of areas where they argue the provisions of CEDAW do not apply nationally; these reservations include a number of recurring, significant issues including:

• Where CEDAW would conflict with Shariah law and its different interpretations (e.g. Bahrain, Bangladesh, Iraq, Libya, Morocco, Oman, Pakistan, Syria etc.) or with other religious issues (e.g. Israel and the issue of women judges in religious courts etc.)

• Where the provisions of CEDAW contradict national legislation as regards family laws or codes (e.g. as regards a child's nationality being decided by that of the father rather than the mother, where all marriages must be registered or where the right of women to freely choose

Continued on page 94 →

CEDAW: A SUMMARY

ARTICLE 1 DEFINES DISCRIMINATION:

'...*any distinction, exclusion or restriction made on the basis of sex which has the effect or purpose of impairing or nullifying the recognition, enjoyment or exercise by women, irrespective of their marital status, on a basis of equality of men and women, of human rights and fundamental freedoms in the political, economic, social, cultural, civil or any other field*'.

ARTICLE 2: DUTY OF THE STATE – The state must ensure the elimination of discrimination in laws, policies and practices nationally.

ARTICLE 3: EQUALITY –The state must take measures to uphold women's equality in all fields.

ARTICLE 4: TEMPORARY MEASURES – States are allowed to implement temporary measures, if this means the acceleration of women's equality.

ARTICLE 5: CULTURE – States must abolish discriminatory cultural practices or traditions.

ARTICLE 6: TRAFFICKING – States must take the appropriate steps to suppress the exploitation involved in prostitution and in the trafficking of women.

ARTICLE 7: POLITICAL AND PUBLIC LIFE – Women must have equal rights to vote, hold public office, and participate in civil society.

ARTICLE 8: GOVERNMENTAL REPRESENTATION – Women must be allowed to work and represent their governments internationally.

ARTICLE 9: NATIONALITY – Women have the right to acquire, retain or even change their nationality as well as that of their children.

ARTICLE 10: EDUCATION – Women have equal rights with men with regard to education.

ARTICLE 11: EMPLOYMENT – Women have equal rights with men in employment (equal pay, healthy working conditions etc.)

ARTICLE 12: HEALTH – Women have equal rights to health care with an emphasis on reproductive health services.

ARTICLE 13: ECONOMIC AND SOCIAL LIFE – Women have equal rights to family benefits, financial credit and equality in recreational activities.

ARTICLE 14: RURAL WOMEN – Rural women must have the right to adequate living conditions, participation in development planning and access to healthcare and education.

ARTICLE 15: EQUALITY BEFORE THE LAW – Women and men must be seen as equals before the law, have the legal right to own property and choose their place of residence.

ARTICLE 16: MARRIAGE AND FAMILY – Women have equal rights with men within marriage, including family planning.

ARTICLE 17-24: refer to the functioning and role of the Committee of CEDAW and reporting procedures.

ARTICLE 25-30: refer to the administration of the Convention.

CEDAW was adopted by the UN General Assembly on the 18th of December 1979, was signed by 64 countries in 1980 and came into force on the 3rd of September 1981.

- 186 countries have ratified the Convention and 99 have ratified the Optional Protocol which recognises and describes the role of the Committee on the Elimination of Discrimination against Women (the committee that monitors States compliance with the Convention) to receive and consider complaints from individuals or groups.

- The 8 countries to have not yet ratified CEDAW are the United States of America, Sudan, Iran, Nauru, Palau, Qatar, Tonga and Somalia.

- States which are party to the Convention must report every 4 years their domicile is restricted etc. Algeria, Egypt, India, Kuwait, Morocco etc.)

- Where governments insist that national legislation is more favourable to women than under CEDAW (France, UK, Ireland etc.)

- Where family legislation confers more favourable rights on women than men (e.g. Ireland, UK etc.) or on men rather than women (e.g. Malta on women's income – seen as part of that of the man or on social welfare being paid to the head of household deemed to be the man)

- Where CEDAW challenges traditional customs and practices which take time to change and which are unlikely to change as a result of direct legislation (e.g. Niger)

- Many states reject the jurisdiction of the Court of International Justice in disputes between states parties outlined in article 29

 **MORE INFO**

For a more detailed discussion of these reservations by state, see www.un.org/womenwatch/daw/cedaw/reservations-country.htm

THE CASE OF THE UNITED STATES AND CEDAW

The United States is the only developed country and democracy that has not ratified CEDAW despite it being endorsed by three of the last four US Presidents, the US State Department and many political, social and religious organisations. The main opposition to the Convention in the US comes from a coalition of conservative political groups and the religious right – the concerns and reservations they express relate not just to the Convention itself but also to the interpretations and statements of the monitoring CEDAW Committee. For these groups, the US should not ratify the Convention

because, in their view, it could/would:

› negate US family law and undermine perceived traditional family values

› force the U.S. to pay men and women the same for 'work of equal value' – this, they believe would undermine the free-market system operating in the US

› ensure access to abortion and contraception, something they oppose

› force the US to allow same-sex marriages

› legalise prostitution

› promote gender re-education along negative lines

› undermine parental rights

As with many other critics of the UN in the US, these groups argue that the Convention would ultimately undermine US sovereignty and lead to 'outsiders' interfering in US affairs (an argument echoed by many other states internationally which have reservations regarding Article 29). They object to the CEDAW Committee overseeing its provisions and as Concerned Women for America argue:
'This, in essence, places the welfare and well being of American women and families at the mercy of 23 individuals, among whom the United States might not even have a voice.'

A coalition of over 190 religious, health, political and educational US associations argue strongly that the US should ratify CEDAW; included in this coalition are the American Bar Association and Amnesty International who have challenged many of the arguments of those opposed to CEDAW.

In response to those who argue that ratification of CEDAW would:

- Give too much power to the international community (with international law superseding U.S. federal and state law, they argue that treaties adopted in the US are not 'self-executing' meaning that that legislation to implement any treaty would come before the House and Senate as any other bill does. They also point to the ability of countries to state reservations to the Convention; where differences do exist, CEDAW calls on states to take measures to progressively promote the principle of non-discrimination thus upholding US law

- Be used to destroy the traditional family structure in the U.S. by redefining 'family' and the roles of men and women in the family or could undermine the rights and role of parents in child rearing, they argue that both CEDAW and the U.S. Constitution recognise the constraints on authorities attempting to interfere with an individual's decisions regarding family. CEDAW simply urges Parties *'to adopt education and public information programs, which will eliminate prejudices and current practices that hinder the full operation of the principle of the social equality of women'.* CEDAW simply calls for the recognition of the *'common responsibility of men and women in the upbringing and development of their children'* and maintains *'the parents' common responsibility (is) to promote what is in the best interest of the child'.*

- Promote and support (and possibly enforce) abortion through its promotion of access to 'family planning', they argue that the Convention does not address abortion and, according to the U.S. State Department, is 'abortion neutral' pointing out that many countries in which abortion is illegal - such as Ireland, Burkina Faso, and Rwanda - have ratified the Convention.

- Force the U.S. to sanction same-sex marriages; they argue that CEDAW is not aimed at all sex-based discrimination, but only at discrimination that is directed specifically against women; there is no provision in the treaty that would compel the U.S. Congress to pass same-sex marriage laws in order to comply with a same sex-marriage claim.

The election of President Obama signalled the possibility that CEDAW might yet be ratified in the USA as his administration has identified the treaty as a priority. Opposition remains strong however, and the arguments against CEDAW in the US mirror many similar arguments elsewhere throughout the world.

WORLDWIDE

"no country treats its women the same as its men"

The startling reality is that there is not a single country in the entire world today where women have achieved equality with men.

Despite the fact that very significant progress has been made in a number of key areas such as education and health, women remain hugely disadvantaged in areas such as economics and politics and, in some countries, they experience widespread and systematic abuse and discrimination.

Gender equity is far from becoming a reality, for example in the period 2008 – 2010:

› Almost 70% of the world's poor are women
› Women make up only 18.6% of the world's parliamentarians
› On average, women get paid less for undertaking precisely the same jobs as men
› Of nearly one billion illiterate adults, two-thirds are women
› There is only one woman for every nine men in senior managerial positions
› An estimated 1 in 5 women worldwide will be a victim of rape or attempted rape in her lifetime and 1 in 3 will have been beaten, forced into sex or abused most usually by a family member or an acquaintance

Afghanista
Sierra Leone
Central Africa
Angola
Zambia
Zimbabwe
Swaziland
Lesotho

1 in 4,300   0.02% | INDUSTRIALISED COUNTRIES

1 in 1,700   0.05% | CSS/CIS & BALTIC STATES

1 in 480
0.20% | LATIN AMERICA & THE CARIBBEAN

1 in 190
0.50% | MIDDLE EAST & NORTH AFRICA

1 in 140
0.07% | THE WHOLE WORLD

1 in 120
0.83% | DEVELOPING COUNTRIES

1 in 110
0.90% | SOUTH ASIA

1 in 37
2.70% | LEAST DEVELOPED COUNTRIES

1 in 31
3.20% | SUB-SAHARAN AFRICA

LIFETIME RISK OF MATERNAL DEATH, 2008

One of the largest and most glaring differences between poor and rich countries is highlighted in the case of maternal deaths. A woman in sub-Saharan Africa has a 1 in 31 chance of dying in pregnancy or childbirth, compared to 1 in 4,300 generally in a developed country. 99% of all maternal deaths occur in poorer countries with only 1% occurring in rich countries indicating quite clearly that such deaths can be avoided with proper health services.

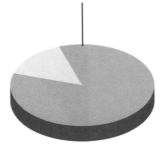

OVER 80% OF ALL MATERNAL DEATHS OCCUR IN SUB-SAHARAN AFRICA AND INDIA

"Most causes of maternal mortality are preventable: why is it then that women continue to be poorly nourished and to die during childbirth? Is child birth the cause of death, or is it the failure to diagnose, prevent and treat the reasons for maternal mortality (all connected with women's status and the access to food, education and health care) responsible?"

Vandana Shiva, 1994

World's lowest life expectancies at birth by sex, 2005–2010

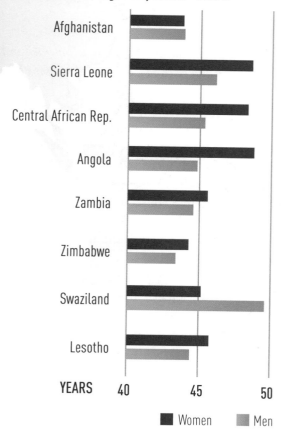

YEARS

■ Women ■ Men

▼ ADULT LITERACY RATES – MEN AND WOMEN

"Literacy gives women a voice – in their families, in political life and on the world stage. It is a first step towards personal freedom and broader prosperity. When women are literate, it is all society that gains."

Irina Bokova, Director-General of UNESCO, 2010

ADULT LITERACY RATE
FEMALES AS A PERCENTAGE OF MALES, 2008

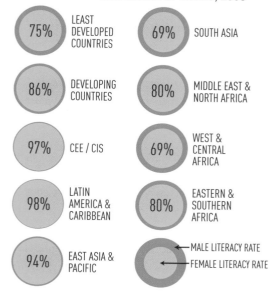

75% LEAST DEVELOPED COUNTRIES

69% SOUTH ASIA

86% DEVELOPING COUNTRIES

80% MIDDLE EAST & NORTH AFRICA

97% CEE / CIS

69% WEST & CENTRAL AFRICA

98% LATIN AMERICA & CARIBBEAN

80% EASTERN & SOUTHERN AFRICA

94% EAST ASIA & PACIFIC

MALE LITERACY RATE
FEMALE LITERACY RATE

▼ WOMEN IN PARLIAMENT 2010

"Where women's voices are heard, policy better reflects their lives. Where under-representation persists, women's interests are repeatedly ignored.'

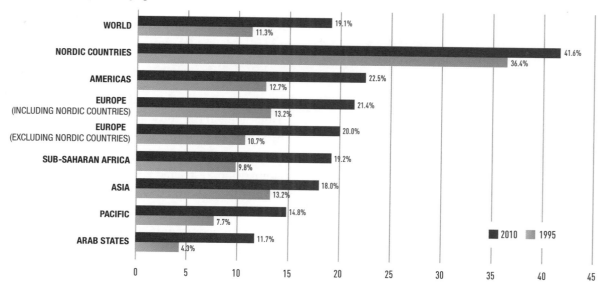

	2010	1995
WORLD	19.1%	11.3%
NORDIC COUNTRIES	41.6%	36.4%
AMERICAS	22.5%	12.7%
EUROPE (INCLUDING NORDIC COUNTRIES)	21.4%	13.2%
EUROPE (EXCLUDING NORDIC COUNTRIES)	20.0%	10.7%
SUB-SAHARAN AFRICA	19.2%	9.8%
ASIA	18.0%	13.2%
PACIFIC	14.8%	7.7%
ARAB STATES	11.7%	4.3%

Sources: UNDP Human Development Report 2010; UNICEF State of the World's Children 2011 and UN Women Progress of the World's Women 2011 – 2012; World Health Organisation and International Parliamentary Union Women in Parliament in 2010.

GENDER BASED VIOLENCE

'Gender based violence (GBV) is any act or threat of harm inflicted on a person because of their gender. It is rooted in gender inequality; therefore, women are primarily affected. Gender based violence refers to an act that results in, or is likely to result in, physical, sexual and psychological harm or suffering, including threats of such acts, coercion or arbitrary deprivation of liberty, whether occurring in public or private life. It encompasses sexual violence, domestic violence, sex trafficking, harmful practices (such as female genital mutilation/cutting), forced/early marriage, forced prostitution, sexual harassment and sexual exploitation, to name but a few.'

Keeping Gender on the Agenda: Gender Based Violence, Poverty and Development: An Issues Paper from the Irish Joint Consortium on Gender Based Violence, 2009

In a 2009 briefing paper on gender based violence (GBV), an Irish consortium of organisations described the issue as *'a phenomenon of epidemic proportions prevalent in many families, communities, societies and cultures across the globe'*. They argue that many women and girls, (and to a lesser extent men and boys) directly experience or live with the results of such violence in their lives and note that GBV presents itself in different forms involving a wide range of perpetrators including intimate partners (*'the most pervasive form'*) and family members, to strangers including the police, teachers and soldiers.

The consortium argues that:

'While gender based violence is a universal problem, it is a problem of extreme magnitude in less developed countries. A recent study in Uganda and Bangladesh reported that more than 80% and 94% of women surveyed respectively had experienced physical, sexual or psychological violence at some point in their marriage/intimate relationship. Gender based violence is exacerbated by war and is increasingly a feature of

conflicts. Widespread rape has been documented in the DRC, Bosnia and Rwanda, leaving a legacy of violence long after peace treaties have been signed'

A range of studies have demonstrated the close links between GBV, poverty and overall development – recessions and increased poverty can trigger an increase in violence while such violence has important negative consequences for productivity, health and general well-being which, in turn, can result in increased poverty and an undermining of development.

Gender based violence has immediate costs for households and communities; at the level of the household, violence often results in *'out of pocket'* expenditure in accessing health services, the police, courts or informal resolution bodies. For example, in Uganda, the average out of pocket cost for services as a result of intimate partner violence in 2009 was $5 - equivalent to three quarters of the average weekly household income. GBV also drains household incomes as women and men often miss paid work and household work is neglected and this drain can have a direct impact on hunger. An inability to work and the potential desertion by the male partner can mean that household members literally starve as daily food purchases are not made and children are left in the care of extended family or neighbours.

For the consortium, gender based violence has serious long-term consequences which are almost impossible to quantify, such as the reduced physical and mental health of women, increased child malnutrition, restricted education of girls and boys, weakened social capital within communities and an overall reduction in the well-being of women, families and communities. Crucially they argue:

'...gender based violence results in lowered participation of women as agents of development which has disastrous implications for realising safer communities and sustainable livelihoods... Gender based violence is an

abuse of human rights and failure to address it amounts to complicity. It is also unquestionably a critical development issue that needs to be addressed for the effectiveness of poverty reduction plans and strategies.'

GBV EXPLORED: HONOUR KILLINGS

Honour killing continues to be practised in countries such as Pakistan, Jordan, Palestine, Israel, the Balkans, Lebanon and Turkey and amounts to the murder of a woman by a male relative because he or his family believe that their 'honour' has been damaged by the perceived or actual *'immoral'* behaviour of the woman. According to writer Syed Kamram Mirza:

'Anything from speaking with an unrelated man, to rumored pre-marital loss of virginity, to an extra-marital affair, refuses forced marriage; marry according to their will; or even women and girls who have been raped—can stain or destroy the family honor. Therefore, family members (parents, brothers, or sisters) kill the victim in order to remove the stain or maintain, and protect the honor of the family.'

- In Jordan and Lebanon, 70 to 75% of the perpetrators of 'honour killings' are the brothers of those killed
- According to a government report, 4,000 women and men were killed in Pakistan in the name of honour between 1998 and 2003, the number of women being more than double that of men
- In a study of female deaths in Alexandria, Egypt, 47% of the women were killed by a relative after the woman had been raped
- Brazil is cited as a case in point, where killing is justified to defend the honour of the husband in the case of a wife's adultery

The UN estimates that every year about 5,000 women are murdered by family members in the name of honour.

Continued on page 100 →

SELECTED FACTS ABOUT SEXUAL AND GENDER-BASED VIOLENCE

- A 2009 study in Uganda and Bangladesh reported that more than 80% and 94% of women surveyed respectively had experienced physical, sexual or psychological violence at some point in their marriage/intimate relationship
- Between 250,000 and 500,000 women were raped in the 1994 genocide in Rwanda
- Worldwide, an estimated 40-70% of homicides of women are committed by intimate partners
- A commonly accepted estimate is that one in every three women has been beaten, coerced into sex or otherwise abused in her lifetime
- The International Organisation for Migration estimates that up to 2 million women are trafficked across borders each year
- More than 90 million African women and girls have undergone female genital cutting
- At least 60 million girls are 'missing' worldwide as a result of sex selective abortions, infanticide or neglect

MORE INFO

- Working towards the elimination of crimes against women committed in the name of honour, Report of the Secretary-General to the General Assembly, July 2002 (A/57/169)
- Gendercide Watch http:www.gendercide.org/case_honour.html
- Amnesty International USA http://www.gendercide.org/case_honour.html

GBV AS A WEAPON OF WAR

The use of systematic sexual violence has become commonplace – a weapon of war – in many conflicts:

- An estimated 250,000–500,000 women and girls were raped during the 1994 genocide in Rwanda
- An estimated 20,000–50,000 women and girls were raped during the war in Bosnia-Herzegovina in the early 1990s
- An estimated 50,000–64,000 internally displaced women in Sierra Leone were sexually attacked by combatants
- An average of 40 women and girls are being raped every day in South Kivu in the Democratic Republic of Congo and it is estimated that more than 200,000 women and children have been raped over more than a decade of that country's conflict
- Out of 300 peace agreements for 45 conflict situations in the 20 years since the end of the Cold War, 18 have addressed sexual violence in 10 conflict situations (Burundi, Aceh, DRC, Sudan/Nuba Mountains, Sudan/Darfur, Philippines, Nepal, Uganda, Guatemala, and Chiapas)

According to Medecins Sans Frontieres:

'The medical consequences of sexual violence are devastating. The physical injuries can be life threatening and many rape victims are at risk of contracting sexually transmitted diseases, including HIV/Aids. This risk is significantly increased during rape because forced sexual intercourse results in injuries and bleeding, thereby facilitating transmission of the (HIV) virus…

…Being raped leads to long-lasting trauma and suffering. Sometimes this takes the shape of mental health disorders whereas at other times it surfaces in less obvious ways such as shame, guilt, sleeping problems, difficulties in daily functioning and withdrawal. Many women report ongoing fear, anxiety, intrusive memories and flashbacks, which are rooted in their experience of the fear of being killed or mutilated…'

> **MORE INFO**
> - http://www.doctorswithoutborders.org/publications/article.cfm?id=1388&cat=ideas-opinions
> - http://www.unfpa.org/emergencies/symposium06/docs/finalbrusselsbriefingpaper.pdf

FEMALE GENITAL MUTILATION

According to the World Health Organisation:

- Female genital mutilation (FGM) includes procedures that intentionally alter or injure female genital organs for non-medical reasons
- The procedure has no health benefits for girls and women
- Procedures can cause severe bleeding and problems urinating, and later, potential childbirth complications and newborn deaths
- An estimated 100 to 140 million girls and women worldwide are currently living with the consequences of FGM
- It is mostly carried out on young girls sometime between infancy and age 15 years
- In Africa an estimated 92 million girls from 10 years of age and above have undergone FGM
- FGM is internationally recognised as a violation of the human rights of girls and women
- The practice is most common in the western, eastern, and north-eastern regions of Africa, in some countries in Asia and the Middle East, and among certain immigrant communities in North America, Europe, Australia and New Zealand.

'It is unacceptable that the international community remains passive in the name of a distorted vision of multiculturalism. Human behavior and cultural values, however senseless or destructive they may appear from the personal and cultural standpoints of others, have meaning and fulfill a function for those who practice them. However, culture is not static but is in constant flux, adapting and reforming. People will change their behavior when they understand the hazards and indignity of harmful practices and when they realize that it is possible to give up harmful practices without giving up meaningful aspects of their culture.'

Joint statement by the World Health Organisation, UN Children's Fund, and UN Population Fund

---

 MORE INFO

· http://www.who.int/mediacentre/factsheets/fs241/en/
· http://topics.nytimes.com/topics/news/health/diseasesconditionsandhealthtopics/femalegenitalmutilation/index.html for a useful collection of articles on FGM

WOMEN CHALLENGING TRADITIONAL LAW AND CUSTOM IN KENYA – THE NTUTU DECISION

In Kenya, most customary or traditional law (predominantly Kikuyu customary law) recognises an unmarried daughter as a son and allows her equal rights alongside her brothers to inherit land.

The daughters of the late Lerionka Ole Ntutu filed an objection to the proposed distribution of his estate as their brothers argued that this estate was governed by the Masai customary law which does not recognise the inheritance rights of daughters. It was for the court to decide which was applicable to this case – Masai customary law, or Law of Succession Act (an act which does not discriminate between male and female children of the deceased).

Continued on next page →

CASE STUDY: WOMEN BLOGGING IN IRAN

The Supreme Council for Cultural Revolution in Iran, a conservative dominated body, declared in 2001 that Internet Service Providers (ISPs) were to remove any anti-government or anti-Islamic websites from their service and that all ISPs were to be placed under the control of the State. Since 2001, more than 70,000 blogs have been set up in Iran mostly by women discussing a range of subjects such as their personal dreams or aspirations for their future, or taboo subjects such as sex, the compulsory wearing of the Hijab and even questioning the behaviour and policies of the state. By 2004, there were over 200,000 registered Farsi language websites. The internet has become an alternative means of communication for the opposition movement in Iran and despite the fact that only 10% of Iranians have access to the internet, Iran now has the third largest blogging community in the world.

Women's rights are an intensely sensitive issue in Iran, especially in more recent years, and those expressing alternative or opposing views to those of the ruling elite have been regularly accused of blasphemy, threatening national security and inciting public unrest. Website activists and individual bloggers have been targeted, arrested and charged and are quite often kept in solitary confinement. In some cases they have been tortured and made provide false confessions. In 2007, over 40 women were arrested for fighting for freedom of expression in Iran by using the internet, 32 of whom were either journalists or bloggers. They were arrested for demonstrating for women's rights in Tehran, and continuing their campaign online as 'cyber-feminists'. Blogging allows people, especially women, to express themselves while also avoiding personal censorship and control of the media within this strict code of Islam.

The judge used the case of Rono vs Rono as an example during his deliberations. In this case, the court agreed that the application of customary law takes pride of place, provided that *'it is applicable and is not repugnant to justice and morality or inconsistant with any written law...'* The court then applied international law, covenants and treaties which had been ratified by Kenya. This included the Declaration of Human Rights, the Covenant of Economic, Social and Cultural Rights, the Covenant on Civil and Political Rights and the Convention of the Elimination of Discrimination Against Women (CEDAW), which stipulates *'any distinction, exclusion or restriction made on the basis of sex which has the effect or purpose of impairing or nullifying the recognition, enjoyment or exercise by women irrespective of their marital status, on the basis of equality of men and women, of human rights and fundamental freedoms in the political economic, social cultural, civil or any other field'.*

Because Kenya had ratified all the international conventions and treaties, the Kenyan Constitution was then amended to include the prohibition of discrimination on the basis of sex. The judge found that these amendments have been made so as not to deprive any person of their legal right only on the basis of sex. He stated that 'finding otherwise would be derogatory to human dignity and equality amongst sex universally applied. I shall add that taking the view otherwise shall definitely create imbalance and absurd situtation'.

Due to this evidence, the court ruled that the daughters were entitled to inherit from the assets of their fathers land.

See: http://www.kenyalaw.org

CASE STUDY: *VIOLENCE AGAINST WOMEN IN SOUTH AFRICA*

A shadow report prepared in March 2010 by the organisations People Opposing Women Abuse and the AIDS Legal Network on behalf of the One in Nine Campaign and the Coalition for African Lesbians highlighted the nature and extent of violence against women in South Africa – a country with one of the highest reported rates of violence and rape in the world. The report, entitled Criminal Injustice: Violence Against Women in South Africa, Shadow Report on Beijing +15:

'In South Africa violence against women has reached epidemic proportions, one of the highest rates in the world of countries collecting such data. It exists in millions of households, in every community, in every institution, in both public and private spaces. VAW cuts across race, class, ethnicity, religion and geographic location. It is enmeshed with particularly violent histories of slavery, imperialism, colonialism and apartheid, and as well entangled in their corollary, the struggles for self determination. All these have left in their wake social and gender relations of a militarised society that has nurtured extremely violent masculinities to the detriment of women.' (2010:7)

The report chronicles some of the realities of this violence:

- As a result of apartheid, many black women live in areas which place them at risk of violence as they depend on walking or public transport at night in poorly lit areas; one study reported that 29% of women interviewed had been gang raped.

- Black people, in particular black women remain overrepresented amongst the poor and are the most vulnerable in terms of various forms of violence and often do not have access to legal services or redress.

- The exact scale of violence against women in South Africa is unknown as official police statistics are seriously flawed (with under-reporting, corruption, difficulties with the classification of crime and over-generalised data) and which provide only a fragment of the overall picture. Even with these limitations, South African Police statistics for reported rape were 69,117 in 2004/5, 68,076 in 2005/6, 65,201 in 2006/7, 63,818 in 2007/8 and 71,500 in 2008/9.

- The Medical Research Council (MRC) estimated that in 2002, some 88% of rape cases went 'unreported'.

- Additional MRC research suggested that the actual levels of violence are much higher than those reported by the police, for example, more than 40% of women interviewed in a Cape Town study had experienced one sexual assault; 45% of women aged 14 – 24 described their first sexual experience to have been one where they had been coerced, persuaded, tricked, forced or raped; 27.6% of men interviewed in another MRC study admitted having raped a woman, while 14.3% had raped a current or ex-girlfriend or wife. Nearly half of the men who said they had raped, had raped more than one woman or girl.

- The levels of domestic violence reported in different research studies also highlight the scale of the problem - 31% of pregnant women surveyed in KwaZulu Natal reported domestic violence; one study in three South African provinces found that 27% of women in the Eastern Cape, 28% in Mpumalanga and 19% in the Northern Province had been physically abused in their lifetimes by a current or ex-partner and 51% of women in the Eastern Cape, 50% in Mpumalanga and 40% in Northern Province had experienced emotional and financial abuse in the year prior to the study.

- Interviews with 1,394 men working for three Cape Town municipalities found that approximately 44% were willing to admit that they abused their female partners.

- Research on intimate femicide (where a woman is killed by her partner) found that every six hours a woman is killed by her male partner.

The report also documented the links between violence against women and increased risk of HIV and AIDS; the particular vulnerabilities of migrant and undocumented women; the impact of violence on children and the issue of homophobic violence. Significantly, the authors of the report state:

'We remain concerned about the inadequate measures taken to ensure the protection of women subjected to violence; the lack of women's access to just and effective remedies, the lack of access to the mechanisms and processes of justice, including compensation. Most significantly, we are concerned about the lack of adequate access, for most survivors of sexual violence, to health and other essential services...'

The report argues that state services including the current Office of the Presidency, not only routinely failed to condemn violence against women, but has, in some cases, invoked African customs and traditions to justify and facilitate the violation of women's rights:

'Increasingly, cultural practices that not only discriminate against women but are often extremely brutal, the practice of ukuthwala (forced marriages of women and under-age girls to older men through abduction), common mainly in the rural Eastern Cape Province, and the recently revived practice of virginity testing of girls and young women in KwaZulu-Natal Province, exist alongside the Constitution and legislation that prohibits such practices or that protects women and girl-children in law but not in practice in these situations...Another contradiction lies in the Recognition of Customary Marriages Act (2000) which entrenches the practice of polygamy, allowing a man to marry more than one wife.

The report acknowledges that South Africa has one of the most progressive and inclusive Constitutions in the world and that the equality clause guarantees non-discrimination on numerous grounds, including gender, sex and sexual orientation but that there are many serious inadequacies in policy and legislation as well as enforcement.

 MORE INFO

- For a more detailed discussion MORE INFO www.powa.co.za/files/ SouthAfricaShadowReportMarch2010.pdf

WOMEN'S RIGHTS ACTIVISTS

Shirin Ebadi is an Iranian lawyer and former judge. She won the Nobel Peace Prize in 2003 for her efforts in the promotion of democracy and human rights, particularly for women and children. She was the first Iranian and the first Muslim woman to receive the award.

Wangari Maathai was one of the first women in East and Central Africa to receive/earn a doctoral degree. She was active within the National Council of Women in Kenya, and was its chairwoman for 6 years. While there she developed the idea of a grassroots organisation, whose main focus was planting trees with women's groups in order to conserve the environment, as well as improve their quality of life.

Hissa Hilal, a Saudi woman, challenges convention every day in her country just by being not only a mother and a wife, but a journalist also. She participated in a reality TV show televised across the Arab world called The Million Poets where she recited her poetry which condemns the excessive conservativeness of Arab society. She has received a number of death threats because of this. She came second overall in the competition.

Ghada Jamsheer is a women's rights activist in Bahrain. In 2001 she launched a campaign for women's rights in the Gulf. She founded and leads the Women's Petition Committee, a grassroots organisation which lobbies for a law which will shift jurisdiction over family and women's affairs from the Islamic Sharia Court to the civil courts.

Mao Hengfeng actively defends women's reproductive rights in China by overtly protesting against the one child policy. She protested against the forced abortion of her second child and was subsequently jailed for two and a half years. The reason stated for her imprisonment was 'intentional damage of property'. She broke two table lamps while in custody.

Mary Robinson was the first ever female President of Ireland. She has since co-founded the Council of Women World Leaders, whose mission is to promote good governance and to enhance the experience of democracy globally by increasing the number, effectiveness and visibility of women who lead at the highest levels in their countries. She was also UN High Commissioner for Human Rights now leads the Mary Robinson Foundation for Climate Justice.

25 year old **Anarkali Honaryar** is a dentist from Afghanistan. She has been awarded Radio Free Person of the year award for her work to promote women's rights in Afghanistan. She works with Afghanistan's Independent Human Rights Commission and travels the country raising women's awareness of the rights they are entitled to.

Miriam Makeba was a legendary South African singer who died in 2008 at the age of 76. Known as Mama Africa and the Empress of African Song, she was one of the most visual and outspoken opponents of South Africa's Apartheid regime, a proponent of civil rights and black power, in addition to being a world famous musician.

For additional profiles, see www.womendeliver.org/knowledge-center/publications/women-deliver-100/

READING

Nalini Visvanathan, Lynn Duggan, Nan Wiegersma and Laurie Nisonoff (2nd edition, 2011) The Women, Gender and Development Reader, London, Zed Books

Irene Tinker (1990) Persistent Inequalities: Women and World Development, Oxford University Press

Lucy Moyoyeta (2005) Women, Gender and Development, Bray, 80:20 Educating and Acting for a Better World

Martha Nussbaum (2001) Women and Human Development: the Capabilities Approach, 2001, Cambridge University Press

World Bank (2012) The World Development Report: Gender Equality and Development, New York, World bank (available to download from the World Bank website)

UNDP (2005) The Arab Human Development Report 2005: The rise of women in the Arab World, New York, UNDP

MORE INFORMATION & DEBATE

http://www.un.org/womenwatch/daw/cedaw/reservations-country.htm

http://www.gendercide.org/case_honour.html

http://www.unfpa.org/emergencies/symposium06/docs/finalbrusselsbriefingpaper.pdf

http://www.who.int/mediacentre/factsheets/fs241/en/

http://topics.nytimes.com/topics/news/health/diseasesconditionsandhealthtopics/femalegenitalmutilation/index.html

www.powa.co.za/files/SouthAfricaShadowReportMarch2010.pdf

Gender and Development is the only journal published which focuses specifically on Gender issues within international development (http://www.genderanddevelopment.org)

http://www.un.org/womenwatch/daw/cedaw/

[*'All that is valuable in human society depends upon the opportunity for development accorded the individual.'*]

Albert Einstein

Chapter 6

PARTICIPATION, EMPOWERMENT AND DEVELOPMENT

Frik de Beer

This chapter reviews the core debates involved in discussions of participation, empowerment and agency in human development. It focuses significantly on the 'capabilities' approach as outlined by Amartya Sen and others and the debates and tensions between the 'liberal' and 'radical' approaches, including case studies from South Africa and Brazil. Finally, the chapter discusses the emerging importance of social movements in the development agenda.

KEYWORDS

Participation, **empowerment,** **agency,** **capabilities,** **ownership,** **social movements,** **South** **Africa,** **Brazil,** **power,** **functionings,** **mobilisation**

INTRODUCTION

Human beings have a need to understand the circumstances – the context – in which they live in order to change it. In the Third World, those circumstances are frequently determined by poverty, illness and conflict. Understanding the context is one thing; finding a way to change it is something quite different. One way of changing the context is to participate in one's own development. By participating in development, individuals, communities and societies gain in self-confidence, knowledge and power and are thus better able to influence their own lives and futures. In short, they optimise their functionings and capabilities – to the benefit of human development as a whole.

Participation can be expressed as *'agency'*; this refers to action or accomplishment. In Africa the debate on agency is found in two discussions; the one argues that aid must be provided to civil society at the local level while the other argues that development requires the combined effort of informal creativity which has to be fed into the formal economy. On the debate about agency in Africa, political scientist Patrick Chabal says:

'The subtext was clear: help yourself and the world will help you. The new discourse was predicated on the force of agency: directed African action would, with the help of the outside world, bring forth greater political accountability, more development and a reduction in poverty.'

Patrick Chabal (2009) Africa: the Politics of Suffering and Smiling, London: Zed Press

Agency or self-help can be an expression of the capabilities people have. Indian economist and Nobel Prize winner Amartya Sen developed the capabilities approach to development; this approach consists of two core concepts: functionings and capabilities. Functionings refer to a person's state of being and doing and this approach sees development as a process of expanding people's human capabilities. This, then, is an approach that places human development and not, for instance, economic development, at the centre. Human beings and their development becomes an end in itself and not a means to other ends.

'Capabilities', in this approach, refer to real or effective opportunities to achieve functionings. Put differently: in the field of development, capabilities do not refer to income, resources, goods, emotions or the satisfaction of preferences. It refers to what philosopher Ingrid Robeyns describes as *'what people are effectively able to do and be' or the freedom 'to enjoy valuable beings and doings'.* Sen does not provide a list of specific capabilities; what he does is to stress the role of activity and freedom in people's ability to make their own choices.

Using own capabilities to live a full life as a human being is based on the assumption that people should be free to shape their life as free beings in cooperation with others in the community and society; their own agency directs their own development. Martha Nussbaum identified and listed the following as examples of capabilities:

- **Life:** To live a full life and not die prematurely
- **Bodily health:** To have access to nourishing food and safe shelter
- **Bodily integrity:** To be free from violation of the body by criminal acts
- **Senses, imagination and thought:** To have freedom to use imagination, thinking and reasoning
- **Emotions:** To experience feelings towards others and objects outside yourself

- **Practical reason:** To be able to reflect and engage in discussion freely
- **Affiliation:** To be part of groups and in relationships with others as determined by the self
- **Other species:** To live conscious of the environment and of the role of humans towards caring for it
- **Play:** To relax and take part in sport, social and other activities.

Source: M. Nussbaum (2003) 'Capabilities as Fundamental Entitlements: Sen and Social Justice', Feminist Economics, 9, 2:3

Functional capabilities – or substantive freedoms as Sen calls it – includes the ability to engage in economic transactions, live to old age and participate in political activities.

The concept of participation in relation to development is in danger of being romanticised, and naively interpreted as always practised for the common good of the community and society. At its core, participation – the expression of own agency - allows people to compete for power to allocate resources. For participation to be meaningful for people in their quest to optimise their functional capabilities a *'level playing field'* must exist and people must participate as equals in deciding and in doing actions that they regard as meaningful.

WHAT IS PARTICIPATION?

People are mobilised to participate in development efforts or projects, but we must have a clear view of what *'participation'* really means. Some, especially governments (and NGOs) mobilise people for a limited and prescribed role and view this as participation. Participation does not mean passive involvement in something (a project!) planned and led by outsiders. When people are mobilised to participate, they should do so fully, and get involved in all aspects of the project. Then they become part of the project's decision-making and planning process. They also become part of the project's implementation and evaluation process. And, if need be, they decide on the adaptations that need to be made to keep a project on track; in short, they participate fully in making decisions on the nature of development and the *management* of the project. Their own agency should be the driving force of development.

Participation may be defined *actively* or *passively*. In the passive sense, it means taking part in - simply playing a role defined by someone else. In the active sense it means to initiate or lead. The passive definition of participation is based on what is called the liberal or weak view (of participation). The active definition of participation represents the radical or strong view. Below we briefly distinguish between these two approaches.

The liberal view of participation regards participation as good, especially if it is organised and orderly. The liberal view emphasises two points. The first is that, through participation, a solid, local knowledge base is used for development. Local people, who, for years, have lived in deprivation, surviving the hardships of their poverty, have a certain ingrained knowledge outsiders do not have. Their *'common sense'* knowledge of environmental dynamics can be of immense value to development efforts. Developers who do not use this knowledge base to the full are placing limitations on projects. Secondly, people who do not participate in their own development have no affinity for development efforts and their results. According to the liberal view of participation, the problem of sustaining development and maintaining facilities can be solved by making sure the local people are present.

In the radical view, participation becomes a way of ensuring equity. In many development projects and programmes, the poorest of the poor do not get their fair share of the fruits of development.

This shortcoming can be corrected by actively facilitating their inclusion, through participation, in development decision-making and activities. Furthermore, we must realise that it is the democratic right of people to participate in matters that affect their future. The wealthy (middle class and elite) in the developed and developing world usually have access to power that obtains and protects their wealth through participation in political processes. Every adult, including the relatively poor, the poor, or the poorest of the poor, has a right to be part of the decision-making mechanism that influences his or her development. This means that, when people are mobilised to participate in a project, they are not just there to make them feel part of the project; they are not present so that they can contribute their local practical knowledge; they are not there to do the physical work. They are there because it is their democratic right to be there and to make decisions regarding the project because it involves their future. The guiding principle is quite clear: don't mobilise people to play a minor role in a project and to fill a subordinate position in relation to professionals, bureaucrats and donors. If the people are not the main role-players, there is something wrong with their participation.

With all said and done a note of warning about the prospect of participation in development must be added. Practitioners and academics in the field of human development are often romantic and hold naïve views about participation. Power and power relations at the local level are complex; not all people entitled to participate are allowed because of social and political dynamics. In this way the poorest often are excluded from participation.

What do people in a community do when they participate? In what do they participate? Their participation may not be limited to an advisory role. If this is so, if the local people act only as advisors to the planners and decision-makers, they are participating in the project in the passive or liberal sense of the word. If, however, power accompanies participation, they participate in the project in the active or radical sense of the word. Such participation sets them free to build on their own agency and let their capabilities become the foundation of a development process that originates within individuals and communities. This only happens when participation is accompanied by empowerment.

WHAT IS EMPOWERMENT?

Empowerment refers to political power whereas mobilising people to participate in projects decided on (and managed by) outsiders – a government department, NGOs - leads to tokenism. 'Tokenism' means that people are mobilised/involved/placed in projects and on committees simply so that these committees look representative. In some cases, people are *mobilised* simply to carry out some form of physical work, or are taught various skills so that they can do such work; this is then called *empowerment*. But empowerment does not mean simply having certain skills or being the token representative on some committee. Empowerment means that one has decision-making power. Of course, certain skills are needed before one can make decisions, but these skills are only tools of enablement, and are not to be confused with empowerment itself.

Related to the skills needed for decision-making is the fact that people can only make enlightened decisions if they have accurate information. Empowerment is therefore a mixture between the right to make decisions and the ability to make those decisions. The guiding principle is that mobilisation must aim at giving the people the power or the right to make decisions, but then go further, and continue a supportive function by providing the community or individual with the information needed to make good decision-making possible.

SOME INDICATORS OF PARTICIPATION AND EMPOWERMENT

1. **AWARENESS, ABILITY TO REFLECT AND TAKE ACTION:** At the initial stage, *'awareness'* arises on the need to solve the common problems faced, to overcome the feelings of exploitation and alienation, and to meet the felt common needs.

2. **CAPACITY TO EXERCISE OWN ABILITIES:** Normally, at the beginning of establishing a group or initiating an activity only one or two individuals may be involved. These are the local activists and group or project leaders who have demonstrated their leadership capabilities to mobilise, organise, facilitate and influence their friends to participate in a group with a view to achieving group goals.

3. **GAINING CONTROL OVER THEIR LIVES:** The people's capabilities to think about their problems and needs and to act upon them. This includes establishing various types of community groups, conducting and sustaining group activities, and pressuring and negotiating with the relevant authorities in order to solve their problems and meet their needs. All these things are part of the process by which people come to gain control over their lives.

4. **DEVELOPING AND ENHANCING CONFIDENCE, SKILLS AND KNOWLEDGE:** In the process of participation, individuals learn. This leads to increase in confidence, skills and knowledge which, in turn, further improve people's existing abilities to organise, solve problems, initiate action, and manage group activities.

5. **GAINING AND EXERCISING POWER OVER ANOTHER PARTY:** An examination of people's empowerment should not be viewed only as individuals' abilities to forward their efforts by working together in a self-help fashion to meet their goals within the sphere where they live. People living in one community, at the micro level, also interact with the outside system, at the macro level, in the development process. The linkage between the two levels lies in the context of their living environment itself, for example, in a planned village settlement, a state-sponsored community development scheme, which directly brings the community into the state patronising structure.

6. **SELF-EVALUATION:** Empowerment does not end when people achieve their group goals. Motivated individuals who possess the characteristics of empowerment begin to evaluate their activities. This self-evaluation process, facilitated by an empowering research approach, enables individuals to reconsider changing the dynamics of their group process in order to maximise its benefits. They can see possibilities – that is, ways of improving the activity, the group and the whole working process – which can promote members' involvement.

Source: adapted from Asnarulkhadi Abu Samah & Fariborz Aref, People's Participation in Community Development: A Case Study in a Planned Village Settlement in Malaysia in World Rural Observations 2009;1(2):45-54.

Photo: Gareth Bentley

Photo: Garry Walsh

" A COUNTRY'S SUCCESS, OR AN INDIVIDUAL'S WELL-BEING, CANNOT BE EVALUATED BY MONEY ALONE; LIFE EXPECTANCY, HEALTH, EDUCATION AND FREEDOM ARE VITALLY IMPORTANT "

Photo: Gareth Bentley

Photo: Gareth Bentley

Photo: Gareth Bentley

Photo: Gareth Bentley

"IN SOME BASIC RESPECTS THE WORLD IS A MUCH BETTER PLACE TODAY THAN IT WAS IN 1990... MANY PEOPLE AROUND THE WORLD HAVE EXPERIENCED DRAMATIC IMPROVEMENTS... THEY ARE HEALTHIER, MORE EDUCATED AND WEALTHIER AND HAVE MORE POWER... NONETHELESS, NEARLY 7 BILLION PEOPLE NOW INHABIT THE EARTH. SOME LIVE IN EXTREME POVERTY—OTHERS IN GRACIOUS LUXURY"

Photo: Garry Walsh

Photo: Garry Walsh

WHY SHOULD PEOPLE PARTICIPATE IN DEVELOPMENT?

People should participate in their own development because it is their human right to do so. From a justice perspective, people must be allowed to determine their own road to development and must be allowed to discover and improve their own capabilities. Development is not just economic development: it entails development of the human being in totality. Participation in development therefore promotes the dignity and self-worth of individuals – expanding their capabilities to achieve things they value. By promoting participation in development on the part of individuals and communities themselves, societies are more likely to become engines of self-development, societies that care for people and the environment.

Participation in the active, radical sense described above is the only way to ensure that people take ownership of their development. Acceptance of ownership is probably the most important result of participation and empowerment. Development projects should not be the property of an NGO or a government department. If the project or its results do not belong to the people, no true participation or empowerment took place. A development activity or project only belongs to people who have the 'freedom [to] enjoy valuable beings and doings' as argued by Ingrid Robeyns. The question that must always guide development support and action (and ultimately its ownership) is: who decides? Without participation in the sense of decision-making, ownership is seldom taken or accepted by a group or community. Being left out of decision-making usually leads to apathy or even rebellion; ownership is simply not accepted by the so-called beneficiary. A group of people or a community that participates makes decisions and takes ownership of their own development illustrates their agency.

CASE STUDY: TAKING OWNERSHIP

A community development worker was responsible for the launching of refuse removal projects in informal settlements in and around the metropolitan area of Johannesburg. These projects did two things. First, it helped clean up the area and second, it created a small income for a group of people who would otherwise be unemployed. She came back one day from one of these areas much earlier than usual. When asked why she was not working that morning in that certain township, she reacted by saying: 'This morning when I arrived at Township X before I could drive into the area I was stopped by a group of people I know well. They are the steering committee of the refuse removal project that we have launched. They stopped me and showed me that they wanted to talk to me. They said to me that they are very thankful for what I have done and for my help to get a project going, but they added that they have decided that they can go it alone now. They don't need me anymore.' The question put to her was: 'Do you feel sad that the people have taken the project?' Her answer: 'No, heavens. This is a fantastic thing! The people have accepted the ownership of a project that was always theirs.'

Adapted from H Swanepoel and F de Beer. Community Development: Breaking the Cycle of Poverty. Cape Town: Juta. (2012 forthcoming)

WHO SHOULD PARTICIPATE IN DEVELOPMENT?

While patriarchal practices in much of the world place decision-making power in the hands of men, participation in development must also include children, youth, older persons and, of course, women. Women in the developing world in general bear the burden of the traditional role of housekeeper and also that of breadwinner. In spite of being numerically the largest section of the population, women remain excluded, to a large extent, from professions, decision-making positions and influential roles. Furthermore, in rural areas, traditional practices generally exclude women from equal participation in the rural economy and politics. Access to land, credit and extension services remain largely the privilege of men. Yet, internationally, women rise to and make use of - or create – opportunities to fully participate in democracy and development. In the box below the story is told of a domestic worker in Brazil who becomes a politician.

HOW DO PEOPLE PARTICIPATE IN DEVELOPMENT?

We distinguish between three types of participation:

- in democratic elections at local and national levels
- as members of social movements
- as members of communities.

The obvious answer to the question 'how do people participate in development?' is that they participate through democratic elections at local and national level. Yet not all countries have democratic constitutions and not all allow elections to take place. Even in those countries that have regular, democratic elections, casting a vote does not necessarily mean participation in development – the time span between elections may be too long and manipulation of the process by the political elite might make elections meaningless as a means of participating in development.

Continued on page 116 →

CASE STUDY: FROM DOMESTIC WORK TO ELECTION CAMPAIGNING

Creuza Oliveira is one of Brazil's 9.1 million domestic workers, 95% of whom are women and 60% are black. Many earn less than five dollars a day. She began her career as a domestic worker at the age of ten when her mother put her into unwaged service.

Experiences of racism and sexual abuse, of working long hours without pay and of being treated as if she were sub-human have shaped her activism. She talks of how her employers kept a special bowl for her under the sink, which they would fill with leftovers after she had served them their food.

Creuza became involved in the struggle for domestic workers' rights after hearing a radio DJ interviewing a woman who was running for political office. The woman said she was going to fight for the rights of domestic workers.

Creuza had always felt that it was not right that she was treated the way she was, but it was not until she heard this woman speak that she realised there was something she could do about it. She joined a group of domestic workers who formed their own association, and, when the dictatorship that gripped Brazil for 20 years ended and trades unions were allowed again, they turned their association into a union.

Source: http://www.ids.ac.uk

Politicians visit their constituencies, make promises and often disappear from the constituency soon after the election. In South Africa, people in townships demonstrated against local government's poor service delivery in the 18-24 months period before the 2011 local government elections.

People also participate in development by taking part in development-related activities organised and promoted by social movements. A social movement is a type of informal or formal group action. It may consist of individuals and/or organisations that share a common concern. A social movement may gain enough power to coerce people with power (usually governments) to take decisions or make changes in favour of the cause promoted by the social movement. Social movements may, for instance, promote issues of health, economic development or environmental conservation. Social movements and their activities represent what Paul Ekins (1992) refers to as 'Another Development'.

An example of a social movement is the Sarvodaya Shramadana Movement of Sri Lanka. The movement was initiated in the late 1950s, initially as a self-help movement, but it currently focuses on nonviolent social transformation. In Burkina Faso, the Six S Association, based on the indigenous NAAM movement (a traditional village body of young people doing various activities), is a social movement that creates cereal banks, dams, wells and income-generating projects. Further examples of social movements are the Bangladesh Rural Advancement Committee (BRAC); Working Women's Forum of India and the Seikatsu Club Consumers' Cooperative of Japan.

Third World governments very often pursue policies and strategies that, ostensibly, support participation by people in their own development. A word of warning about these strategies must be voiced, however; governments and politicians make words such as participation, community development and empowerment part of their political language and speeches, and part of government policies and strategies. This they do by giving such words a meaning that suits their agendas – a meaning that may differ significantly from the original meaning of the concepts. It is then up to social movements to organise and pressurise government to take action when politicians fail the people. Chapter 16 summarises the history and achievements of the social movement, the Treatment Action Campaign, in South Africa.

While social movements operate regionally, nationally and even internationally, local initiatives for development are found in communities all over the Third World. If supported in a way that empowers people's active participation, communities take ownership and create developments that address their needs. The case study opposite shows how very poor people in a remote rural area participated in their own development and empowerment and took ownership of development projects. The people in this case study practised their freedom to convert goods and services into functionings (according to the capabilities approach), according to their personal, social and environmental circumstances. In short: they used their agency to bring about development.

CASE STUDY: MOTHERS OF THE MOUNTAIN COMMUNITY

The superintendent of a hospital in a rural area in the foot hills of a mountainous area realised that the same women would bring their children with kwashiorkor to the hospital for treatment. The child would be hospitalised for two weeks and the mother would receive information on a balanced and healthy diet for her child. However, within a month or two she would be back with the same child needing treatment again. The superintendent decided to speak to these women about the feeding of their children.

In his discussions with these women he soon realised that they knew what constitutes a healthy diet, but that they simply did not have the means to provide the right food to their children. The superintendent's discovery led to a discussion between him and the mothers about a food garden to supplement their children's diet. After a while, a group of twenty mothers declared themselves willing to start a garden. They acquired a piece of land in the hospital grounds (with ample water). Their garden was an instant success, so much so that more mothers wanted to join them. When this one project reached its capacity, women started their own gardens on land acquired from the tribal chief. Some women had no feeling for gardening and they decided to start with a small poultry farm where they would raise broilers. Again, their endeavour was met with instant success.

The efforts of the women in food gardening and poultry farming caught the eye of the tribal authority and the service providers in the area. Because of the mountainous terrain there were many springs in the area. Through the offices of the authorities and a few NGOs, groups were created to develop those springs. These springs just had to be developed and the water piped from them to tanks in the various villages. When the first efforts to develop the springs proved to be a fairly easy task, a number of groups emerged with this in mind and a large number of villages got water in this way. As water became easier to use more food gardens appeared.

At this stage an NGO appointed a project manager who got a small flat inside the hospital where she could reside.

With the help of the new project manager, a group of women identified the harvesting and selling of the thatch grass that covered a large portion of the area. A buyer of the thatch was found in Johannesburg. They arranged with the transport services to park a railway truck at the siding which the women would then fill with thatch grass. This project was a great success and really brought prosperity to the area.

Other successful projects launched by the community included the building of three primary schools, a food garden in every school, and a fund for poor children who could not afford their own books.

After about two years since the first project (began by the mothers with kwashiorkor babies), there were about 200 projects in that area and the local people ran these projects with minimal help (from the project manager and a few NGOs).

Source: Adapted from H.J Swanepoel, HJ and F.C. De Beer, 2006, Community Development. Breaking the cycle of poverty, Juta, Cape Town.

CONCLUSION

In the radical interpretation of participation as empowerment, the role of the government, parastatals and NGOs in development should be an enabling and supportive one. However, this understanding of the goal of participation has been discredited by the enthusiastic support it was (and is) given by the international development community and national governments. Consequently, Taylor (2001: 136) argues that participation has never been radical: *"[T]o be genuinely radical ... it is necessary for there to be a 'challenge from below' and a spontaneous coming together of different individuals and groups ... ".* This challenge can come from social movements and community-groups as illustrated by the case studies in this chapter. When this challenge comes people will take empowerment and use their agency to build on their own capabilities and achieve the functionings they decide upon.

READING

Patrick Chabal (2009) Africa: the Politics of Suffering and Smiling, London, Zed Press

Amartya Sen (1999) Development as Freedom, New York, Albert Knopf

Ingrid Robeyns (2005) The Capability Approach: a theoretical survey, Journal of Human Development 6 (1)

Martha Nussbaum (2000) Women and Human Development; the Capabilities Approach, Cambridge University Press, Cambridge

HJ Swanepoel, and FC De Beer (2006) Community Development. Breaking the Cycle of Poverty, Juta, Cape Town

P Ekins (1992) A New World Order: Grassroots Movements for Social Change, London, Routledge

MORE INFORMATION & DEBATE

http://www.sarvodaya.org/about - website for Sarvodaya

http://www.moibrahimfoundation.org/en/section/the-ibrahim-index - explores democracy and governance issues

http://www.sp2.upenn.edu/restes/praxis.html - very useful University of Pennsylvania site with substantial onwards referencing

http://ctb.ku.edu/en/default.aspx - Community Tool Box, a University of Kansas site with practical suggestions and ideas related to community development

http://www.socialwatch.org/node/13749 - site for the Basic capabilities Index of Uruguay-based NGO Socialwatch

http://www.moibrahimfoundation.org/en/section/the-ibrahim-index - explores democracy and governance issues

http://www.sp2.upenn.edu/restes/praxis.html - very useful University of Pennsylvania site with substantial onwards referencing

http://ctb.ku.edu/en/default.aspx - Community Tool Box, a University of Kansas site with practical suggestions and ideas related to community development

http://www.socialwatch.org/node/13749 - site for the Basic capabilities Index of Uruguay-based NGO Socialwatch

['...*wealth, prosperity, influence and future promise for the few; poverty, exclusion, voicelessness, and stagnant hopelessness for the many*']

Michael Kelly SJ

Chapter 7

JUSTICE AND DEVELOPMENT 'AN ILLUSION OF INNOCENCE...?'

Colm Regan

This chapter sketches out many of the key dimensions of the justice perspective on development and underdevelopment. Having posed the question as to whether poverty in the world today is a matter of duty or charity, the chapter surveys the arguments and debates posed by philosophers and activists concerning 'our place in the world'. Case studies explore the challenging issues of HIV and AIDS, sustainable development and global social and political priorities. The chapter concludes with a brief discussion of the 'alternative' ideas of a 'Global Resource Dividend', the 5:10:5:10 movement and a review of the debates around the role of the internet.

KEYWORDS

Positive and negative duties, **poverty and human rights**, HIV and AIDS, **welfarism**, culpability, rectificatory, **distributive** and **regulative** justice, the global sink, **global resource dividend**, role of the internet

How do you see and understand these facts – inevitable or not; avoidable or not; a natural part of the order of things or a creation of human action and inaction?

TODAY, 1.4 BILLION LIVE IN EXTREME POVERTY (ON LESS THAN $1.25 PER DAY)

Is there a relationship between the wealth and welfare of some and the poverty and ill-fare of others and if yes, what is that relationship?

THE AVERAGE LIFE EXPECTANCY FOR A NORWEGIAN IS 81 YEARS IN AFGHANISTAN IT IS 44.6 YEARS

Is the co-existence of the (over) development of one part of the world alongside the (under)development of other parts side by side and in full public view a matter of concern? Why?

IN 2005, 82% OF THE WORLD'S WEALTH WAS CONCENTRATED IN HIGH INCOME OCED COUNTRIES

Are these issues a matter of essentially, providing assistance and support to those in need or one of challenging the gap between rich and poor? Are they a matter of charity or of justice, why?

915 IS THE AVERAGE MATERNAL MORTALITY RATE (PER 100,000 LIVE BIRTHS) IN THE 'BOTTOM 20 COUNTRIES' – THE GLOBAL AVERAGE IS 273

What are our duties with regard to such issues; are we, in any way, obliged to respond and to what extent? Where should our response begin ... and end?

INTRODUCTION

Questions such as those opposite have long underpinned and challenged discussion of development and aid and allied areas for decades. While many of the dominant debates appear to be essentially technical (e.g. how can we improve current international development models and initiatives, in what ways are international economic relations fair or unfair ? etc.), underpinning them has been a set of ethical, philosophical and political questions that go right to the heart of today's world. How do we understand why the world is so unequal; how did (does) this state of affairs happen; how do we fit in and what are our duties (if any) to those in desperate need? Is there a direct connection between wealth and poverty?

In recent years, these and similar questions have been the focus of vigorous debate and disagreement among analysts, commentators and activists, and the debates have raised very fundamental and uncomfortable questions about our place in the world – especially for those of us living in the 'West' or those elsewhere in the world living an affluent lifestyle. Why is it that, while many millions continue to suffer from the most extreme forms of poverty, hunger, ill-health and exclusion, the more affluent members of the world's population (wherever located, including in the 'emerging economies' of the developing world) allow it to continue and why do we apparently do so little to eradicate it? Faced with world hunger and inequality, do we have a moral duty to respond or is it a matter of personal choice and preference? In what ways are we implicated in the current order of things? What begins as a discussion on how to assist those in need is rapidly transformed into a far more challenging, complex and disturbing debate.

EXPLORING HUMAN SUFFERING AND DUTY IN THE WORLD

Australian philosopher and activist Peter Singer has been pre-eminent in addressing many of these questions and concludes that we do indeed have a 'duty' to respond to human suffering in the world for a number of reasons:

- One, such suffering is evil
- Two, people 'must' respond to relieve suffering whenever they are in a position to do so and not to do so is morally wrong
- Three, affluence places very considerable moral demands on the affluent to continue to respond up to the point where they would sacrifice something of equal moral importance if they were to give more.

Arguing that the world requires (and possesses the possibility for) a *new ethics* to match an ever increasing interconnectedness associated with globalisation, Singer goes even further arguing that we are obliged to give as much as we can spare without damaging our own health, our children's education or whatever else is deemed to be *'morally important'*..

'... going out to nice restaurants, buying new clothes because the old ones are no longer stylish, vacationing at beach resorts – so much of our income is spent on things not essential to the preservation of our lives and health. Donated to one of a number of charitable agencies, that money could mean the difference between life and death for children in need.'

Singer insists that our obligations are not limited to those people in need in our own area or community, those that we, in some way *'know'* (our *'immediate circle of concern'*) but extend equally to those we do not know and those whom we will never meet and who exist thousands of miles away; his essential focus is on those living in absolute and avoidable poverty:

'In the present situation we have duties to foreigners that override duties to our fellow citizens. For even if inequality is often negative, the state of absolute poverty that has already been described is a state of poverty that is not relative to someone else's wealth. Reducing the number of human beings living in absolute poverty is surely a more urgent priority than reducing the relative poverty caused by some people living in palaces while others live in houses that are merely adequate.'

He argues that because the cost of eliminating extreme hunger and poverty is very small in the grand scheme of things and makes up such a small percentage of our overall income and wealth, that there is no moral *'get out'* clause that we can appeal to. Equally, he argues that in recent times, we cannot claim ignorance of the plight of others (modern communications has ended that argument) and, as distinct from the past, we now have the mechanisms and structures in place to deliver assistance via an extensive network of official and voluntary aid and development agencies. Singer, echoing the work of Amartya Sen and others, argues that affluent individuals should donate a minimum of 1% of their income to the elimination of absolute poverty – not to do so is *'morally wrong'*:

'Those who do not meet this standard should be seen as failing to meet their fair share of global responsibility and, therefore, as doing something that is morally wrong ... There is something to be said for seeing a 1 percent donation of annual income to overcome world poverty as the minimum (in practice, Singer argues for a much higher level of giving than this) *that one must do to lead a morally decent life. To give that amount requires no moral heroics. To fail to give it shows indifference to the indefinite continuation of dire poverty and avoidable poverty-related deaths'*

Peter Singer (2002) One World: the Ethics of Globalisation, New Haven and London, Yale University Press: 194

For philosophers such as Singer and others (including Peter Unger who challenges *'our illusion of innocence'* and Gareth Cullity who insists that we have both individual and collective responsibilities), the fundamental argument is that if it is within our power to prevent something bad or evil from happening, at little cost to ourselves, then we have a moral obligation to do it. According to these commentators, international issues such as poverty, hunger and aid fall into this framework and arguments about who is ultimately responsible for such situations or those about the effectiveness of aid or the realities of international and local corruption do not, in any way, diminish our obligations. Ultimately, they insist that it is immoral and unjust to allow world poverty and hunger to continue to exist alongside immense and extensive global affluence that is capable of ending such immense human suffering.

For Amartya Sen, our obligations to uphold justice and challenge injustice arise from our innate (and intuitive) human capacity to reason and our sense of what is *'reasonable'* in any given situation – he insists that we have a duty to act *'reasonably'* and that our obligations go way beyond those immediately close neighbours to embrace people throughout the world as, in the twenty-first century we have few *'non-neighbours'* left, our obligations in justice are now inextricably global.

'We could have been creatures incapable of sympathy, unmoved by the pain and humiliation of others, uncaring of freedom, and – no less significant – unable to reason, argue, disagree and concur. The strong presence of these features in human lives does not tell us a great deal about what particular theory of justice should be chosen, but it does indicate that the general pursuit of justice might be hard to eradicate in human society, even though we can go about that pursuit in different ways.'

Amartya Sen (2009) The Idea of Justice: 414-415

For these commentators, the challenge of underdevelopment and its most significant human consequences should not be the focus of a *'welfarist'* approach, one that places the primary emphasis on providing *'assistance'* to the poor and hungry (an approach characterised by Paulo Freire as *'assistencialism'*) but rather should be driven by a relentless pursuit of justice and the rejection of all forms of injustice. Even if we cannot achieve a state of an absolutely *just society* (however defined), we are still obliged to create as just a situation as is reasonably possible in any given circumstance.

Characteristic of this approach is also the view, championed by Thomas Pogge, that we have two sets of duties – positive and negative. While there is considerable debate about the scale of individual and collective positive duties (to act positively to reduce human suffering), there is almost unanimous agreement that we have negative duties (the principle of, at least, *doing no harm*). While we continue to argue about the scale and nature of our duties in the world, we do not have the *right* to continue to act unjustly and that the current world order is profoundly unjust to the very significant advantage of the affluent (the minority) and the disadvantage of the poor (the majority).

A key argument in discussions of our moral duties and obligations in the world revolves around whether they are primarily individual or primarily collective, and around the scale and degree of moral responsibility affluence places on individuals as well as on society.

Singer places primary emphasis on the duty of the individual and argues for what is described as the *'extreme demand'* approach – every individual is morally obliged to respond to the maximum extent regardless of whether others do likewise – each case of need must be judged on its own merits. This approach is described as *'iterative'*. On the other hand, Cullity prefers an *'aggregative'* approach – where the moral demands of affluence are essentially collective (this does not, however, absolve individual obligation) where the demands on the individual, while still significant, are more moderate, and where the demands on society as a whole are greater. Globally, justice and injustice need to be approached collectively. Cullity has also criticised Singer for placing too high a level of demand on affluent individuals, arguing that there are justifiable limits for limiting individual contributions (although he agrees such limits are infinitely greater than what is the current norm, he accepts that there exist *'moral requirements of great self-sacrifice'*).

For those who critically engage with the arguments and analysis of philosophers such as Singer, Cullity and Pogge, there are a number of significant issues to be addressed. These include (but are by no means limited to) the following:

- On what basis should the *'culpability'* or otherwise of the world's affluent citizens be calculated – on the basis of past wrongs from which they have benefitted (slavery, colonial plunder, resource pillaging etc.); participation in the current unjust world economic structures (cheap products, resource use, environmental damage etc.) or simply shared characteristics, such as membership of the human species? How should we assess the degree and nature of culpability and the corresponding duties?

- Which causal factors should be given most weight in our assessment and what are the implications of this – are they individual, collective, institutional?

- How do their analyses relate to the *'rights and responsibilities'* associated with other, more *'minimalist'* views of duty and rights in the world? Are some *'individual rights'* to be *'over-ruled'* by other duties - private property vs. collective need, individual rights vs. environmental duty?

- The *'solutions'* offered by, for example, Singer and Pogge are simplistic (Singer's emphasis on the value and impact of charities and aid organisations or Pogge's Global Resource Dividend - see below) and unworkable.

- They are *'extreme'* in the duties they place on ordinary citizens who, themselves, are *'victims'* (albeit at an entirely different level) of the international political and financial system.

Note: for an excellent introduction to some of these debates, see Alison Jaggar (ed., 2010) Thomas Pogge and His Critics, Cambridge, Polity Press and the Ethics and International Affairs Journal, 19.1 (Spring 2005).

DEBATE IT: Duty and Responsibility in the World

ARGUMENT

RESPONSE

We cannot be held accountable for what happened in the past through slavery, colonial plunder etc.; what happened in the past cannot be undone

If this is true, then we have no right to the benefits of the past and must make recompense. We do have responsibility for what happens now

If we listened to philosophers such as Peter Singer and Thomas Pogge and acted on their suggestions, western economies would be weakened and undermined and this would have negative consequences for everyone including the world's poor

The current western model of development is based on harming others, especially the poor in the Third World and this is unacceptable – we do not have the right to harm others. Anyway, western development is unsustainable in the longer term anyway and has to change

The emphasis on charity and on giving is discredited because aid no longer works, even if once it did

Even if aid doesn't work, we have obligations to help others and we recognise this intuitively in very many ways. Our duty is to ensure that aid does work, much more effectively that is the case today

If I take my duties seriously and act on them and others do not, then it will have no impact and, anyway, I cannot change the system

There are duties that can only be realised collectively e.g. climate justice but this does not absolve us from our own individual duties – to say that it does is merely a self-serving excuse

Who should decide what is morally correct in today's world, there are no agreed international moral codes. So long as I do not deliberately harm someone else, I have done my duty

We all accept a variety of internationally recognised codes of different types around genocide, murder, airline travel, dumping at sea etc., we need these for our own protection as much as for that of others. All major world religions and philosophies accept we have duties to others – even those we may not know

I have rights and responsibilities, especially to my own family, my community and my country – they come first

Agreed but duties do not cease at the end of the street or at our airports, otherwise our own well-being would be compromised

DEBATING THE LINKS BETWEEN (IN) JUSTICE AND (UNDER)DEVELOPMENT

In his analysis of the moral demands of affluence in the world, Australian philosopher Garrett Cullity identifies three key approaches to justice and to our collective responsibilities, all of which have significant implications for current approaches to development (and underdevelopment).

- **Rectificatory justice:** we are collectively responsible for assisting the world's poor because we are collectively responsible for creating and sustaining their poverty. This gives them – the poor, at least, a moral claim on us – the rich. The challenge is to identify and act upon what we ought to do collectively to rectify the injustice. Individually, the challenge is to identify what each person ought to contribute as a 'fair share', regardless of what others might do (or not do). Challenges to such an approach include the following: since most people are neither colonialists nor international financiers, in what ways are we collectively responsible; is responsibility divisible so that shares can be divided individually; can shares of responsibility be discharged individually independently of others; what if others are not discharging their responsibilities etc.? Cullity argues that nothing in these criticisms undermines the principle of individual responsibility.

- **Distributive justice:** this approach argues that it is simply a matter of fact that the world's resources are inequitably distributed, rather than offering an explanation of how that distribution came about, the reality of sustained and extensive inequality in the distribution of resources implies a duty to change the situation. Therefore, it is morally wrong for the affluent across the world not to discharge a fair share of this collective duty. Challenges to this view again include the debate as to what is an appropriate individual share of responsibility and how to discharge it; how are priorities decided and by whom; is my share diminished by others not doing their fair share etc.? And, again, Cullity argues that these questions do not undermine the fundamental one of responsibility.

- **Regulative justice:** this perspective essentially objects to the rules that currently govern international trade and financial accountability; it objects to the rules themselves, rather than the unequal distributions that result from their application. There is compelling and documented evidence that such rules unjustly enforce the poverty of others to the advantage of the affluent and that, therefore, we are collectively responsible for reforming them; and, again, each individual has a role to play in such reform. Many of the same challenges described above apply to this approach also.

As noted earlier, Cullity places primary emphasis on our collective responsibilities rather than simply on our individual responsibilities. The collective or aggregative approach does not undermine individual accountability but it does insist that there are reasonable (and morally defensible) limits to what can be expected of the individual. However, the current situation worldwide does require that we change '… *our practice, in a demanding direction.*' Any justice-based perspective insists that '*we combine a proper recognition of the desperate needs of other people with a full engagement with the goods that provide us with our own interests*'. He challenges the '*comfortable vindication of moral 'common sense' as revealed in the practical attitudes of most affluent people*' and concludes '*We are left with a serious practical challenge; but not, I think, one that is beyond us.*'

The remainder of this chapter explores further many of the principles, ideas and alternatives outlined above with respect to five issues or case studies:

- HIV and AIDS
- The debate on the 'Global Commons'
- The 'Global Resource Dividend'
- Current global social and political priorities
- The 5:10:5:10 movement

CASE STUDY 1: HIV AND AIDS

'AIDS is a justice issue, not primarily a sex issue. ... Perhaps an even more basic issue than economic and gender relations in the countries most affected by AIDS is the justice of the interlocking local and global economic systems that disrupt traditional societies, displace economic and educational infrastructures, and cut off access to kinds of prevention and treatment of disease whose efficacy in Europe and North America is well established'

Lisa Sowle Cahill

Against the backdrop of many decades of activism in Zambia and Southern Africa (where HIV & AIDS remain a core justice issue), researcher, teacher and activist Michael Kelly SJ argues that the virus is driven by four forces: poverty, gender disparities and power structures, stigma and discrimination, and exploitative global socio-economic structures and practices:

'The more these thrive, the more HIV and AIDS will flourish. Equally, the more HIV and AIDS prosper, the greater the likelihood that poverty, gender disparities and power structures, stigma and discrimination, and disruptive socio-economic structures and practices will flourish and ensure the continuation of the epidemic.'

These forces constitute *'a network of domination, oppression and abuse'* that excludes millions of human beings from sharing in, building up and enjoying a more just and equal world; he criticises the two current dominant responses to the virus, one a biomedical and pharmaceutical response and the other which places emphasis on human behaviour practices for their concentration on the immediate causes and effects of HIV/AIDS and their failure to deal with the underlying and structural causes of the epidemic. Although there is global agreement that prevention should be the mainstay of the response to the epidemic, the policy directions currently being

promoted seem destined to remain a never-ending struggle with the immediate causes of the epidemic - sexual behaviour, mother-to child transmission, blood supplies, and injecting drug use.

'But because initiatives do not directly or sufficiently concern themselves with issues of poverty, inequalities in society, gender disempowerment, or north-south relations, the epidemic is likely to maintain the upper hand and perpetuate its unjust outcomes.'

The heavy emphasis of recent years on the behaviour change response is criticised by Kelly for its unspoken assumption that different patterns of behaviour are real options for individuals and for its failure to address the social factors that shape behaviour.

In his analysis, Kelly emphasises a number of key 'justice' elements mediating the issue:

- Central to effective treatment are good nutrition, proper medication for common illnesses and opportunistic infections, the availability and accessibility of antiretroviral drugs (ARVs), the medical and social infrastructure that can deliver and monitor treatment and supportive, human care. However, it is extremely difficult, and in too many cases almost impossible for poor households to satisfy these requirements. Recent economic policies and strategies which place primary emphasis on economic reforms have negatively impacted on the nutritional status of the poor and on the ability of health services to respond to people's most basic needs.

- Antiretroviral drugs and therapy provide vital and effective treatment for those affected but today, according to Kelly, ART reaches less than one in five of those in need; the rest are left to die and *'notwithstanding its enormous* benefits ... *the treatment of AIDS bristles with still unanswered justice, equity, ethical and practical questions'* regarding the costs and sustainability of ART provision throughout the life of an infected person (currently the lives of these people depend on political and economic decisions made in the developed countries). It should also be noted that the recent decision to significantly reduce support for the Global AIDS Fund will make the situation even more unjust and indefensible.

- To be poor and female increases vulnerability; as has already been explored in detail in Chapter 10. On physiological grounds alone, women and girls are at greater risk of infection than men and boys and women's vulnerability is substantially increased by a range of social, cultural, economic and legal factors - all embedded in extensive gender inequality. For Michael Kelly: *'Several established practices in society also have the twofold outcome of demeaning women and enhancing their risk of HIV infection ... sexual violence, indulgence towards men who take sexual liberties, the practice of older married men of having a 'girlfriend' on the side, early marriage, widow inheritance, ritual cleansing, and dry sex ...*

'For the greater part, this stalking of women by HIV and AIDS arises from society's unjust allocation to them of an inferior status. Were it not for the unjust treatment and exploitation that women experience, the epidemic would not have its current worldwide grip. It would not have its current stranglehold on southern Africa. Fewer men would be infected. Far fewer women would be infected, and because this would reduce the incidence of parent–to–child transmission, fewer children would be infected.'

- Kelly argues that globalisation has resulted in *'wealth, prosperity, influence and future promise for the few; poverty, exclusion, voicelessness, and stagnant hopelessness for the many'*. This situation has significantly increased the vulnerability of countries, communities and individuals to HIV and AIDS, especially in circumstances where poverty and inequity, working together, provide a fertile breeding ground for the epidemic. He argues that the World Bank and the International Monetary Fund (IMF) have given priority to economic stability, far ahead of every social need and human right.

Currently, international trade, intellectual property rights issues and the repayment of debt (see Chapter 8) do not favour poor countries in Africa or elsewhere in the world; instead they are heavily weighted in favour of the wealthier countries, while creating barriers to the market access of goods from poorer countries; this reality maintains countries in their poverty thus impacting directly on HIV and AIDS.

In conclusion, Kelly reiterates three key justice principles:

- There is strong synergy between the AIDS epidemic and four basic root causes: poverty; gender disparities and power structures; stigma and discrimination; and exploitative global economic structures and practices

- Responding to HIV and AIDS is intimately connected with the practice of justice

- AIDS and justice issues are so intimately linked that action on behalf of justice will almost automatically be action against the epidemic. Equally, action against the epidemic will be action on behalf of justice.

Note: For a more extended outline of Michael Kelly's argument, see http://www.jctr.org.zm/downloads/hivkelly.pdf

CASE STUDY 2: THE DEBATE ON THE 'GLOBAL COMMONS'

Philosopher Peter Singer provides an additional illustration of a justice perspective in the context of sustainable development issues with a case study of what he terms the 'global sink'. He begins his discussion in the following way:

'Imagine that we live in a village in which everyone puts their waste down a giant sink. No one quite knows what happens to the wastes after they go down the sink, since they disappear and have no adverse impact on anyone, no one worries about it. Some people consume a lot, while others with more limited means, have barely any, but the capacity of the sink seems so limitless that no one worries about the difference. As long as that situation continues, it is reasonable to believe that, in putting waste down the sink, we are leaving 'enough and as good' for others, because no matter how much we put down the sink, others can also put as much as they want, without the sink overflowing.'

The phrase *'enough and as good'* is derived from the arguments of John Locke made in 1690 in his justification for the existence of private property. Locke argued that the earth and all its contents *'belong to mankind in common'* and that the use of such *'commons'* for private gain is acceptable so long as there is *'enough and as good'* left in common for others. Singer draws on Locke to argue that the sink analogy is relevant if its capacity is, or appears to be, limitless so that everyone can put in as much waste as desired so long as this does not hinder others' capacity to do likewise.

However, Singer argues that if the capacity of the sink is reduced or is found not to be limitless the argument is seen to be fallacious. There are common consequences if the sink's capacity is fully used up – smells arise, local water holes (where children may swim) become unusable and many begin to fear that if the use of the sink is not reduced, the village water supplies may become contaminated. If we continue to dump waste in the sink we no longer leave *'enough and as good'* for others and hence *'our right to unchecked waste disposal becomes questionable'* as others right to use the sink is reduced or eliminated if we wish to avoid the unwanted results. This he describes as the *'tragedy of the commons'* in that the sink needs to be used equitably but how? This challenge raises fundamental questions of distributive justice. Singer draws a parallel between such a local sink and the global atmosphere where its capacity to absorb our gases without harmful consequences has been, at best, severely restricted and that the benefits and goods generated by the *'global sink'* have been so unevenly distributed internationally.

Singer goes further:

'The average American, by driving a car, eating a diet rich in the products of industrialised farming, keeping cool in summer and warm in winter, and consuming products at hitherto unknown rates, uses more than fifteen times as much of the global atmospheric sink as the average Indian. Thus, Americans, along with Australians, Canadians, and to a lesser degree Europeans, effectively deprive those living in poor countries of the opportunity to develop along the lines that the rich ones themselves have taken. If the poor were to behave as the rich now do, global warming would accelerate and almost certainly bring widespread catastrophe.'

He argues that since the wealth of the industrialised world is inextricably linked to their extensive use of carbon fuels, it is a small step to conclude that the present global distribution of wealth is the result of the wrongful expropriation by a small fraction of the world's population of a resource that belongs to all human beings *'in common'*. Furthermore, if we ask when the level of greenhouse gases contributed by each individual in the developing world will equal those currently contributed by each individual in the developed world, this will only occur towards the end of this century – so, the argument is simple if, as far as the atmosphere is concerned, the developed world *'broke it'* then the onus is on those nations to *'fix it'* – the polluter pays principle.

For Singer, there are fundamental justice issues implicated in current debates about the global atmosphere and the many technical and political objections and reservations about proposals for change do not alter the fundamental moral and justice issues involved. The solution favoured by Singer is based upon a combination of an equal per capita entitlement share to future global sink capacity tied to current UN population projections and a system of global emissions trading (where countries can buy or sell shares internationally). There are, however, many ongoing debates and disagreements regarding such solutions.

Whatever the objections to suggested solutions there might be, for Singer, there is no ethical basis for the present distribution of the atmosphere's capacity to absorb greenhouse gases without drastic climate change.

'If we, as citizens of the industrialised nations, do not understand what would be a fair solution to global warming, then we cannot understand how flagrantly self-serving the position of those opposed to signing even the Kyoto protocol is.'

Note: for a more extended discussion of the issues, see Peter Singer (2002) One World: the Ethics of Globalisation, New Haven and London, Yale University Press; Cristovam Buarque (1993) The End of Economics?: Ethics and the Disorder of Progress, London, Zed Press – for an interesting 'alternate' perspective on the debate on the Amazon region, also see Buarque's argument on http://www.diaplous.org/amazo.htm).

CASE STUDY 3: THE GLOBAL RESOURCE DIVIDEND

German philosopher and co-founder of Academics Stand Against Poverty, Thomas Pogge argues in World Poverty and Human Rights (2nd edition, 2008) that one of the greatest challenges to any morally sensitive person today is the extent of global poverty; he argues that there are two ways of understanding how such poverty is a major moral challenge:

- one, we may be failing to fulfil our positive duty to help others in acute distress
- two, we may be failing to fulfil the more strict negative duty not to uphold, contribute to, or profit from the unjust impoverishment of others

He argues that some people believe that the existence of *'radical inequality'* in the world is a sufficient condition to violate our negative duties to others.

He summarise five key conditions that define radical inequality:

- The worse-off are very badly off in absolute terms
- They are also very badly off in relative terms – very much worse off than many others
- The inequality is impervious: it is difficult or impossible for the worse-off substantially to improve their lot; and most of the better off never experience life at the bottom for even a few months and have no vivid idea of what it is like to live in that way
- The inequality is pervasive: it concerns not merely some aspects of life, such as the climate or access to natural beauty or high culture, but most aspects or all
- The inequality is avoidable: the better off can improve the circumstances of the worse-off without becoming badly off themselves

These conditions alone are, he argues, insufficient to prove a violation of negative duties in the context of global poverty today and adds a further five conditions:

- There is a shared institutional order that is shaped by the better off and imposed on the worse off
- This institutional order is implicated in the reproduction of radical inequality in that there is a feasible institutional alternative under which severe and extensive poverty would not persist
- The radical inequality cannot be traced to extra-social factors (such as genetic handicaps or natural disasters) which, as such, affect different human beings differentially
- The better-off enjoy significant advantages in the use of a single natural resource base from whose benefits the worse-off are largely, and without compensation, excluded
- The social starting positions of the worse-off and the better-off have emerged from a single historical process that was pervaded by massive, grievous wrongs

These conditions lead Pogge to argue that:

'... the existing radical inequality is unjust, that coercively upholding it violates a negative duty, and that we have a moral reason to eradicate world poverty.'

This analysis leads Pogge to make a *'modest'* proposal for reform in the shape of a Global Resources Dividend (GRD) through which '... *those who make more extensive use of our planet's resources should compensate those who, involuntarily, use very little.'*

He proposes a modest GRD of approximately $300 billion which could radically alter the situation of approximately 2,533 million people experiencing severe poverty while only amounting to half the defence budget of the US alone; half the 'peace dividend' of the high income countries in 2007 as a result of the ending of the Cold War and about one-seventh of the market value of current oil production. This GRD is modest and could be developed further depending on agreed objectives and timeframes.

Pogge concludes *'It is clearly possible – without major changes to our global economic order – to eradicate world hunger within a few years by raising a sufficient revenue stream from a limited number of resources and pollutants.'*

Thomas Pogge (2008) World Poverty and Human Rights, 204 -212

CASE STUDY 4
CURRENT GLOBAL SOCIAL AND POLITICAL PRIORITIES

$1.29 TRILLION – ESTIMATED COST OF THE WARS IN IRAQ AND AFGHANISTAN BY END 2011
$1.63 TRILLION – WORLD MILITARY SPENDING, 2010
$8.44 TRILLION – OUTFLOW OF ILLICIT RESOURCES FROM DEVELOPING COUNTRIES 2000 – 2009
$12 – 13 TRILLION – IMF 2009 LOWER ESTIMATE OF THE COST OF BANK BAILOUTS FOLLOWING THE FINANCIAL CHAOS OF THE PREVIOUS TWO YEARS

Sources: US Center for Defence Information 2011, Stockholm International Peace Research Institute 2011 and Global Financial Integrity Report 2011

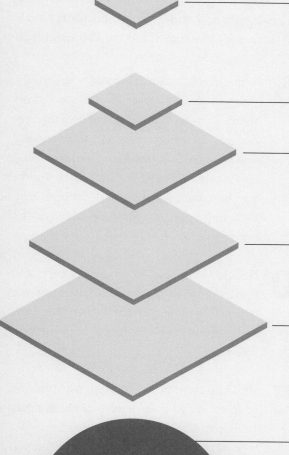

$1.2 BILLION

Estimated annual cost of saving the lives of the 529,000 women who die annually from complications during pregnancy, childbirth or immediately following it - nearly all of these deaths occur in developing countries (source: World Bank 2011).

$1.5 BILLION

Cost of one US Stealth Bomber

$7 BILLION

Estimated yearly cost of supplying adequate clean water and effective sanitation for all worldwide (source: UN Millennium Task Force 7 Report 2004)

$7.5 BILLION

Yearly cost of scaling up health investments to reduce child mortality rates by 2/3rds (WHO Commission on Macroeconomics and Health 2002)

$12.4 BILLION

US spend on cosmetic surgery (breast augmentation, liposuction, nose reshaping, eyelid surgery and tummy tuck) in 2007 (source: Aesthetic Medicine News, 2012)

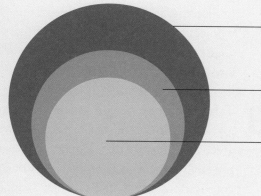

$51 BILLION

estimated cost of eliminating malaria worldwide by 2020 (source: World Health Organisation

$30 BILLION

cost of eradicating world hunger (FAO report)

$20 BILLION

Minimum estimated cost of eliminating malaria worldwide by 2020 (source: World Health Organisation

$150 BILLION

estimated annual amount of extra aid needed to achieve all the MDGs

Oxfam

$160 BILLION

estimated cost of tax evasion to developing countries by Multinational Companies

Action Aid Report 2009

$167 BILLION

estimated amount the US Department of Energy plans to spend on maintaining the U.S. nuclear weapons stockpile

Institute for Policy Studies, 2011

CASE STUDY 5 – FAIRSHARE INTERNATIONAL AND THE 5:10:5:10 PROPOSAL

FairShare International, which began in Australia, is an international movement of individuals, families and businesses who choose to take a stand for a fair share for *'everyone, always'*.

FairShare argue:

'…We are against the unjust distribution and misuse of the world's resources – money, water, energy and minerals. We are for a fair go for all people.'

'Of the earth's water, energy and minerals used each day, 80% is consumed by 20% the earth's wealthy population (about a billion of us). Leaving almost nothing for the majority of people and compromising the existence of future generations, this consumption is neither fair nor sustainable. By joining the FairShare community, you can help achieve:

- *decreasing the gap between rich and poor*
- *better use of the earth's resources*
- *more ethical connections with others*
- *conservation of the natural environment and threatened species'*

✎ MORE INFO

- See www.fairshareinternational.org and also see Alex Evans, (2011) Resource scarcity, fair shares and development. WWF-UK / Oxfam Discussion Paper (http://assets.wwf.org.uk/downloads/wwf_oxfam_scarcityfairsharesdev2011.pdf)

FairShare proposes a practical formula - 5.10.5.10 - to put these four goals to work daily and asserts that such goals offer new and ethical standards for life in the 21st century world of affluence. The formula is:

5 Redistribute your personal wealth by giving away at least 5% of your gross annual income to projects and programmes that provide direct assistance to financially disadvantaged individuals, families and communities 'anywhere on earth, and/or for the care of the natural environment, including conservation of the earth's resources'.

10 Reduce your use of water, energy and minerals by at least 10%, based on the national average per capita consumption, and sustain such reduced consumption forever.

5 Build up community through contributing at least 5% of your leisure time annually in direct, 'face to face' assistance to people who could use your help or with others in your community who are tackling social or environmental challenges.

10 Take democratic action at least ten times a year to correct practices associated with greed and injustice that damage people and the environment (e.g. letters to politicians, corporations and the media; support campaigns etc.).

MEASURING SOCIAL JUSTICE – EXCLUSION OR INCLUSION?

'Political actors – as well as individual citizens – must therefore act to uphold the principles of a sustainable and socially just order in which economic strength and social justice do not undermine but complement and facilitate each other.'

This is one of the final conclusions of the 2011 study entitled Social Justice in the OECD – How Do the Member States Compare? undertaken by German foundation Bertelsman Stiftung. The report seeks to outline how the concept of social justice can be *'operationalised'* and then measured through the use of a series of quantitative and qualitative indicators to produce a Social Justice Index.

The report recognises that social justice is a contested concept that is difficult to define and then agree and is even more difficult to measure. However, building on the work of others such as John Rawls, Amartya Sen and John Roemer (and evidence collated from the Stiftung's own research on governance indicators), the report develops the Index and its 'measurement' of social justice in 31 OECD countries.

Report author Daniel Schraad-Tischler argues that the modern concept of social justice refers to the aim of realising equal opportunities and life chances and offers a vision or ideal capable of obtaining the social consensus needed and argues that establishing social justice depends less on compensating individuals and groups for *'exclusion'* and more on investing in *'inclusion'*. Such a conception of justice is concerned with guaranteeing each person *'genuinely'* equal opportunity through focused investment in individual capabilities.

The Justice Index focuses on six specific areas - poverty prevention, access to education, labour market inclusion, social cohesion and non-discrimination, health and intergenerational justice; and overall, the index has 21 quantitative and 8 qualitative indicators. Not all dimensions are accorded equal 'weight' and the first 3 carry the greatest weight.

SOCIAL JUSTICE INDEX

Poverty prevention (triple weight)	Access to education (double weight)	Labour market inclusion (double weight)	Social cohesion and non-discrimination (normal weight)	Health (normal weight)	Intergenerational justice (normal weight)
• Poverty rate • Child poverty • Senior citizen • Poverty	• Education policy (qualitative) • Socioeconomic background and student performance • Pre-primary education	• Employment rate • Older employment • Foreign-born to native employment • Employment rates by gender women/men • Unemployment rate • Long-term unemployment • Youth unemployment • Low-skilled unemployment ratio	• Social inclusion policy (qualitative) • Gini-coefficient • Non-discrimination policy (qualitative) • Income inequalities women/men • Integration policy (qualitative)	• Health policy (qualitative) • Infant mortality • Healthy life expectancy • Perceived health in relation to income levels	• Family policy (qualitative) • Pension policy (qualitative) • Environmental policy (qualitative) • CO2-Emissions • Research and development • National debt level

Key Conclusions from the 2011 Index

Iceland and **Norway** are the most *'socially just'* countries in the 31 countries surveyed while **Turkey** (which ranks in the bottom five in each of the six targeted dimensions) is the least so

The north European states - **Iceland**, **Norway**, **Denmark**, **Sweden** and **Finland** score highly on access to education, social cohesion and intergenerational justice and lead the index by some distance (yet **Sweden** has a youth unemployment rate three times higher than the general rate)

Most central and north western European states rank in the upper midrange with the **Netherlands**, **Switzerland** and **France** ranking higher than **Germany** and the **UK**

Ireland, Hungary, Poland and **Slovakia** rank in the lower midrange together with their southern European neighbours and the **Czech Republic** ranks highly because of its very low poverty levels

All southern European countries lie considerably below the OECD average, with **Turkey** and **Greece** at the bottom; in both these countries, fair access to education and intergenerational justice are particularly underdeveloped

While **Canada** is the top performer among the non-European OECD states, **Australia** despite its relatively inclusive labour market struggles with problems in poverty prevention and educational justice performs poorly

Japan and **South Korea,** where income poverty is relatively widespread, fail to rank above the bottom third and **Japan** receives particularly low marks for intergenerational justice

The **United States** lies near the bottom with *'alarming poverty levels'* ranking only slightly better than **Mexico** and **Chile**.

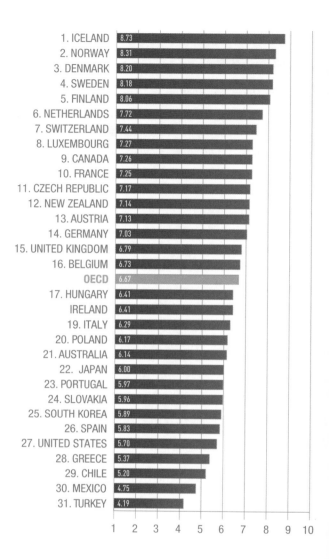

1. ICELAND	8.73
2. NORWAY	8.31
3. DENMARK	8.20
4. SWEDEN	8.18
5. FINLAND	8.06
6. NETHERLANDS	7.72
7. SWITZERLAND	7.44
8. LUXEMBOURG	7.27
9. CANADA	7.26
10. FRANCE	7.25
11. CZECH REPUBLIC	7.17
12. NEW ZEALAND	7.14
13. AUSTRIA	7.13
14. GERMANY	7.03
15. UNITED KINGDOM	6.79
16. BELGIUM	6.73
OECD	6.67
17. HUNGARY	6.41
IRELAND	6.41
19. ITALY	6.29
20. POLAND	6.17
21. AUSTRALIA	6.14
22. JAPAN	6.00
23. PORTUGAL	5.97
24. SLOVAKIA	5.96
25. SOUTH KOREA	5.89
26. SPAIN	5.83
27. UNITED STATES	5.70
28. GREECE	5.37
29. CHILE	5.20
30. MEXICO	4.75
31. TURKEY	4.19

Scores for Ireland, the UK and the US (of 31 countries)

Dimension	Ireland	U.K.	U.S.
Poverty Prevention	15	19	29
Child Poverty	16	19	28
Access to Education	29	21	20
Public Expenditure on Early Childhood Education	31	23	22
Labour Market Inclusion	23	15	16
Unemployment and long-term unemployment	29	15	22

Source: Daniel Schraad-Tischler (2011) Social Justice in the OECD – How Do the Member States Compare? Sustainable Governance Indicators 2011, Gütersloh, Bertelsmann Stiftung

READING

Garrett Cullity (2004) The Moral Demands of Affluence, Oxford, Oxford University Press

Valerie Duffy and Ciara Regan (2010) This Is What Has Happened...HIV and AIDS, women and vulnerability in Zambia, Bray, 80:20 Educating and Acting for a Better World

Alex Evans, (2011). Resource scarcity, fair shares and development. WWF-UK / Oxfam Discussion Paper

Thomas Pogge (2008 2nd ed.) World Poverty and Human Rights, New York, Polity Press

Michael Sandel (2009) Justice: What's the Right Thing to Do? London, Allen Lane

Amartya Sen (2009) The Idea of Justice, London, Allen Lane

Peter Singer (1993) A Companion to Ethics, London, Blackwell

Peter Unger (1996) Living High and Letting Die: Our Illusion of Innocence, Oxford, Oxford University Press

MORE INFORMATION & DEBATE

http://www.oxfam.org/en - information and campaigns on economic issues

http://www.eurodad.org - site for 54 NGOs working together on debt, finance and poverty issues; see also http://www.debtireland.org

http://www.tradejusticemovement.org.uk - UK based focus on trade issues by 60 member organisations

['*...by far the main beneficiaries of trade liberalization have been the industrial countries*']

Chapter 8

5:50:500
THE ECONOMICS OF INEQUALITY

Bertrand Borg and Colm Regan

This chapter explores the various international economic, financial and trade related mechanisms and structures that fuel and support the current international economic system to the advantage of the rich and the disadvantage of the poor. It presents a simple illustrative equation – 5:50:500 to capture the core parameters of this reality. Issues such as the brain drain, the EU's Common Agricultural Policy, trade protectionism and capital flight are also briefly reviewed. The chapter concludes with three case studies of campaigning organisations seeking fundamental reform of this system.

KEYWORDS

Financial flows from poor to rich, **debt repayments**, trade protectionism, **tariff escalation**, subsidies, **rules of origin**, the 'Washington Consensus', **EU Common Agricultural Policy**, trade related intellectual property rights, **brain drain**, corruption, **WTO, IMF**

INTRODUCTION

Casual readers of a very official looking report by the Secretary General of the UN to the General Assembly at its 65th session on July 30th, 2010 (Item 18 (b) of the provisional agenda, Macroeconomic Policy Questions) entitled International Financial System and Development would be forgiven for thinking it contained a serious misprint. The Report noted that developing countries as a group continued to provide net financial resources to developed countries in 2009, amounting to $US513 billion! Surely this could not be true – the Poor World transferred $US513 billion to the Rich World? In one year? Alas, it was not a typo but a statement of fact.

The Report went on to note (without irony) that:

'...while still substantial, this amount is notably lower than the record high of $883 billion reached in 2008... The decrease reflects the transitory narrowing of global imbalances as a consequence of the global economic and financial crisis. The structure of flows underlying the reversal of the increase in financial transfers in 2009 indicates, for the most part, a disorderly unwinding of accumulated global imbalances'.

However, the Secretary General's Report also noted that such *'transfers'* were expected to revert to previous levels with an *'...expected increase in outward resource transfer from developing countries in 2010'*. The estimated figure for 2010 was $US641.2 billion.

The Annex to the Report calculated the yearly net transfer to developed economies from developing economies from 1998 to 2010 (estimate) which amounted to a staggering $US5, 740.6 billion (see page 152).

Source: United Nations General Assembly A/65/189, July 2010 International financial system and development, Report of the Secretary-General

We often hear talk of how much we give to the world's poor, inundated as we are with billboards, TV advertisements, newspaper articles and fundraising campaigns all calling for our financial generosity. As a result, the dominant perspective in the West is that the relationship between the Rich and Poor worlds is one where *'we give and they get'* – it is a measure of the success of the dominant *'story'* in development and its associated *'communications'* that the public in the West believes to be true the exact opposite of what the facts indicate.

Our worldview is significantly mediated through the lens of charity and *'assistencialism'* (to use the phrase of Brazilian educator Paulo Freire). Even the title in the Annex referred to above in the UN Secretary General's Report is titled *'Net transfer of financial resources* **to** *developing economies and economies in transition, 1998-2010'*, when the evidence presented highlights the opposite and should read *'Net transfer of financial resources* **from** *developing economies and economies in transition, 1998-2010'*.

This chapter describes and explores how this reality operates and is based around a simple equation 5:50:50 as described below.

5:50:500 IN SUMMARY

The basic idea behind 5:50:500 is simple (and is based on OECD and IMF figures):

- **Every year, for the past 10 years aid given to the Developing World by non-governmental agencies (voluntary aid) has amounted to at least $5 billion. Estimates vary from a low of $5 billion to a high of $15 billion, so we have chosen the lower figure**

- **Every year for the past 10 years aid given to the Developing World by governments (official aid) has amounted to at least $50 billion. The actual** figures vary considerably year by year from $50 billion in 2000 to $105 billion in 2007, so we have chosen the lower figure

- Every year over the past 10 years, the Developing World loses an average of $500 billion as a result of the operation of the current unjust international economic system. The figure for 2007 was $792 billion, that for 2008 $895 billion and the estimate for 2010 $641 billion, so we have chosen the lower figure

So, in 2007, the actual ratio was 8:105:792 and we have chosen to highlight the ratio 5:50:500 because it is on the conservative side of the estimates and because it graphically captures the degrees of magnitude involved in what can be fairly described as the unjust relationship between the 'Rich' and 'Poor' worlds.

HOW 5:50:500 WORKS

Each year:

- Interest payments on total Third World debt of some $375 billion = $34 billion minimum
- The cost to the Developing World of unjust trade barriers = $130 billion minimum
- The cost of 'trade related intellectual property' issues = $40 billion minimum
- The cost of corruption and capital flight = $193 billion minimum
- The cost of the brain drain = $100 billion minimum

So, in short, for every $55 billion transferred to the Poor World by the Rich World, $500 billion is transferred in the opposite direction.

Note: all these figures, together with a number of others, can be found within the IMF's 2008 World Economic Outlook database at http://www.imf.org/external/ns/cs.aspx?id=28. Alternatively, the United Nations General Assembly has issued a report on The International Financial System and Development, which can be found at http://www.un.org/Docs/journal/asp/ws.asp?m=A/63/96

The remainder of this chapter explores the key dimensions of this equation and tries to illustrate the following realities behind the equation.

- Walk into a supermarket in Lusaka, Zambia, and in many sections you'll find European products (butter, tomatoes etc.) being sold cheaper than locally-produced, Zambian products

- The vast majority of the world's cocoa is grown in West Africa, and yet you never see any West African chocolate bars

- The EU's single most expensive endeavour, its Common Agricultural Policy, costs EU citizens €55 billion every year in subsidies to our farmers. But what does it cost farmers in Kenya, Bolivia or Bangladesh?

- Have you ever stopped to wonder where the various Filipino nurses and Indian doctors who staff our hospitals were educated, or who bankrolled their studies?

- How is it that pharmaceutical multinationals make billion-dollar profits while millions die every year due to easily curable ailments?

The 500 in our ratio is concerned with all these questions (many of the issues relating to the 5 and the 50 in our equation are addressed in Chapter 9 below). This chapter is based, in part, on a more detailed resource by Bertrand Borg, Dylan Creane and Colm Regan (2009), 5:50:500 or how the world rewards the rich at the expense of the poor and where we fit in!, (see the reading at the end of this chapter).

DEBT REPAYMENTS

A variety of grassroots and NGO-driven campaigns have sprung up over recent years around the debt agenda. Initiatives such as the Jubilee Debt Campaign and massive events like the Live 8 concert have focused public attention on debt repayments and called for broad (at times even across-the-board) cancellation of debt. And, listening to Bono or Bob Geldof's statements, you'd be forgiven for thinking that debt repayments are a thing of the past.

Unfortunately, despite world leaders' pledges and celebrities' euphoria, debt repayment continues to cripple the world's poor. Only 10% of the debt owed by the world's poorest countries has been cancelled. According to 2010 World Bank figures (Global Development Indicators 2010), the combined debt of Low and Middle Income Countries (the poorest worldwide) amounted to $3.5 trillion in 2009. In the majority of cases, these countries can barely pay off the interest accrued on these debts. In 2009, Low and Middle Income Countries paid $536.6 billion in interest alone.

For example, in 2009, Africa's external debt stood at around US$300 billion with African countries spending about 16% of the continent's export earnings on servicing their external debt. While this is significantly less than during the height of the African debt crisis of the 1990s, it still diverts scarce financial resources from key areas of spending for development and growth. Interest payments accounted for more than 14% of total expenses in 15 countries for 2009, including Jamaica, Ghana, Bangladesh, India, Egypt, Brazil and Columbia. As a result, Africa's external debt has *impeded economic growth and sustainable human development in numerous countries* (Office of the UN Special Advisor on Africa, 2010).

CLOTHING, INDIA'S SECOND LARGEST EXPORT TO THE U.S., IS TAXED AT 19%. IMPORTS FROM COUNTRIES SUCH AS FRANCE, JAPAN AND GERMANY ARE CHARGED AT BETWEEN ZERO & 1%

IMPORT TAXES IMPOSED ON GOODS FROM DEVELOPING COUNTRIES ARE, ON AVERAGE, 4 TO 5 TIMES HIGHER THAN TARIFFS APPLIED ON TRADE BETWEEN RICH NATIONS.

A SHIRT MADE BY A WORKER IN BANGLADESH ATTRACTS 20 TIMES MORE IMPORT TAX WHEN IT ENTERS THE U.S. THAN ONE IMPORTED FROM BRITAIN.

FOR DUTCH GOODS IT IS JUST 1%, WHICH MEANS VIETNAM - A COUNTRY WITH 81 MILLION PEOPLE LIVING IN POVERTY - PAYS MORE IN US CUSTOMS DUTIES THAN THE NETHERLANDS, WHICH EXPORTS FOUR TIMES AS MUCH TO THE U.S.

BARRIERS TO TRADE

In 2001, the Secretary General of the UN received a report from the 'High Level Panel on Financing for Development' which noted that, after eight rounds of multilateral trade negotiations:

'.... by far the main beneficiaries of trade liberalization have been the industrial countries. Developing countries' products continue to face significant impediments in rich country markets. Basic products in which developing countries are highly competitive are precisely the ones that carry the highest protection in the most advanced countries. These include not only agricultural products, which still face pernicious protection, but also many industrial products subject to tariff and non-tariff barriers.'

Report of the High-Level Panel on Financing for Development 2001:7 http://www.un.org/reports/financing/full_report.pdf

The UN Panel also estimated that the Third World was losing US$130 billion each year as a result of trade barriers, and reckoned that even a 50% tariff cut on imports from developing countries could generate up to US$155 billion in extra revenue every year. Trade barriers take various forms, and range from straightforward taxation-based measures such as import tariffs to hidden costs to trade, such as overly-stringent health and safety regulations.

Barriers to trade, such as import tariffs, are often discussed, but *'hidden'* trade barriers receive much less attention, often because their protectionist nature is veiled by pretexts unrelated to trade. Trade barriers (be they overt or hidden) are always to the detriment of the exporting nation. This does not make them inherently bad. The problem for developing nations is that international trade regulations are skewed to favour the rich and powerful – trade has been effectively forced open with respect to the items that rich countries are good at manufacturing (technology and services) and have remained firmly closed in the items where the rich are not so competitive – agriculture and textiles. Reciprocal tariff reductions are reciprocal in name only, as subsidies, hidden trade barriers and the sheer financial muscle of large-scale corporations squeeze out smaller-scale operations from the developing world. Such double standards are inherent throughout the international trading system.

Tariff escalation is one of the European Union's favoured tariff barriers. *'Tariff escalation'* is the raising of tariffs in line with a product's level of processing. In other words, unrefined products (commodities) such as raw fruit and vegetables are allowed into EU markets tax-free, but processed variants of these, such as fruit juices or canned foods, are taxed.

Moving up the product refinement cycle is in every country's interest, since higher-processed goods command a price premium, and do not suffer the degree of price volatility which plagues commodities. Think of the cost of a raw leather hide, and then compare it to the price you pay for a pair of new boots in any high street, and it becomes clear that product refinement is a far more lucrative business than simply harvesting commodities.

Escalating tariffs discourage Third World countries from refining their export commodities, ensuring that this lucrative business is left to developed Western states and leaving the poor where they started – exporting low-value commodities which are extremely price volatile. For example, Nigeria drills and exports unrefined oil, only to re-import refined oil; Zambia exports copper only to re-import pots and pans, and Peru exports leather hides and then re-imports leather jackets and shoes.

Sometimes it is seemingly innocuous things that are *de facto* hidden barriers to trade. In January 2009 the EU declared a number of insecticides unsafe, and announced that products carrying any traces would be refused importation – despite the science behind the ban being highly contested. These insecticides, which the World Health Organisation (WHO) recommends for use in high-risk malarial regions, have now been arbitrarily declared unsafe by the European Commission, leaving a number of African states with a difficult choice: stop using the insecticides in malaria infested regions and endanger the lives of those who live there, or continue to protect their citizens from malaria and forego the lucrative EU export market. See, for example, a discussion on the challenges facing Thailand as a result of the ban at http://www.businessreportthailand.com/thailand-eu-vegetable-ban-12654

In some cases, regulatory standards are set so high that only the richest corporations can afford to gain market access. In mid-2008 the EU announced that, by December 2010 all clothing manufacturers importing into the EU will have to comply with its REACH regulatory standards, which require manufacturers to identify and quantify the chemicals present within the garment to an accuracy of 0.1%. The EU itself has calculated that meeting these new regulations is likely to cost EU companies alone €2.3 billion over 11 years (see 'EU metals producers fret over cost of new REACH law', A. Stablum, Reuters 12 Feb 2008). Needless to remark, small-scale firms within the developing world stand no chance of clearing this regulatory hurdle.

Similarly, EU 'Rules of Origin' restrictions are often used as a hidden tariff barrier. Rules of Origin are used to determine where a product originates, for the purposes of international trade. A product may be classified in a preferential tariff band, or conversely be penalised with higher tariffs, depending on its Rules of Origin classification. The EU's Rules of Origin system is sometimes so

RAW COCOA BEANS (No tax)

↓

COCOA BUTTER +10% TAX

↓

COCOA POWDER +10% +5% TAX

↓

CHOCOLATE OVER 20% TAX

THE MORE VALUE THE POOR ADD TO THEIR GOODS, THE MORE IMPORT TARIFFS INCREASE

THIS EXPLAINS WHY GERMANY PROCESSES MORE COCOA THAN IVORY COAST, THE WORLD'S LARGEST PRODUCER; AND WHY BRITAIN GRINDS MORE COCOA THAN GHANA

restrictive as to verge on the farcical. To illustrate: fish caught in Fijian seas, canned in a Fijian cannery and exported by a Fijian company was not considered Fijian fish if the vessel or crew were not Fijian or European, and could therefore not gain duty-free access to EU markets (see 'Partnership or Power Play?' Oxfam Briefing Paper April 2008). Similarly, pineapple juice made with Ghanaian pineapples and juiced, bottled and exported by Ghanaian companies could not be considered Ghanaian (for Rules of Origin purposes) if the sugar used in the juice came from elsewhere. The result, once again, is the juice being denied duty-free entry into the EU.

A further example of the arbitrary nature of EU regulations is highlighted by the ban on camel milk from being imported or sold, for fear of a foot-and-mouth disease crisis. This despite the camel being immune to foot-and-mouth disease, its milk being lower in fat than cow's milk, and its suitability for sufferers from lactose intolerance (see 'Camel Milk on the Menu?' R. Pagones, Los Angeles Times 25 June 2007).

It can be very difficult to argue against such hidden barriers to trade, especially because they are often dressed up as health and safety regulations – and it takes a brave woman (or man) to argue against health and safety regulations. But all too often, EU health and safety regulations adopt what is known as the 'precautionary principle', meaning that precautions should be taken even if there is no scientifically proven causal relationship between the product and the risk.

SUBSIDIES

A subsidy is a form of financial assistance given by governments to specific businesses or business sectors, generally to make the latter more competitive or prevent them from going bankrupt. Governments often subsidise a great deal of public services, education and health services being two good examples: although taxes cover part of their cost, governments often cover the significant shortfall through subsidies.

Within market (capitalist) economies, subsidies are governments' equivalent to a Monopoly game's 'Get out of jail' card. If a country's exports are too expensive to be competitive, a government can subsidise them, thereby rendering them cheaper. Conversely, a government may want to protect its own producers from cheaper foreign imports: it therefore subsidises local manufacturers, allowing them to drive their product costs down

to competitive (sometimes artificially so) prices. In some cases a government may want to encourage the development of a particular industry or business sector. In order to do so, it offers subsidies as an incentive to encourage growth (thereby stimulating supply of the good/s in question. This should theoretically lead to a fall in prices as supply matches, and sometimes overtakes, demand).

The merits and demerits of subsidies are constantly debated by economists and public policy experts. Those from the *laissez-faire school* believe that subsidies artificially skew global markets and sacrifice long-term efficiency gains on the altar of short-term protectionism. On the other hand, economists from the Keynesian tradition - who believe the state has an active role to play in stimulating growth – tend to see subsidies as a *'necessary evil'* which help protect jobs and entire industries. Whether subsidies do more harm than good remains a moot point, and is best left to professional economists to debate. The problem with subsidies *vis-a-vis* the Third World is that, all too often, it's a case of two weights and two measures, a case of *'Do as I say, not as I do'.*

SUGAR

FARMERS IN EUROPE ARE GUARANTEED A PRICE THREE TIMES HIGHER THAN THE WORLD PRICE. MOZAMBIQUE LOSES MORE THAN £70 MILLION A YEAR DUE TO RESTRICTIONS ON IMPORTING INTO EUROPE AND THE DUMPING OF CHEAP EXPORTS AT ITS DOOR.

CASE STUDY 1: THE CASE OF THE MALAWIAN 'STARTER PACK'

A tragic example of the impact of subsidies is that of the Malawian famine of 2001. In the immediate aftermath of the famine, with farmers struggling to achieve decent crop yields, the government of Malawi initiated a *'Starter Pack'* agricultural programme. Through this programme, farmers would receive one sack of corn seed and two bags of fertiliser free of charge, in order to kick-start their recovery following the famine. The IMF vetoed the Starter Pack programme and prohibited Malawi from distributing the free corn and fertiliser as it considered them an agricultural subsidy. WTO members had previously agreed to cap their agricultural subsidies at their existing levels and not raise them any further, and the IMF believed that by offering this Starter Pack, Malawi was in breach of its WTO obligations.

The IMF's position failed to take into account two issues: first, the extraordinary circumstances in which this Starter Pack was being offered, coming as it did on the back of a nationwide famine. Second, and perhaps more importantly, the IMF didn't take into account Malawi's existing agricultural subsidy levels. When in 1990 WTO members had agreed to cap agricultural subsidy levels, they had started on uneven territory: the EU and its Common Agricultural Policy subsidised European farmers to the tune of over €40 billion a year, while farmers in Malawi got nothing. Introducing the Starter Pack in the aftermath of the famine therefore not only made sense given the dire food situation in Malawi, but also in a broader sense – by supplying some form of government-financed help to farmers, Malawi was actually evening out the subsidy playing field. But WTO rules were firm: subsidies had been capped at 1990 levels, and if that meant that Europe and the USA got to keep their subsidies and countries like Malawi weren't allowed to supply agricultural Starter Packs – tough.

It would be simplistic and false to solely blame the IMF for the Malawian famine and the subsequent problems. But the Malawi Starter Pack example is useful because it illustrates the *'economy-before-ethics'* approach that organisations such as the IMF have pursued for the past couple of decades. When Malawi blamed the IMF for its food shortages, the IMF was moved to admit that it had *'no expertise in food security policy'*, and while admitting that it had ill-advised Malawi on how to manage its grain stocks, *'it was not the responsibility of the Fund to implement the advice'* (see Francis Ng'Ambi, *'Who Caused the Malawi Famine?'* African Business, Jan 2003).

For an organisation with as much clout and influence as the IMF to claim that Malawi could have opted to ignore its advice is disingenuous at best. Developing countries rely on international financial institutions such as the IMF and World Bank in order to secure much-needed loans, foreign direct investment and other such injections of capital. To such countries, receiving a seal of approval and positive credit rating from the IMF is vital: a negative country report automatically makes it less attractive to investors and leads to a worsening credit rating. A country's credit rating determines how 'expensive' it is for its government to borrow money (i.e. at what rate of interest). Much the same as an individual given a bad credit rating may find it impossible to get a mortgage, a country with a low credit rating will find it increasingly difficult to raise money through public borrowing.

THE WASHINGTON CONSENSUS

In the late 1980s, the IMF and World Bank adopted what came to be known as the Washington Consensus. This *'consensus'* amounted to a set of policy prescriptions that impoverished countries were to adopt and which would, the prevailing western worldview argued, stimulate economic growth. The Washington Consensus could be summed up in three words: *'stabilise, privatise, liberalise'*. The free market, according to the IMF, was key - government involvement only hindered growth, and the more liberalised an economy, the more efficiently it would function; protectionist measures (such as subsidies) were to be discouraged.

Developing countries, therefore, were presented with the ultimate Hobson's choice: follow the IMF's instructions, liberalise their economies and obtain the loans they so badly needed; or ignore the IMF's suggestions, get a negative credit rating and become investment pariahs. Left with little choice but to follow these Structural Adjustment Programmes (as the IMF named them), the governments of developing states quickly privatised their main industries and steered clear of interventionist trade policies – leading to the financial meltdown of the Asian 'Tiger' economies, the collapse of the Argentinean economy and several other countries becoming ensnared in a poverty cycle (as countries were forced to borrow even more money to cope with the damage wrought by previous loan conditions).

And all along, while developing countries were being liberalised to near death, affluent states continued to protect their own economies using the exact same measures they were ordering developing states to abandon.

Today, the European Union spends approximately 45% of its annual budget on agricultural subsidies. The Common Agricultural Policy (CAP) costs the EU a massive US$665 billion every year, with large-scale agribusiness pocketing 6- and 7-figure subsidies in the process. Thanks to CAP subsidies, each cow within the EU receives US$2.60 per day in subsidies – which is more than what over 2 billion people across the developing world have to live on. The USA has its own share of distorting subsidies, with cotton perhaps being the most notorious. The USA is the world's third largest cotton producer, mainly due to its tradition of heavily subsidising cotton farming, thereby encouraging overproduction. Between 1996 and 2006 the USA paid out an average of US$2-3 billion per year in subsidies to cotton farmers. The injection of such massive sums of money into the cotton sector naturally comes at a price: Oxfam estimates that US cotton subsidies have led to an artificial collapse in the global price of cotton of between 6% - 14% and resulted in cotton farmers across West Africa earning up to 20% less than they would, were they on an equal playing field with their American competitors (see Alston, J. Sumner, D. & Brunke, H. *'Impacts of Reductions in US Cotton Subsidies on West African Cotton Producers'*, Oxfam America (2007) http://www.oxfamamerica. org/files/Paying_the_Price.pdf) .

Here, at least, there have been signs of improvement. A number of developing states, tired of the *'do as I say, not as I do'* attitude prevalent at high-level trade talks, brought the World Trade Organisation's (WTO) Doha trade round to a halt in 2008, with agricultural subsidies being the main bone of contention. The discussions were lengthy and convoluted. In a nutshell: the US refused to lower its cotton subsidies unless the EU lowered its own agricultural subsidies. The EU, in turn, insisted that the US scrap its cotton subsidies if negotiations were to even proceed. Developing nations, spearheaded by Brazil's leadership, argued that the subsidies were not only unfair, but also unlawful. As yet, no resolution has been reached with WTO Director Pascal Lamy noting, according to a BBC news report of January 28th, 2011, that the next stage of the Doha talks would be the first one *'where some developing countries will pay a price in terms of market access in manufacturing, in agriculture, in services'* (http://www.bbc.co.uk/news/business-12309484).

For more on the debate on the nature and impact of the Washington Consensus, see the World Bank's discussion of the issue at: http://web.worldbank.org

TRIPS

TRIPS (the Agreement on Trade-Related Aspects of Intellectual Property Rights) is one of the World Trade Organisation's most controversial policies. TRIPS deals with the copyrighting and patenting of intellectual property, with the aim of protecting the creative rights of innovators (be they of an artistic, scientific, engineering or other nature) and encourage further ingenuity.

Intellectual property rights and human development, however, often do not make for the best bedfellows as TRIPS rewards innovators with a time-limited monopoly over their creation/s, effectively allowing them to control the sales, price and distribution of their creation as they please. Granting an author a monopoly over their words is uncontroversial, but when patents and exclusivity are placed on life-saving medicines, or indigenous peoples are told that they must now pay for their centuries-old medicinal remedies, a number of ethical questions arise.

Perhaps the most well-known TRIPS and development issue relates to anti-retrovirals (ARVs), which are drugs used to treat HIV. Pharmaceutical companies who patented their ARV creations and sold them at extortionate rates prevented developing nations with large-scale HIV epidemics from manufacturing generic 'copies' of the ARVs by invoking TRIPS*. Developing nations, left with no choice, subsidised the ARVs (with costs of up to US $15,000 per year for every patient) as best they

could. By 2001, a number of Indian and Brazilian pharmaceutical companies were producing generic ARV 'copies' and selling them at a fraction of the price – in blatant disregard of TRIPS. As a result, TRIPS was amended in 2001 to state that its regulations were supportive of *'access to medicines for all'* and ARV prices continued to plummet as generics flooded the market in the developing world. Nowadays, the most commonly used generic ARV treatment costs a patient just US$87 per year.

TRIPS does not distinguish between traditional, community-based knowledge and industry generated knowledge, with no provision for the patent-less protection of traditional herbal knowledge. As a result, large numbers of western companies have piggybacked on the traditional knowledge of the world's poor in order to profit. RiceTec, a US firm, has a patent on a variety of basmati rice, effectively preventing Thai and Indian farmers from growing it themselves; the Cameroonian pygeum tree, which has been used for centuries by locals as an anti-inflammatory, is now patented by a European pharmaceutical and sold as Tadenan, raking in US$150 million a year in profits; in India, a US company patented an insecticide that locals had used for thousands of years. In Argentina, Brazil, Mexico, India and Taiwan alone, the welfare loss caused by patented medicines has been calculated at US$11 billion, and studies have shown that drugs are up to 41 times costlier in countries with patent protection. According to NGO, Unipharma, in 2010 alone, the world's five largest pharmaceutical companies had estimated total revenues of US$248 billion.

*HAART, the first mass-produced ARV, cost US$10,000 per person per year when it was first released in 1996. Mass production of generic ARVs began in 2000/1, and by the end of 2001 an HIV+ patient could get an annual course of ARVs for just US$295. http://www.avert.org/generic.htm

CAPITAL FLIGHT

Capital flight, as its name implies, is the movement of private (non-business) capital out of countries and into others, usually away from Third World economies and into their western counterparts. Capital flight occurs largely for three reasons. First, rich investors may fear or dislike the political and economic system of the country of origin, and wish to invest it elsewhere. Bear in mind that this applies to the multi-millionaire investor, not to an honest Third World employee who seeks to protect his savings against inflation. Second, capital flight is sometimes the by-product of tax evasion, where large sums of money are transferred out of a country and into a secretive banking system in order to avoid paying tax.

The third kind of capital flow is not only the most significant in terms of volume; it is also the most evil and illegal. This is the export of massive fortunes acquired by illegal means – the fortunes of dictators and despots, of corrupt mega-businesspeople and their cronies – to collusive secretive banks in Switzerland, Luxembourg and the UK, where their millions can lie hidden, unknown to anyone but the individual and his or her bank manager.

This latter wealth is generally derived from activities such as corruption, embezzlement, elaborate tax avoidance schemes and other criminal activity (including drug dealing, extortion and arms dealing). Official statistics on these illicit capital flows do not exist, since, by definition, they escape detection. Nevertheless, a number of investigations into illegal capital flight have been undertaken, and the results are staggering.

A 2011 study by the Global Financial Integrity Program (see next page) calculated that in 2008 alone, developing countries lost an estimated US$1.26 trillion in illegal financial flows and that, on average, developing countries lost between US$725 billion to US$810 billion per year over the nine-year period 2000- 2008 (see *Illicit Financial Flows from Developing Countries: 2000-2009*, Dev Kar and Karly Curcio, Global Financial Integrity).

Transparency International, the anti-corruption NGO, has calculated that *'only'* US$140 billion of swindled cash could be retrieved and argues that it is *'immoral'* for Western governments to allow such funds to freely circulate in their respective countries while Africa sinks under the weight of debt and poverty. It is very difficult to effectively assess the amount and impact of such corruption (and the capital flight associated with it) but it is clear that the losses to poorer countries are immense while the benefits to the elites of such countries and to rich countries (or more accurately, to Rich World corporations) are very significant.

This aspect of the issue has been described in the following terms by Peter Eigen, founder of Transparency International:

'... if you pay, say, a $5 million bribe to a minister in order to promote a project which has no use, which is even harmful to the economy, say a huge power dam or a huge pipeline that costs $500 million, then the damage to the economy is close to $500 million rather than the $5 million.'

Eigen also argues that western countries not only turn a blind eye to such corruption, they often actually encourage it:

'... Therefore, many of these countries not only allowed their exporters to bribe, but they even subsidised the systematic bribery through generous tax write-offs, by support, by export financing agencies, export guarantee agencies. And what we found: a system in which even the most respectable companies had no compunction to systematically bribe, to pay huge amounts of money,

$5-$10-$20 million, to the decision makers in other countries, in Indonesia, in Nigeria, in India, in order to get billions in contracts in these countries.'

Montgomery, M. 'The Costs of Corruption: Interview excerpts with Peter Eigen', American Radioworks http://americanradioworks. publicradio.org/features/corruption/eigen.html

A study of capital flight from 25 highly-indebted, low income countries in sub-Saharan Africa during the period 1970 to 1996 by University of Massachusetts researchers concluded that the value of such flight amounted to a total of more than US$193 billion (in 1996 dollars) while the external debt of these countries stood at $178 billion in 1996. The study concluded:

'... taking capital flight as a measure of private external assets, and calculating net external assets as private external assets minus public external debts, sub-Saharan Africa thus appears to be a net creditor vis-à-vis the rest of the world'.

Is Africa a Net Creditor? New Estimates of Capital Flight from Severely Indebted Sub-Saharan African Countries, 1970-1996 (2000) by James K. Boyce and Léonce Ndikumana, University of Massachusetts

'STAGGERING AS THIS AMOUNT IS...' - ILLICIT FINANCIAL FLOWS FROM DEVELOPING COUNTRIES

In January 2011, US based, non-profit organisation Global Financial Integrity published a study on illicit financial flows from developing countries for the period 2000 to 2009 and estimated that such flows amounted to between $858 billion and $1.06 trillion per year. The top 10 countries identified in the report with the highest outflows were as follows:

- China: $2.18 trillion
- Russia: $427 billion
- Mexico: $416 billon
- Saudi Arabia: $302 billion
- Malaysia: $291 billion
- United Arab Emirates: $276 billion
- Kuwait: $242 billion
- Venezuela: $157 billion
- Qatar: $138 billion
- Nigeria: $130 billion

The Report identified many of the mechanisms and trends associated with such outflows:

- The mispricing of trade accounts for an average of 54.7% of cumulative illicit flows over the period 2000-2008 and is the major channel for the transfer of illicit capital from China.
- Bribery, theft, kickbacks, and tax evasion were the greatest means characteristic of the major exporters of oil - Kuwait, Nigeria, Qatar, Russia, Saudi Arabia, the United Arab Emirates, and Venezuela.
- Oil exporters Russia, the United Arab Emirates, Kuwait, and Nigeria, are becoming more important as sources of illicit capital.
- Asia accounted for the largest volume of illicit outflows with half a trillion in 2008 alone; over the nine-year period examined, an average of 89.3% of total flows from Asia was transferred abroad through trade mispricing.
- In real terms, illicit outflows through trade mispricing grew faster in the case of Africa (28.8% per annum) than anywhere else.
- The report estimated that for 2009, such flows would amount to $1.30 trillion with the reduction on the 2008 figure due to a decline in trade mispricing as a result of the slowdown in world trade following the global financial crisis.

Global Financial Integrity published another report in 2010 on *Illicit Financial Flows from Africa: Hidden Resource for Development* in which authors Dev Kar and Devon Cartwright-Smith studied the 39-year period from 1970 through 2008 and concluded that during that period illicit flows reached a conservative estimate of some US$854 billion and speculated that the real figures could amount to some US$1.8 trillion.

This amounts to some US$998.50 per person, outpaces Official Development Assistance to the region by a factor of 2:1 and amounts to sufficient funds to pay the entire cost of the Millennium Development Goals with cash to spare. The countries identified as topping the list in Africa are Nigeria, Egypt, Algeria, Morocco and South Africa.

While funds generated by illicit cross-border flows accounted for about 3% of the global total and criminal proceeds from drug trafficking, racketeering, counterfeiting and more equalled about 30-35%, the proceeds of commercial tax evasion, mainly through trade mispricing, are by far the largest element at 60-65%. The report concludes that the figures for Africa are likely to be of the same magnitude. The authors also argue that the massive flow of illicit money out of Africa is:

'...facilitated by a global shadow financial system comprising tax havens, secrecy jurisdictions, disguised corporations, anonymous trust accounts, fake foundations, trade mispricing, and money laundering techniques and that the impact on Africa is 'staggering' and has its greatest impact on those at the bottom of income scales in their countries, removing resources that could otherwise be used for poverty alleviation and economic growth.'

The report notes that over the period 1970-2008, Africa lost US$854 billion in cumulative capital flight – this amounts to enough to not only wipe out the region's total outstanding external debt but leave US$600 billion for poverty alleviation and economic growth and concludes:

'Staggering as this amount is, the estimate still does not include illicit flows generated due to smuggling, trade in narcotics and contraband, violations of intellectual property rights, human trafficking, sex trade, and other illegal activities.'

Sources: Dev Kar and Karly Curcio (2011) Illicit Financial Flows from Developing Countries 2000-2009, Washington, Global Financial Integrity

Dev Kar & Devon Cartwright-Smith (2010) Illicit Financial Flows from Africa: Hidden Resource for Development, Washington, Global Financial Integrity

THE BRAIN DRAIN

The Brain Drain is essentially the emigration of skilled workers and professionals to countries of greater prosperity than their own. Kenyan doctors, Pakistani IT programmers and Filipino and Zambian nurses all fulfil important roles within our society, but this comes at a significant cost to their respective countries. Take Ghana, where life expectancy is 59, only 57% of the population is literate and 37% of the population is undernourished, 47% of all its university graduates live within OECD (developed world) countries. In Guyana, a staggering 89% of all its University graduates live overseas (statistics derived from the 2007 UNDP Human Development Report).

This migration pattern is hardly new. In the early 1960s, brain drain from East to West Germany reached such heights that the former erected the Berlin Wall to stem it. Similarly, Ireland experienced significant brain drain before the Celtic Tiger economic boom encouraged the Irish to seek their fortunes back home. Neither is the brain drain restricted to the most impoverished states: Lithuania lost some 200,000 young, well-educated citizens to more prosperous European countries (Ireland in particular) between 1994 and 2004 and Malta currently has a brain drain rate of 56% with doctors and dentists especially susceptible. Countries like Malta and Lithuania, however, are EU members and

THE NET TRANSFER OF FINANCIAL RESOURCES FROM DEVELOPING ECONOMIES BETWEEN 1998 AND 2010 WAS
$US5,740.6 BILLION

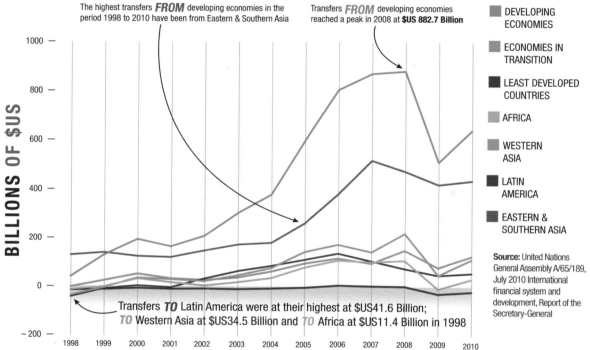

The highest transfers **FROM** developing economies in the period 1998 to 2010 have been from Eastern & Southern Asia

Transfers **FROM** developing economies reached a peak in 2008 at **$US 882.7 Billion**

Transfers **TO** Latin America were at their highest at $US41.6 Billion; **TO** Western Asia at $US34.5 Billion and **TO** Africa at $US11.4 Billion in 1998

BILLIONS OF $US

1000 — 800 — 600 — 400 — 200 — 0 — −200 —

1998 1999 2000 2001 2002 2003 2004 2005 2006 2007 2008 2009 2010

- DEVELOPING ECONOMIES
- ECONOMIES IN TRANSITION
- LEAST DEVELOPED COUNTRIES
- AFRICA
- WESTERN ASIA
- LATIN AMERICA
- EASTERN & SOUTHERN ASIA

Source: United Nations General Assembly A/65/189, July 2010 International financial system and development, Report of the Secretary-General

ILLICIT FINANCIAL FLOWS
THE TOP 10

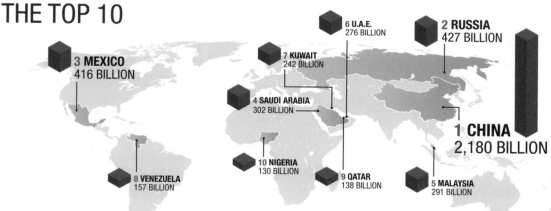

6 **U.A.E.** 276 BILLION

2 **RUSSIA** 427 BILLION

3 **MEXICO** 416 BILLION

7 **KUWAIT** 242 BILLION

4 **SAUDI ARABIA** 302 BILLION

1 **CHINA** 2,180 BILLION

10 **NIGERIA** 130 BILLION

8 **VENEZUELA** 157 BILLION

9 **QATAR** 138 BILLION

5 **MALAYSIA** 291 BILLION

Sources:
Dev Kar and Karly Curcio (2011) Illicit Financial Flows from Developing Countries 2000-2009, Washington, Global Financial Integrity
Dev Kar & Devon Cartwright-Smith (2010) Illicit Financial Flows from Africa: Hidden Resource for Development, Washington, Global Financial Integrity

middle-income countries. Although brain drain is a problematic issue for their respective governments, they enjoy a safety net of (relative) prosperity and free access to the world's largest common market. Third World nations do not.

The Kenyan government, for example, spends over US$40,000 on each medical student within the country, and yet in 2006 there were 167 Kenyan born and trained doctors working in OECD countries. Research undertaken in order to calculate the cumulative cost of losing doctors to the West has shown that for every doctor that emigrates, Kenya loses over US$500,000. Multiply that by the 167 doctors known to have left by 2006 and we get the total cost of the Kenyan medical doctor brain drain – US$87 million (J. Kirigia, et al. *'The Cost of Health Professionals' Brain Drain in Kenya'* BMC Health Services Research, Vol.6 2006).

The costs of the brain drain are threefold:

- **Skilled workers and professionals cost money to educate**, as illustrated by the Kenyan doctors example above. When state-subsidised students are lured to greener pastures, governments of LDCs (already hard-pressed to invest in high-level education), see no return on their investment and are discouraged from pressing on with educational reform.

- **When skilled workers emigrate, they leave gaps in the workforce** that must be somehow filled. The UNDP estimates that, in order to fill the human resource gap created by brain drain, 'Africa employs up to 150,000 expatriate professionals at a cost of US$4 billion a year' (see A. Roisin, 'Brain Drain: Challenges and Opportunities for Development' UN Chronicle April 2007).

- **The human development cost.** When a doctor or nurse emigrates from Kenya to the EU, the most devastating cost is not financial, it is medical. In Malawi, there are only 2 doctors for every 100,000 people. The result is a 46-year life expectancy and one of the highest infant mortality rates in the world. When an engineer leaves Namibia to work in the USA, it is Namibian technical expertise that pays the biggest price.

Many commentators refer also to the *'brain gain'* where remittances are sent 'home' by migrants working abroad. In 2005, the World Bank estimated that such remittances amounted to a total of US$167 billion (*'Global Economic Prospects 2006: Economic Implications of Remittances and Migration'* World Bank). However, the Bank also points out that between 30% and 45% are remittances from one developing country to another.

THREE CAMPAIGNS ON TRADE AND FINANCE

Focus on the Global South (Focus) is a non-governmental organisation working in Thailand, the Philippines and India. Focus was established in Bangkok in 1995 and is affiliated with the Chulalongkorn University Social Research Institute. Focus combines policy research, advocacy, activism and grassroots capacity-building in order to generate critical analysis and encourage debates on national and international policies related to corporate-led globalisation, neo-liberalism and militarisation. Its major campaigns focus on four key areas – Reclaiming the Commons; Peace and Democracy: Climate Justice and deglobalisation. See: http://www.focusweb.org/content/who-we-are

Tax Justice Network

Tax is the most important, the most beneficial, and the most sustainable source of finance for development. Tax revenue in Africa, for example, is worth 10 times the value of foreign aid. The long-term goal of poor countries must be to replace foreign aid dependency with tax self-reliance.

The Tax Justice Network is an international independent organisation launched in the UK in

2003 with the objectives of mapping, analysing and explaining the role of taxation and the harmful impacts of tax evasion, tax avoidance, tax competition and tax havens, especially for developing countries. The Network includes academics; accountants; development organisations and NGOs; faith groups; journalists, trade unions and others. The Network has published a variety of research reports and proposals for tax reform.

See http://www.taxjustice.net

The Ethical Trading Initiative

The Ethical Trading Initiative is an alliance of companies, trade unions and voluntary organisations working in partnership to improve the lives of workers worldwide who make or grow consumer goods – *'everything from tea to T-shirts, from flowers to footballs'*. EFI was established to challenge that reality that millions of people around the world experience inadequate, often shocking, conditions at work and to ensure that member companies adhere to an agreed code under which products they make or sell are ethical.

Ethical trade means that retailers, brands and their suppliers take responsibility for improving the working conditions of the people who make the products they sell, ensure that supplier companies respect and protect worker's right and labour law. The basic code which member companies adopt addresses issues such as wages, hours of work, health and safety and the right to join free trade unions.

Check out EFI's basic code on http://www. ethicaltrade.org/eti-base-code to identify what are the key questions to ask.

Also see an interesting case study on how Tesco initiated a programme to apply the principles set out by Professor Ruggie (the United Nations Secretary General's Special Representative on the issue of human rights and transnational corporations and other business enterprises in 2008) in South Africa (http://www.ethicaltrade.org). On the Report itself, see http://www.business-humanrights.org/SpecialRepPortal/Home.

READING

Bertrand Borg, Dylan Creane and Colm Regan (2009), 5:50:500 or how the world rewards the rich at the expense of the poor and where we fit in!, Bray, 80:20 Educating and Acting for a Better World)

Peter Stalker (2009) No Nonsense Guide to Global Finance, New Internationalist, London

Oxfam (2002) Rigged Rules and Double Standards: trade, globalisation and the fight against poverty, Oxford

Christian Aid (2005) For Richer or Poorer: transforming economic partnership agreements between Europe and Africa, London

Nicola Cantore, Sheila Page and Dirk Willem te Velde (2011) Making the EU's Common Agricultural Policy coherent with development goals, London, Overseas Development Institute

UNDP (2005) Human Development Report: International cooperation at a crossroads: Aid, trade and security in an unequal world, New York

Graham Dunkley (2004) Free Trade: Myth, Reality and Alternatives, London, Zed Books

MORE INFORMATION & DEBATE

http://www.afrodad.org - the African Forum and Network on Debt and Development

http://www.taxjustice.net – site for the Tax Justice network

http://www.gfintegrity.org - a programme of the US based Centre for International Policy

http://www.debtireland.org - Irish Debt and Development Coalition

http://www.eurodad.org - European Network on Debt and Development

http://www.oxfam.org/en/campaigns/trade - Oxfam site related to trade issues

www.unctad.org - United Nations Conference on Trade and Development site

http://www.odi.org.uk - the Overseas Development Institute, UK

SUMMARISING BASIC NEEDS

BEATRICE MAPHOSA

At the core of many of the development issues and debates addressed in *80:20 Development in an Unequal World* is the priority agenda of resolving the immediate and urgent needs of a large segment of humanity. Addressing the basic needs of the world's poorest is, to a very significant degree, the litmus test of our intent and resolve. This brief section summarises key dimensions of those basic needs as a backdrop to much else in the book.

Essentially, those who have consistently focused on a 'basic needs approach' argue that certain key human needs are more immediately urgent than others and cannot await (and should not await) more general economic and social development for their realisation. Basic needs should be addressed directly through both targeted and specific interventions and not indirectly via broader development strategies and plans. Basic needs cannot wait for general development and can be effectively and efficiently addressed with currently available resources. Addressing basic needs immediately will also assist broader development goals and strategies and will be less costly than if delayed or approached indirectly. The focus on basic needs emerged in development thinking and practice as part of the need to 'dethrone' GNP as the dominant measurement tool and for an explicit focus on the needs of the poor rather than solely the needs of an economy. As an approach and a theory, it became associated with the International Labour Organisation (ILO) and influenced considerable work in the 1970's and early 80's before the debt crisis and changes in the dominant models of development in western thinking and strategy effectively ended its influence. Nonetheless, basic needs approaches heavily shaped and influenced the emergence of the human development approach already addressed in chapters 1 and 2.

HUNGER

"Starvation is death by deprivation; the absence of one of the essential elements of life. It's not the result of an accident or a spasm of violence, the ravages of diseases or the inevitable decay of old age. It occurs because people are forced to live in the hollow of plenty. For decades, the world has grown enough food to nourish everyone adequately. Satellites can spot budding crop failures; shortages can be avoided. In the modern world, like never before, famine is by and large preventable. When it occurs, it represents civilization's collective failure."

R. Thurow and S. Kilman (2009) *Enough: Why the Poorest Starve in an Age of Plenty*

TOP FIVE 'HUNGER HOTSPOTS'

1. ERITREA 65% **3. HAITI 57%**
2. BURUNDI 62% **4. ZAMBIA 44%**
                    5. ETHIOPIA 41%

% of population suffering undernourishment, 2006-2008

Region	Total Population 2006-08	Numbers Undernourished (% of total)	% Change Since 1990-92
World	6652.5	850 (13)	↓ 19
Developing Countries	5420.2	839.4 (15)	↓ 22
Least Developed Countries		263.8 (33)	↓ 16
All Africa	962.9	223.6 (23)	↓11
Sub-Saharan Africa	801.5	217.5 (27)	↓13
Asia	3884.3	567.8 (15)	↓ 27
Latin America	528.2	38.6 (7)	↓ 35

Source: Food and Agriculture Organisation (2011) State of Food Insecurity in the World, Rome

LITERACY

In the United States, an estimated 30 million people over the age of 16 read no better than the average elementary school child. Worldwide, nearly 800 million adults are illiterate in their native languages; two-thirds of them are women. Yet the ability to read and write is the basis for all other education; literacy is necessary for an individual to understand information that is out of context, whether written or verbal. Literacy is essential if we are to eradicate poverty at home and abroad, improve infant mortality rates, address gender inequality, and create sustainable development. Without literacy skills—the abilities to read, to write, to do math, to solve problems, and to access and use technology—today's adults will struggle to take part in the world around them and fail to reach their full potential as parents, community members, and employees.

Comment by US based organisation Proliteracy

LITERACY RATES BY REGION, 2009

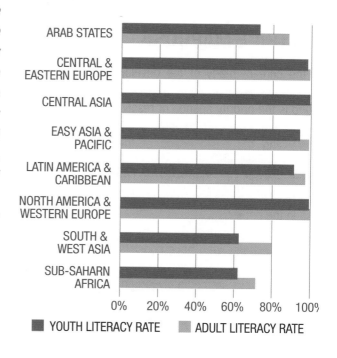

- YOUTH LITERACY RATE
- ADULT LITERACY RATE

Source: UNESCO Institute for Statistics, September 2011.

MATERNAL HEALTH

Comment from the United Nations Family Planning Association:

'Every minute, a woman dies in pregnancy or childbirth... The consequences of losing over half-a million women every year have a ripple effect in families, communities and nations. Children without mothers are less likely to receive proper nutrition, health care, and education. The implications for girls tend to be even greater, leading to a continued cycle of poverty and poor health. And every year, $15 billion in productivity is lost due to maternal and newborn mortality, a huge burden on developing nations.'

ESTIMATED NUMBER OF MATERNAL DEATHS 2008 & RATE OF DECLINE (%) SINCE 1990

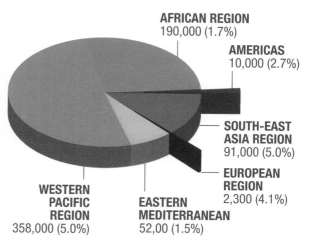

AFRICAN REGION
190,000 (1.7%)

AMERICAS
10,000 (2.7%)

SOUTH-EAST ASIA REGION
91,000 (5.0%)

EUROPEAN REGION
2,300 (4.1%)

WESTERN PACIFIC REGION
358,000 (5.0%)

EASTERN MEDITERRANEAN
52,00 (1.5%)

Source: World Health Organisation (2011) World Health Statistics, Geneva

WATER & SANITATION

Comment by former South African Minister of Water Affairs and Forestry Kader Asmal:

'Those of us who have never been thirsty, truly thirsty, struggle to empathise with those for whom real thirst is a daily torment. Nor do those of us who have not had to carry heavy containers of water, day in and day out for long distances over rough terrain, really understand the real value of having water – safe water – on tap.'

Quoted in Van Vuuren, L (ed.), 2007, Our water, Our culture: A glimpse into the water history of the South African people, Water Research Commission, Gezina, Pretoria, South Africa.

REGIONAL DISTRIBUTION OF THE 2.6 BILLION PEOPLE NOT USING IMPROVED SANITATION FACILITIES
2008, POPULATION (MILLION)

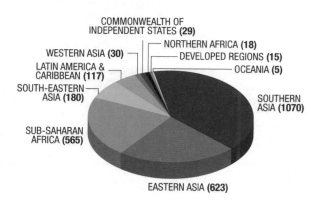

REGIONAL DISTRIBUTION OF THE 884 MILLION PEOPLE NOT USING UNIMPROVED SOURCES FOR DRINKING WATER
2008, POPULATION (MILLION)

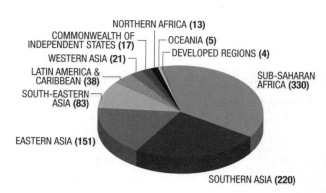

WORK

Comment by US researcher L. Randell Wray, 2009:

'...the burden of joblessness is borne unequally, always concentrated among groups that already face other disadvantages: racial and ethnic minorities, immigrants, younger and older individuals, women, people with disabilities, and those with lower educational attainment'.

EMPLOYMENT SHARES BY SECTOR AND SEX, WORLD AND REGION (%) 2009

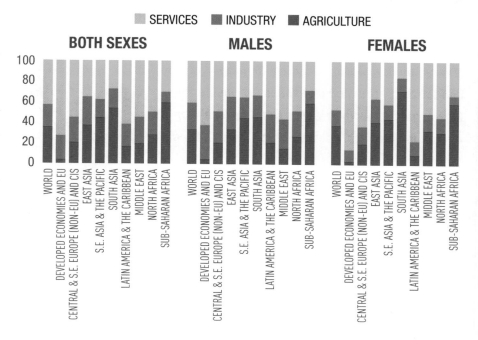

CHILDREN'S HEALTH

Comment by the World health Organisation 2011:

'The improvements and progress are encouraging – but stark disparities persist. Sub-Saharan Africa is still home to the highest rates of child mortality, with one in eight children dying before reaching five – more than 17 times the average for developed regions (1 in 143). Southern Asia has the second highest rates with 1 in 15 children dying before age five.

Under-five deaths are increasingly concentrated in sub-Saharan Africa and Southern Asia. In 1990, 69 per cent of under-five deaths occurred in these two regions – in 2010 that proportion increased to 82 per cent. About half of all under five deaths in the world took place in just five countries in 2010: India, Nigeria, Democratic Republic of Congo, Pakistan and China.'

LEVELS AND TRENDS IN THE UNDER-FIVE MORTALITY RATE, BY MILLENNIUM DEVELOPMENT GOAL REGION
1990–2010 (DEATHS PER 1,000 LIVE BIRTHS)

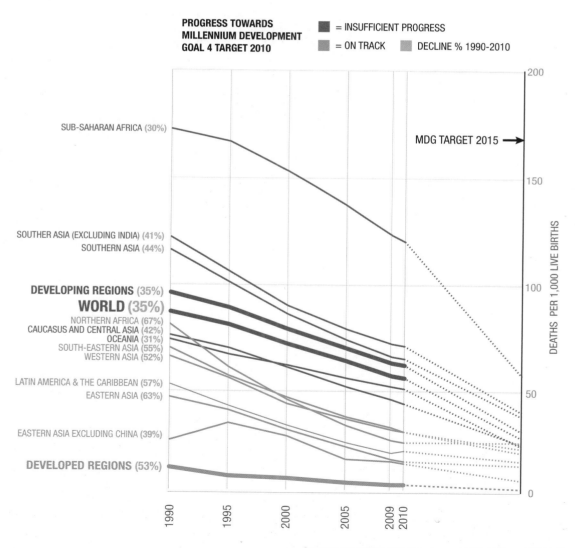

PROGRESS TOWARDS MILLENNIUM DEVELOPMENT GOAL 4 TARGET 2010

= INSUFFICIENT PROGRESS

= ON TRACK   DECLINE % 1990-2010

MDG TARGET 2015 →

SUB-SAHARAN AFRICA (30%)

SOUTHER ASIA (EXCLUDING INDIA) (41%)
SOUTHERN ASIA (44%)

DEVELOPING REGIONS (35%)
WORLD (35%)
NORTHERN AFRICA (67%)
CAUCASUS AND CENTRAL ASIA (42%)
OCEANIA (31%)
SOUTH-EASTERN ASIA (55%)
WESTERN ASIA (52%)

LATIN AMERICA & THE CARIBBEAN (57%)
EASTERN ASIA (63%)

EASTERN ASIA EXCLUDING CHINA (39%)

DEVELOPED REGIONS (53%)

DEATHS PER 1,000 LIVE BIRTHS

200
150
100
50
0

1990   1995   2000   2005   2009 2010

Source: UN Millennium Development Goals Report 2011, New York

159

[*'...where moral values and enlightened self–interest intersect...'*]

Chapter 9

DEBATING AID
DEVELOPMENT COOPERATION TODAY

Mary Rose Costello and Colm Regan

International aid has been the subject of intense debate and vigorous disagreement particularly as regards its effectiveness, its impact and its value in human development terms. Aid has become big business at the level of government, NGO and celebrity culture; it has convinced critics and defenders alike, and remains the subject of much misinformation and myth-making on both sides. This chapter summarises and reviews many of the debates and much of the evidence.

KEYWORDS

Negative and positive duties, option or duty, multilateral and bilateral aid, history of aid, impact of aid, aid critics and aid defenders, NGOs and aid, South-South aid, China and aid

INTRODUCTION

The *'aid'* agenda today has become dominated by a series of questions and debates, truths and half-truths, *'certainties'* and confusions as policymakers, NGOs, journalists and academics survey the aid landscape. And, the public, at least those in donor countries, possess strong opinions and reservations about the entire aid project.

- *Is aid effective; does it do what we intend it to do?*
- *What is it reasonable to expect of aid as it currently exists and what expectations are unrealistic and unfair?*
- *Are NGOs more effective than governments at aid?*
- *To what degree is aid undermined by administrative costs, by corruption or by waste?*
- *Is aid a 'duty', in the modern world or an 'option' and whose responsibility is it?*
- *Where does aid begin, and end and what is its relationship to other 'development issues' e.g. trade, investment, enterprise, human rights etc.?*

At one end of the spectrum of the debates are the recent and popular views of Zambian Dambisa Moyo who argues that:

'One of the most distressing aspects of the whole aid fiasco is that donors, policymakers, governments, academicians, economists and development specialists know, in their heart of heart, that aid doesn't work, hasn't worked and won't work ...'

(Moyo 2009)

As against this view, the analysis of Roger Riddell suggests:

'...aid would appear to have an important gap-filling role to help meet the basic needs of poor people...the evidence presented here confirms that in helping to address these short-term needs, much aid has been successful (between 75 and 90 percent) when it has provided tangible goods needed and used directly by poor people...'

(Riddell 2007)

However, the aid debate is not simply a technical one about whether aid works or not, it is also a moral and ethical debate about our place and role in the world and its associated responsibilities. This chapter explores many of these debates and the evidence underpinning them; it focuses strongly on the key arguments against aid and those in support of aid and it outlines the key patterns associated with aid today*.

COMMON THEMES OR IDEAS UNDERPINNING THE DEBATE ON AID

The literature on aid today is voluminous and detailed; the arguments of those who criticise aid as well as those who defend aid are addressed below but first, it may be useful to identify some of the bigger themes and ideas underpinning these debates including those of Peter Unger (1996), Peter Singer (2002), Thomas Pogge (2008), Garrett Cullity (2004) and Amartya Sen (2009); these include:

- We are all a part of a bigger world picture and we are now connected to others in a bewildering series of ways. These connections have implications (both positive and negative) and it is impossible for us to cut ourselves off from them (even if we wished) or to isolate ourselves from them. As a result, we have duties to others, even others we do not know and are unlikely ever to meet – for example, the duty to acknowledge that our actions (and inactions) in trade, environment etc., routinely impact on others and we do not have the right to ignore such impacts.
- We all have what are often termed *'negative and positive duties'* – our negative duty is to ensure that whatever we do does not harm others

---

* This chapter is based upon a more extended essay Debating Aid written by the current authors with Bertrand Borg (2010)

(especially the poor and the weak) and if harm has been done, we have a positive duty to make reparation for it. Given the current economic, political, environmental and social order which clearly benefits the rich world while hugely disadvantaging the poor, we are all morally and causally deeply involved. The *long arm of injustice* must have an equally *'long reach of justice'*.

- The rights which we proclaim for ourselves imply corresponding duties such as the responsibility that such rights apply to all, without exception. Therefore, we have a duty to promote the rights of all; either rights apply to everyone or the very concept of rights is meaningless.

- Some commentators argue that the traditional stricture *'love thy neighbour as thyself'* means literally what it proclaims. Human thought, rationality and emotion go beyond the merely *'individual'* and embrace some element of a *'universal'* or bigger picture. This implies that we should give equal weight to the interests and needs of others as we would give to our own. While *'self-interest'* is a key motivating and ethical framework by which to live one's life, it is, nonetheless insufficient as all of us recognise interests and needs outside those of pure self-interest.

- Part of our condition as human beings, in fact the essence of being human, is our capacity to reason and to act *'reasonably'* and to search for and constantly re-assess what *'reasonableness'* means at any point in time or in any context. This essence places us under an obligation to others – we have a duty to 'go as far as we can reasonably go'.

While the implications of these debates are explored more fully in Chapter 7, it may be useful in the context of this chapter to summarise a number of important implications for the aid debate *per se*.

- Some philosophers argue that not only do we have a duty to support those in need but this duty is *'absolute'* – development aid, they argue, is

a way of countering international injustice, just as the welfare state seeks to redress local injustices. For philosophers such as Emmanuel Kant, humankind's moral choices had to uphold what he termed a 'categorical imperative' – an absolute and universal moral standard which all human beings have an obligation to uphold.

- Such a perspective has significant implications – if charity is a duty, then 'giving to the poor' is no longer a good-spirited, voluntary deed but rather an issue of social justice with corresponding duties placed upon both the individual and the state. In this context, a rich government which knowingly does nothing to help reduce poverty (and not just at home) is morally culpable. The priority to assist should not be determined solely by nationality but rather by need – a dying child at the other end of the world is no less important than one at the end of our street.

- Many argue that if it is in our power to prevent something bad from happening, at little cost to ourselves, then we have a moral duty to do so. Using arguments about 'bad' aid to dismiss all aid is akin to dismissing all music on the basis of poor musicians.

- Against this, many argue that while people have mutual obligations to each other, such obligations are neither absolute nor universal – they are a matter for individuals and governments since each person/state has a right to dispose of personal property as they see fit.

- Thus, even if helping the poor is morally right, it cannot and should not be enforced and even if an individual is indifferent to the plight of the poor, they should not be forced into assisting or punished for not doing so as an enforced morality is no morality at all.

AID: THE HISTORY OF AN IDEA

1945-1960: a new beginning

Though aid to poorer countries existed in different forms before World War II - mostly via religious and voluntary agencies – it wasn't until the late 1940s that aid giving became part of ongoing government and *development* policy. After the War, the United States provided a relief package known as the *Marshall Plan* to help rebuild Europe and to ensure Europe remained a US ally.

1960-1970: a decade of development

The United Nations branded the 1960s as the *Development Decade* and it became characterised as the 'glory years' for development aid; support for aid was strong and the amount of aid made available was growing. There was evidence to suggest that aid was actually working, with the rates of growth increasing in the economies of some of the poorest countries, including those in Sub-Saharan Africa. Official aid continued to increase and, in 1969, the General Assembly of the UN agreed an international aid target of 0.7% of GNP for donor countries.

1970s-1980s: aid in decline

By the start of the 1970s commitment to official aid was beginning to decline and increases in ODA had come to a halt and aid as a percentage of GNP fell as compared with previous levels. The situation was compounded by an international recession and the oil crisis. Official aid began to increase again towards the end of the decade in real terms to $27 billion by 1980, however, this growth in real terms masked a decline in aid as a percentage of GNP to half of the 0.7% agreed. The 1970s also witnessed the growth in the size and influence of non-governmental organisations.

1980s-1990s: the lost decade of development

The 1980s began with the continued negative consequences of the global recession for development aid, with aid levels stagnating and in many cases declining. A focus on development aid and economic growth became the main theme of the decade and despite recommendations against such policies, there was an increase in the number of *conditions* attached to aid; policies associated with neo-liberal economics led to the introduction of Structural Adjustment Programmes (SAPs)*, which included increasing the role of the *free market* and cutting social expenditure. These policies had direct impact on areas such as health and education expenditure in developing countries with resulting negative impact on the poorest.

1990-2000: the dependency debate

By 1992 debates around the notion of *aid dependency* grew in importance. This debate coincided with an all-time low in ODA as a percentage of GDP; in 1990 ODA was at an average of 0.33% but by the end of the decade it had dropped to just 0.22% and with a worsening situation for the world's poorest the aid debate again moved away from promoting economic growth to directly alleviating poverty.

2000-Present Day: an aid revival?

The new century began with a positive start with the biggest ever meeting of heads of state committing to the 8 Millennium Development Goals (MDGs). Debates moved away from the *does aid work* agenda to that of making aid more effective, there has also been movement towards more recipient-focused development strategies which seek to give recipient countries more control over development and poverty eradication strategies. In 2008, DAC aid amounted to $119.7 billion, however as the economic climate has deteriorated rapidly in recent years it has become clear that aid levels are, once again, under threat despite the positive picture.

---

* Neo-liberal economics - an economic school of thought which considers the free market to be the best, most efficient route to economic growth and which seeks to transfer control of the economy from the public to the private sector. In development, neo-liberal economics were typified by the so-called Washington Consensus (a list of 10 policy proposals) adopted by the IMF and World Bank in the 1990's (see Chapter 8).

AID: THE CURRENT SITUATION

By 2010, net official development assistance from members of the Development Assistance Committee (DAC) of the OECD reached US$ 128.7 billion - the highest real ODA level ever and net ODA as a share of GNI amounted to 0.32%, higher than in any year since 1992. The largest donors by volume were the United States, the United Kingdom, France, Germany and Japan. Denmark, Luxembourg, the Netherlands, Norway and Sweden continued to exceed the agreed target of 0.7% of GNI.

Aid increased from Australia, Austria, Belgium, Canada, Denmark, Finland, France, Japan, Korea, Germany, the Netherlands, Norway, Portugal and the United Kingdom but declined from Greece, Ireland, Italy, Luxembourg, New Zealand, Spain, Sweden (although it continues to allocate approximately 1% of its GNI to ODA) and Switzerland.

Other countries contributing aid in 2010 included the Czech Republic, Estonia, Hungary, Iceland, Israel, Poland, Slovak Republic, Slovenia and Turkey.

In 2005 aid donors made formal commitments to increase their ODA including the 15 DAC members of the EU to a minimum target of 0.51% of their GNI in 2010. While 9 individual countries achieved (or nearly achieved that goal), 6 did not. The combined effect of these increases has been to raise ODA by 37% in real terms since 2004, however, when comparing the 2010 outcome against the promises made there was a shortfall of some US$19 billion. Commitments also made to increase aid to Africa have not been delivered on and the percentage shortfall is higher than the shortfall in overall aid.

DEFINING OFFICIAL DEVELOPMENT ASSISTANCE (OR OFFICIAL AID)

The Organisation for Economic Cooperation and Development (OECD) defines official aid as *grants or loans to countries and territories* on the Development Assistance Committee (DAC) list of recipient countries and to international organisations which are:

- Undertaken by the official (state or international organisation) sector
- With the promotion of economic development and welfare as the main objective
- At concessional financial terms (if the aid is a loan, it must have a grant element of at least 25%)
- In addition to financial flows, technical co-operation is included in aid
- Grants, loans and credits for military purposes are excluded
- Transfer payments to private individuals (e.g. pensions, reparations or insurance payouts) are in general not counted either.

Multilateral aid is made up of contributions to international institutions for use in, or on behalf of, developing countries and in financial aid and technical co-operation by an international institution to developing countries. Bilateral aid is ODA provided on a country-to-country basis.

AID *'governments continue to miss agreed targets'*

KEY ISSUES AND TRENDS

> While aid rose in 2009 as a percentage of GNI (to 0.31%), this was due to a decrease in GNI bringing overall aid to US$119.6 billion. These figures fall far short of commitments made in almost every decade since the 1970's and including those made as recently as 2005.

> 'Real ODA' (see note on opposite page) remains less than half the agreed UN target

> Aid remains urgently in need of radical reform

> Gender issues remain largely 'invisible' in aid

> Aid delivery continues to be dominated by 'external' technical assistance and local ownership of aid programmes remains weak

> Donors continue to short change Sub-Saharan Africa overall and climate change against the pledges announced

> Aid remains significantly 'tied' also with inappropriate 'conditionality' attached

> Aid remains heavily driven by foreign policy concerns rather than human development and has largely failed to prioritise the MDG agenda

OFFICIAL AID RECIPIENTS 2009
BY INCOME GROUP

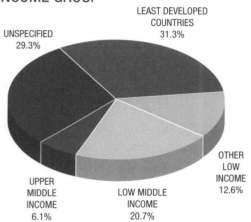

LEAST DEVELOPED COUNTRIES 31.3%

UNSPECIFIED 29.3%

OTHER LOW INCOME 12.6%

UPPER MIDDLE INCOME 6.1%

LOW MIDDLE INCOME 20.7%

OFFICIAL AID RECIPIENTS 2009
TOP 10 COUNTRIES

1. AFGHANISTAN
2. ETHIOPIA
3. VIET NAM
4. PALESTINIAN ADM. AREAS
5. TANZANIA
6. IRAQ
7. PAKISTAN
8. INDIA
9. COTE D'IVOIRE
10. CONGO, DEM. REP.

SOURCES

The Reality of Aid Network (2010), Aid and Development Effectiveness: Towards Human Rights, Social Justice and Democracy, IBON Books, Philippines

OECD DAC (2011) Development Aid at a Glance: Statistics by Region - Developing Countries, Paris

www.oecd.org/dac/stats

HUMANITARIAN AID TO SUB-SAHARAN AFRICA, 1995-2008

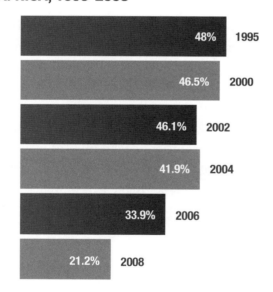

48%	1995
46.5%	2000
46.1%	2002
41.9%	2004
33.9%	2006
21.2%	2008

In 2008, Bilateral humanitarian assistance amounted to US$8.8 billion (up from US$6.3 billion in 2007). Since 2000 an increasing amount of bilateral humanitarian assistance has been directed to Sub-Saharan Africa, reflecting human insecurity there. As a proportion of "real aid" to this region, humanitarian assistance has grown from 9.1% in 2000 to 16.0% in 2008 with just 6 countries accounting for 47% of all bilateral humanitarian assistance in 2008 – Afghanistan, Iraq, Sudan, the Democratic Republic, of the Congo, Ethiopia, and Somalia.

DAC DONOR COUNTRIES 2009 WITH AID AS A % OF GROSS NATIONAL INCOME

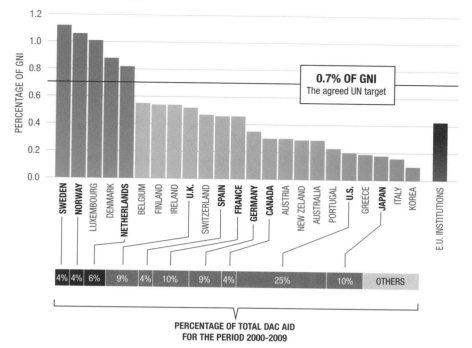

0.7% OF GNI
The agreed UN target

PERCENTAGE OF GNI

SWEDEN · NORWAY · LUXEMBURG · DENMARK · NETHERLANDS · BELGIUM · FINLAND · IRELAND · U.K. · SWITZERLAND · SPAIN · FRANCE · GERMANY · CANADA · AUSTRIA · NEW ZELAND · AUSTRALIA · PORTUGAL · U.S. · GREECE · JAPAN · ITALY · KOREA · E.U. INSTITUTIONS

| 4% | 4% | 6% | 9% | 4% | 10% | 9% | 4% | 25% | 10% | OTHERS |

PERCENTAGE OF TOTAL DAC AID
FOR THE PERIOD 2000-2009

"The financial crisis response shows how countries can work together and support each other in times of difficulty. Rich country governments managed to find astonishing sums of money to spend on bank bailouts and fiscal stimulus to rescue their own economies. Yet the long-term effort to resolve the poverty and environmental crises in Southern countries also requires political attention. Millions of people worldwide have insufficient food to eat, are vulnerable to disease and disaster, and receive minimal income."

The Reality of Aid Report, 2010

Note: Aid figures are routinely 'massaged' by donor governments and the figures often officially quoted include items not related to aid as popularly understood. In contrast, 'Real Aid' is calculated by subtracting items such as debt cancellation and the costs related to of developing country refugees and students arriving in donor countries from reported ODA.

For more information on this issue, see Reality of Aid Network (2010) Aid and Development Effectiveness: Towards Human Rights, Social Justice and Democracy, IBON Books, Philippines.

AID BY SECTOR
ESTIMATED PERCENTAGES BY AID SECTOR, 2008

13% NON-DAC OFFICIAL DONORS

23.6% NGOS

63.4% 'REAL' DAC DEVELOPMENT ASSISTANCE

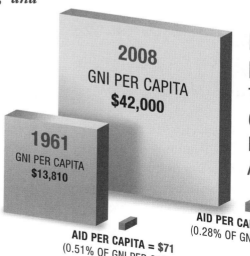

2008
GNI PER CAPITA
$42,000

1961
GNI PER CAPITA
$13,810

AID PER CAPITA = $71
(0.51% OF GNI PER CAPITA)

AID PER CAPITA = $118
(0.28% OF GNI PER CAPITA)

RICHER BUT MEANER
THE GROWING GAP BETWEEN DONOR WEALTH AND 'REAL AID'

THE 2010 REALITY OF AID REPORT EXPLORES AID

The Reality of Aid Report is published every two years and is based on the knowledge, expertise and experience of aid agencies, community-based organisations, governments and academics. It is published by the Reality of Aid Network – an international organisation that brings together 172 member organisations, including over 40 civil society networks from Europe, the Americas, Africa, and in the Asia/Pacific region. The Report provides an independent assessment of aid policies and practices and forms a basis for ongoing discussion between policy makers and civil society at national and international levels.

The Reality of Aid Report for 2010 analysed aid and development cooperation issues and focused on two major areas of criticism of current government practices and programmes:

- The fact that governments consistently fail to achieve agreed and publicised aid targets
- The quality of the aid that is delivered needs to be improved if it is to achieve its stated aims and objectives

The detailed criticism includes the following points:

On failing to reach agreed targets

- With just five years remaining to realise their commitment to the Millennium Development Goals (MDGs), donors continue to fall far short of the pledges they made in 2005.
- While ODA performance as a % of GNI only increased slightly because of a 3.5% decline in collective donor GNI, ODA continues to fall short of the commitments made to 0.7% of GNI – this is now a long standing failure, year upon year.
- Reality of Aid estimates 2009 'real ODA' at US$112.7 billion, which is only 0.29% of GNI, far removed from the UN target of 0.7%. 'Real ODA' is calculated by subtracting debt cancellation and the costs of spending on Southern refugees and on students arriving in donor countries from reported ODA.
- The Report argues that aid commitments are affordable despite the current economic crisis. In 2008, the amount of aid was equivalent to just 1.8% of total donor government revenues which was below the 2% level of 1990. Aid per donor country citizen was only US$118 in 2010.

On the quality of aid

- Despite commitments made to make aid more effective, aid donor countries only marginally improved their performance over that of 2005 as regards targeting human development goals, gender equality and the poorest countries in Africa. In their aid allocations and policies, donors are giving only slightly increased priority to the fundamental issues of poverty reduction and strengthening the rights of the poor with only 42% of 'new aid dollars' being spent directly on human development goals.

  Aid continues to be driven by foreign policy concerns with the majority of funds being allocated to increased support for debt cancellation, support for refugees in donor countries, and to Iraq and Afghanistan.

- Donor governments continue to fail to prioritise global 'public goods' and the MDGs in areas such as education, health, food security, and poverty reduction.

- Country to country humanitarian assistance

continues to grow as a proportion of *'real aid'* amounting to 8.3% in 2008 and with increasing amounts directed to Sub-Saharan Africa. This priority in aid must continue according to the Report.

- Gender equality remains largely *'invisible'* in donor aid activities. Only 4.1% of official aid funding goes to activities where gender equality is stated as a principal objective while only $US411 million of a total aid budget of $US122 billion was allocated to organisations working on women's equality, a fundamental human development issue.

- Donor governments continue to undertake donor controlled *'technical assistance'* programmes which currently take up at least one-third of all DAC bilateral aid, something which undermines local control and ownership.

- Governments continue to *'short-change'* Sub-Saharan Africa - by at least US$14 billion compared to their pledges for 2010. Donors have failed to deliver on their 2005 commitment to provide an additional US$25 billion a year to Sub-Saharan Africa by 2010 – the region containing many of the world's poorest and most vulnerable people. By 2010, total donor aid to Sub-Saharan Africa is expected to be only US$36 billion against a target of US$50 billion.

- Governments are reneging on an earlier pledge that financing for climate change activities and programmes would be *'additional'* to ODA. Climate change financing must focus on the most vulnerable, particularly women, taking account of human rights as well as of development effectiveness principles.

- Donors have not strengthened local country ownership and leadership on bilateral aid programmes and policies despite repeated commitment to this principle.

- Aid regularly remains *'tied'* – the linking of aid to donor country contractors and suppliers despite commitments to significantly reduce the practice.

- Donor governments continue to influence and direct policies in aid-dependent poor countries especially through requiring compliance with International Monetary Fund and World Bank conditions, even though such conditions often undermine human development objectives, especially those of the poor and marginalised. The Report argues that agreements should be based on international law rather than the conditionality of the Fund or the Bank.

- The costs associated with servicing aid programmes by poor countries and local NGO and civil society organisations have significantly increased as the channels for official ODA have increased. This reality has significantly reduced the potential for citizens in the poorest countries to achieve real democratic ownership of locally determined needs and priorities.

For more information and analysis from the Reality of Aid Network, see http://www.realityofaid.org/

THE DEBATE ON AID

THE CRITICS OF AID

'The less aid you have, the better you do? The aid lobby claims the opposite, and exhorts a doubling, tripling of aid. But if it is aid on the old terms, it will not bring progress. Saying that, the world is richer than ever and implying that development can simply be purchased with more of the same resources is a monumental deception. It ignores the primordial need for countries to foster their own capacity to development.'

David Sogge 2006

There are a number of important and fundamental criticisms of aid offered by those who are sceptical, openly critical or downright dismissive of aid as it is practised today – this section examines some of the key arguments but, at the outset it is necessary to note two important points which must be taken into account in considering those critical voices speaking out against aid:

- Most critics are not against all forms of aid. Few argue against humanitarian assistance in times of disaster – almost everyone agrees we have a duty to help those in dire need in times of emergency.

- Most criticism has been directed at official or government aid although there is considerable and growing criticism directed towards NGO aid also (see below).

FOUR KEY CRITICISMS OF AID

1 **Aid does not work because it is not effectively planned or targeted; it doesn't get to those who need it most and is routinely undermined by corruption** – aid is insufficiently focused on human development needs; it does not relate to income levels in recipient countries (the lowest income countries account for 75% of those living in poverty but only 40% of all aid); aid remains volatile and undependable; aid remains driven by donor needs and objectives and aid is frequently focused on non-development objectives (these arguments are made by critics of aid such as Stephen Browne (a former senior UNDP staff member) as well as by the OECD Development Assistance Committee itself). On this critique, see the work of Stephen Browne, 2006, *Aid and Influence: Do Donors Help or Hinder?* (Earthscan, London).

2 **Aid does not achieve its objectives because it is inherently negative** – Zambian Dambisa Moyo argues that aid has not only failed to work in terms of promoting development, it has actually made things worse; for Moyo, aid is not part of the solution – 'aid is the problem' and her solution is for Africa to trade its way to development instead of becoming reliant on volatile aid payments. Moyo, like many before, argues that no country will ever develop itself via handouts (see Dambisa Moyo, 2009, *Dead Aid: Why Aid is Not Working and How There is Another Way for Africa* (Allen Lane, London).

William Easterley, another critic poses a similar question – why the West's charitable work has in fact accomplished *'so much ill and so little good'* and argues that aid fails to get below the surface of underdevelopment; is too focused on the needs of governments; does not address the needs of the poor and is, in effect *'self-serving'* (see William Easterley, 2007, *The White Man's Burden* (Oxford University Press).

3 **Aid cannot overcome a chronically unjust international economic system** - many critics of aid highlight the realities of the international economic and financial system, which systematically operates to the disadvantage of the poor and in favour of the rich. They argue that no amount of aid, however well planned and delivered, could hope to overcome the disparities and that the debate on aid is often a distraction. The main focus should be on radical reform of the international political and economic system (see Johnathan Glennie, 2008, *The Trouble With Aid: Why Less Could Mean more for Africa* (Zed Books, London).

4 **Aid doesn't work because it was never designed to do so** - critics often view aid as part and parcel of the agenda of the West in the Third World and that government aid is an inappropriate mechanism for human development as it is based upon a series of erroneous assumptions – that the poor are poor because they have been 'left out' of development; because they lack key resources which are not available locally; because they lack incentives to increase productivity, etc. (see Teresa Hayter, 1971, *Aid as Imperialism* (Penguin, London).

For Hayter, aid is inappropriate because it is *'…simply another version of the fallacious theory that one can reach the poor by expanding a process controlled by the rich.'*

THE DEFENDERS OF AID

'There is plenty of evidence of impact in terms of skills development, improvement in services, infrastructure, production, income and well-being as well as in specific areas such as education and health (schools, vaccines, basic medicines etc.)…'

Roger Riddell, 2008

The past 30 years has witnessed the growth of an *'industry'* in measuring the impact and benefits of aid, something which is inherently very difficult as aid does not exist in isolation from all the other factors that impinge on, or shape, development including local, national and international policies and practices. The international economic and financial system is broadly *'hostile'* to the needs and interests of the world's poor and aid has to operate in this context. In such a situation, aid alone cannot be blamed for the failure of development. With this in mind, there are a number of key points which have been made in the argument supporting the existence of aid.

FIVE ARGUMENTS IN SUPPORT OF AID

1 **Aid is necessary in crisis situations although the delivery of humanitarian aid could be improved** – few would argue against the need to provide immediate and effective assistance to those affected by different types of disasters or emergencies, especially as they affect the poor and least powerful most directly. For a detailed discussion on this dimension of aid, see the arguments and work of the Humanitarian Practice Network at http://www.odihpn.org

2 **A great many individual aid projects and initiatives do have impact and do promote key dimensions of human development** – analyst and aid activist Roger Riddell has published two of the most exhaustive studies of aid and its impact. He argues that there is significant evidence of the effectiveness of aid in real human development terms – skills, services, welfare, education, health, agriculture, medicines etc.). Riddell holds that aid is at its best when addressing people's short-term basic needs and can be argued to be effective at 75% - 90% levels. But, he argues aid is only one part of the picture and is essentially a *'gap filler'*. See Roger Riddell 2008, *Does Foreign Aid Really Work?* (Oxford University Press) and the UNDP Human Development

Report for 2005. Further information on aid and its effectiveness can be found at the Development Co-operation Directorate of the OECD.

3 **Engaging with aid overall has led to considerable learning about the complexities and contradictions but also the possibilities of the development process itself** – writers and analysts such as Jeffrey Sachs, Paul Collier and philosopher and activist Thomas Pogge provide detailed and pragmatic examples and proposals where aid is not only needed but is also effective within the context of human development today. See Jeffrey Sachs 2005, *The End of Poverty* (Penguin, London), Paul Collier 2007 *The Bottom Billion*, (Oxford University Press) and Thomas Pogge 2008 *World Poverty and Human Rights* (Yale University Press).

4 **Aid on its own is by no means a sufficient condition to promote development – an obvious but nonetheless crucial observation** – arguing that aid has failed to produce development misses the mark by some considerable distance as aid, on its own, could never hope to overcome the many obstacles (local, national and international) to full human development. On this theme, see the work of Thomas Pogge cited above.

5 **Economic, let alone human, development does not equal the sum of the parts of a series of individual aid projects no matter how successful or well designed and managed they might be. Aid alone cannot be blamed for the failure of development** – this position is argued in the UNDP's Human Development Report for 2005 as well as by Roger Riddell (see above).

However, within the analysis offered by those who broadly defend aid, there is an acknowledgement of the numerous problems with aid which urgently need to be addressed - it remains largely short-term; its impact is heavily influenced by other factors like conflict and international trends; and aid volumes and therefore the impact of aid itself remains *'volatile'*. What is not so clear from the evidence available is whether this short-term, practical type of aid contributes to longer-term sustainable development. So, for Roger Riddell the key question is not *'Does aid work?'* but rather *'How can aid to poor countries be made more effective?'* Unlike some critics of aid its defenders believe that the problems visible in the delivery of aid do not mean aid should be cut or abandoned altogether, but that constant advancement must be made to increase aid effectiveness and transparency to ensure it gets to those who need it the most.

NGOS AND AID

While official aid is, by far the largest source of aid to developing countries, NGOs are increasing the scale of their presence and programmes in human development. Non-governmental organisations (NGOs) is a broad term used to describe a variety of independent organisations involved in different types of agendas in aid, development, welfare, education etc. Observers tend to identify three characteristics common to development or aid NGOs:

- they are directly or indirectly involved in development and humanitarian work

- they are separate from government and private for-profit organisations

- their business is of a not-for-profit nature

It includes NGOs based in developed countries and NGOs based in developing countries. The main sources of funds for NGOs are government donations, private or individual donations and money given by private foundations or trusts.

There are three main areas of NGO work:

- The provision of direct support (such as food, medical assistance and shelter) which assists in overcoming the immediate consequences of an international crisis or a natural disaster.
- Short and long-term projects involved with *'local capacity building'* including training local leaders, defending human rights and encouraging effective participation and the *'empowerment'* of people who are directly affected by global poverty and inequality.
- Highlighting the causes and consequences of world inequality within *'developed'* countries in order to stimulate and promote activism for real change. NGOs relate to the public in a number of ways - fundraising, development education and campaigning. NGOs are also involved in advocacy and lobbying for policy change.

Studies show that most NGOs allocate aid and development funding largely according to the needs of their beneficiaries and poverty consistently appears as the main determinant of resource allocation. This contrasts with Official Development Assistance (ODA), which tends to be allocated according to political, economic and strategic considerations. In 1997 an OECD study concluded that 90% of all NGO projects achieved their immediate objectives and in 2003 a 25-year review of EU support to over 8,000 NGO projects found that 60% of projects were rated as either satisfactory or excellent. Despite this, many commentators make the valid point that the overall impact of the work of NGOs is really just a drop in the ocean by comparison with the scale and nature of the challenge.

NGO AID DEBATED

The past decades have been characterised by a major increase in government aid and projects channelled through NGOs, sometimes to the point where the political and financial independence of NGOs has been questioned. With the increasingly diverse and widespread work of NGOs, the debates as to the merits of directing aid through this channel are increasing.

Some advantages

- NGOs often (but by no means always) bypass governments, have more direct contact with local communities; therefore their projects can be more locally focused and effective
- Unlike governments, NGOs often have more access to areas affected by conflict or political strife due to their assumed *'neutrality'*
- The problems associated with *'tied'* aid are easier to avoid with NGOs as their projects are usually smaller in scale, therefore, larger scale, often inappropriate projects can be avoided
- NGOs are often the leaders in openly challenging injustice and inequality at local and international level and in this context play an important role in supporting local NGOs and organisations
- Some studies show that NGO projects are often more cost effective than projects run by local governments

Some disadvantages

- As governments have increased their funding and influence in NGOs, their independence has come into question. While more money can be positive, it can undermine and even reduce NGOs' ability or willingness to speak out or to follow independent policies and strategies.

- Some NGOs still promote the *'starving black baby'* image. This can seriously distort the perception of people in the developing world, reinforcing age-old stereotypes. This undermines understanding and can feed discriminatory attitudes and, sometimes, even racism.

- For fundraising purposes, many NGOs offer simplistic and highly questionable *'solutions'* to world development issues, *'solutions'* that generate very considerable criticism and debate especially from Third world organisations and commentators e.g. donating inappropriate or second hand goods, child sponsorship schemes; volunteering programmes etc.

- There is a view that some NGO's have become too large and over professionalised. Instead of listening to the needs of the people themselves, there is now an increase in using the views of academics to steer policy and projects, reinforcing *'western views'* of development at the expense of local views.

On the role of NGOs in development, see Roger Riddell 2005 *Does Aid Really Work* (Oxford University Press), Ian Smillie 1995 *The Alms Bazaar: Altruism Under Fire – Non-Profit Organisations and International Development* (ITDG Publishing, London) and Daniel Korten 1990 *Getting to the Twenty First Century: Voluntary Action and the Global Agenda* (Kumerian Press, Virginia).

CASE STUDY 1: SOUTH-SOUTH COOPERATION: THE CHANGING FACE OF AID?

International development cooperation is commonly viewed in the context of North-South relations. Traditionally, aid flowed from donor countries in the North to donor-recipient countries in the South. This placed the balance of power in the North leaving impoverished Southern countries with little room for manoeuvre and dependent on Northern aid flows. Recently the aid environment has experienced significant change in this regard; with Northern economies in stagnation, some strong Southern economies have emerged as big players on the world stage. These include countries like Brazil, India, Saudi Arabia, Venezuela and, most notably, China. China has now overtaken Japan as the second largest economy in the world while India's economic output is expected to surpass the gross domestic product (GDP) of Canada by next year. Brazil is the fastest growing economy in all of the Americas. With the rapid growth of development assistance from so-called *'emerging donors'* from the South, there is increasing interest in South-South Cooperation especially within official government circles. Southern donors argue they can bring a feeling of mutual understanding when providing aid assistance and can avoid some of the mistakes made in the past by Northern donors.

A study produced by the United Nations Economic and Social Council in 2009 proposes a definition of South-South ODA as consisting of *'grants and concessional loans provided by one Southern country to another to finance projects, programmes, technical cooperation, debt relief and humanitarian assistance and its contributions to multilateral institutions and regional development banks'*. A 2008 study covering 18 major Southern contributors estimated South-South ODA flows between US\$9.5 billion and US\$12.1 billion disbursements for 2006, or 7.8% to 9.8% of total international development assistance. The growing presence of Southern donors has stimulated debate about both old and new issues in development co-operation – the place of geo-politics in aid; the place of commerce and resource extraction; human rights observance, environmental issues and issues around transparency, accountability and monitoring.

For more information on South-South Development Cooperation, see Reality of Aid Network 2010 *South-South Development Cooperation: A challenge to the aid system?* http://www.realityofaid.org/roa-reports/index/secid/373/South-South-Development-Cooperation-A-challenge-to-the-aid-system

CASE STUDY 2: THE INCREASING PRESENCE OF CHINESE AID IN AFRICA

China is fast becoming a major source of aid in Asia, Latin America and most notably Africa. Estimates vary as to how much aid China is actually giving but speculation places it between US$1 and US$2 billion with up to half going to Africa, although exact figures cannot be accessed. Like many Southern donors, China is reluctant to appear to be reproducing donor-recipient hierarchies, traditionally associated with Western countries.

In theory, Chinese aid comes with no political or economic strings attached, but with little transparency and accountability concerns about issues such as human rights, corruption and environmental responsibility are widespread. It's *'no questions asked'* aid policy makes Chinese aid very attractive to countries with unstable or undemocratic governments, which has raised the suspicions of a number of western commentators. Many argue that China's varied interests in Africa threaten to undermine American and European efforts to promote peaceful societies in the region.

China's involvement in countries like Sudan and Zimbabwe gives substance to the argument that China favours corrupt, unstable or authoritarian regimes as against the more democratic states favoured by the EU and the United States (this argument is hotly disputed as EU and US interests routinely engage undemocratic regimes across the world).

China continues to develop new markets to ensure economic growth and there are now over 800 Chinese companies operating in nearly all African nations. Cheap Chinese goods routinely compete in African economies against African producers, something that western states are also guilty of. Chinese aid is often also linked to the use of Chinese labour and there are consequent fears regarding labour standards and working conditions.

For its part, China has made no bones about its aid hinging upon self-interest. With over 50 nations, the continent of Africa represents more than one-quarter of the United Nations General Assembly – a significant voting bloc - and some commentators argue that gaining power on the international stage is the underlying reason for China's recent increased aid efforts in Africa. Another reason is China's need for natural resources to fuel economic expansion. Africa provides China with 30% of its energy imports, meeting 5% of China's energy needs and rivalling the Middle East as a source of energy.

China also sees Africa as a growing market for its goods and in 2006 Sino-African trade was worth US$56 billion bringing in revenue critical to some of the world's poorest nations. A recent AFRODAD report points out that Chinese firms are building roads, rehabilitating infrastructure, and bringing in wireless communication systems where landlines have not worked, especially in rural areas. It is also worth noting that China's interest in Africa has given African nations more options to negotiate better trade deals with Western competitors, leading to Western corporations and governments now facing competition.

Engaging with corrupt governments and leaders, questionable human rights and poor environmental records, the lack of aid conditions (traditionally applied by Western governments) and increases in economic activity by China in many African countries ensures that this debate will continue.

For more information, see African Forum and Network on Debt and Development, 2010 'Assessing the growing role and developmental impact of China in Africa: An African Perspective' in Reality of Aid Network *South–South Development Cooperation: A challenge to the aid system?*

READING

Bertrand Borg, Mary Rose Costello and Colm Regan (2010) Debating Aid, Bray, 80:20 Educating and Acting for a Better World

Stephen Browne, 2006, Aid and Influence: Do Donors Help or Hinder? London, Earthscan

William Easterley, 2007, The White Man's Burden, Oxford, Oxford University Press

Johnathan Glennie (2008) The Trouble With Aid: Why Less Could Mean more for Africa, London, Zed Books

Daniel Korten (1990) Getting to the Twenty First Century: Voluntary Action and the Global Agenda, Virginia, Kumerian Press

Dambisa Moyo, 2009, Dead Aid: Why Aid is not working and How There is Another Way for Africa, London, Allen Lane

Reality of Aid Network (annual) Reality of Aid Report

Roger Riddell (2005) Does Aid Really Work, Oxford, Oxford University Press

Jeffrey Sachs 2005, The End of Poverty, London, Penguin

Ian Smillie (1995) The Alms Bazaar: Altruism Under Fire – Non-Profit Organisations and International Development, London, ITDG Publishing

MORE INFORMATION & DEBATE

http://www.oecd.org/dac - Development Assistance Committee of the OECD

http://www.realityofaid.org – main NGO site for analysis of aid issues

http://www.aidwatchers.com - The Aid Watch blog maintained by the Development Research Institute, New York University (contains the views of critic William Easterley)

http://www.concordeurope.org - European Confederation of NGOs

http://www.guardian.co.uk/global-development - development section of the Guardian newspaper, London

http://www.odi.org.uk/work/themes/details.asp?id=2&title=aid – UK Overseas Development Institute themed section on aid issues

[*'...HIV and AIDS is not a democratic disease...'*]

Michael Kelly SJ

Chapter 10

WOMEN, VULNERABILITY AND HIV AND AIDS

Valerie Duffy

This chapter focuses directly on the impact of HIV and AIDS on women and reviews the arguments and evidence as to why women are so susceptible to the virus. Focusing on sub-Saharan Africa, South Africa and India, the chapter outlines five key dimensions of the vulnerability of women.

KEYWORDS

'Feminisation of HIV and AIDS', vulnerability of women, HIV and AIDS in sub-Saharan Africa, India and South Africa, biological, economic, educational, socio-cultural and legal-political vulnerability of women

'...Men have too much power. We have been taught to submit to men – we have been told, we cannot say no to sex. If your husband approaches you, whether you are ready or not ready, you have to say yes to sex. Women have no control over their bodies. You do not know how a man is using his body. You do not know how many partners he is meeting. But when he goes to his wife, she cannot refuse him. You have to let him do what he wants. He can bruise you, but you cannot cry out. It is like women are being bought to be sex machines...'

'...A 'no' must be said. We need to have rights, just like men. If a man finds a woman in a relationship outside of marriage, they can sue them. But women cannot if they find a man is doing the same. Men need to be punished for this also. Proper, practical equal rights are needed between men and women, not theoretical ones. Practical ones. If we continue to crawl in front of men, they will continue to look down on us. Stand up and show them that we are making a change. Raise your head and tell them that we are teaching men to make a change. Change has to begin with me, you and him and her, so that others can also change...'

'... The two most important things that need to be done. Firstly, education. We need to share what we know and learn, we need to educate young people. The second is empowerment in order to fight poverty. If people have something to eat, they do not need to sell their bodies in order to feed their family. We need more honesty in the aid that is given to Zambia. There needs to be a 'down up' approach as we need to feed the roots to empower the leaves. If you feed the leaves, everything becomes dry. The government needs to not enrich themselves, but enrich the roots.'

Chieftainess Mwenda, Zambia (pictured on previous page)

INTRODUCTION

(This chapter is extracted from a larger report *This is what has happened... HIV and AIDS, women and vulnerability in Zambia* by V. Duffy and C. Regan 2010, Bray, 80:20 Educating and Acting for a Better World)

It is now undeniable that HIV and AIDS is more than ever, a devastating attack on women, most notably on women in sub-Saharan Africa - the only region in the world where, according to UNAIDS, HIV rates are higher among women than men. Of the 23 million adults currently infected in sub-Saharan Africa, 57% are women: women aged between 15 and 24 years, are three times more likely to become infected than men of a similar age. This increase in the number of women and girls becoming infected at ever younger ages is now referred to as the 'feminisation of HIV and AIDS'.

In the initial stages of the spread of HIV, men appeared to be more infected. More recently though, it is women who have become more vulnerable, especially in countries where the primary transmission is through heterosexual intercourse. The negative impact of the virus on the lives of women is more severe than for men, principally due to their subordinate status in society. In many sub-Saharan African countries, socio-cultural practices and traditions sustain women's unequal status leaving them vulnerable to poverty, discrimination and violence – and ultimately to HIV infection.

This chapter explores each of these vulnerabilities and some of the associated human consequences primarily in the context of Zambia.

Continued on page 183 →

THE PARTICULAR VULNERABILITY OF WOMEN CAN BE HIGHLIGHTED AT FIVE LEVELS:

BIO-MEDICAL VULNERABILITY

Women remain biologically seven times more vulnerable to the transmission of the virus during sexual intercourse than men; cultural practices reinforce this and women's role as the primary care givers also leaves them vulnerable.

ECONOMIC VULNERABILITY

The poverty experienced by women and their economic dependence on men leaves them vulnerable often with little option but to sell themselves in order to survive or to feed their children.

SOCIAL AND CULTURAL VULNERABILITY

Certain cultural practices associated with the subordination of women to men help ensure women's vulnerability to HIV and AIDS. The practice of multiple concurrent sexual partnerships is lethal in this context.

LEGAL VULNERABILITY

While women are seen in theory to be equal in rights to men, the practice often denies this with traditional law as well as constitutional law often discriminating against women and structures and institutions routinely enforce this discrimination.

EDUCATIONAL VULNERABILITY

The ongoing challenge of ensuring female access to, and completion of, education at primary and post-primary levels contributes to the subordination of women.

HIV AND AIDS IN AFRICA

Sub-Saharan Africa is more heavily affected by HIV and AIDS than any other region of the world. An estimated 22.5 million people are living with HIV in the region - around two thirds of the global total. In 2009 around 1.3 million people died from AIDS in sub-Saharan Africa and 1.8 million people became infected with HIV. Since the beginning of the epidemic, 14.8 million children have lost one or both parents to HIV and AIDS.

The social and economic consequences of the AIDS epidemic are widely felt, not only in the health sector but also in education, industry, agriculture, transport, human resources and the economy in general. The AIDS epidemic in sub-Saharan Africa continues to devastate communities, rolling back decades of development progress.

Currently, sub-Saharan Africa faces a triple challenge:

- Providing health care, antiretroviral treatment, and support to a growing population of people with HIV-related illnesses.

- Reducing the annual toll of new HIV infections by enabling individuals to protect themselves and others.

- Coping with the impact of millions of AIDS deaths on orphans and other survivors, communities, and national development.

Both HIV prevalence rates and the numbers of people dying from AIDS vary greatly between African countries.

- In Somalia and Senegal, HIV prevalence is under 1% of the adult population, whereas in Namibia, Zambia and Zimbabwe around 10-15% of adults are infected. Southern Africa is the most impacted by AIDS; in South Africa the HIV prevalence is 17.8% and in three other southern African countries, the national adult HIV prevalence rate now exceeds 20%. These countries are Botswana (24.8%), Lesotho (23.6%) and Swaziland (25.9%).

- West Africa has been less affected by HIV and AIDS, but some countries are experiencing rising HIV prevalence rates. In Cameroon HIV prevalence is now estimated at 5.3% and in Gabon it stands at 5.2%. In Nigeria, HIV prevalence is low (3.6%) compared to the rest of Africa. However, because of its large population (it is the most populous country in sub-Saharan Africa), this equates to around 3.3 million people living with HIV.

- Adult HIV prevalence exceeds 5% in Uganda, Kenya and Tanzania.

Overall, rates of new HIV infections in sub-Saharan Africa appear to have peaked in the late 1990s, and HIV prevalence seems to have declined slightly, although it remains at an extremely high level.

Sources: UNAIDS World AIDS Day Report | 2011 and http://www. avert.org/aids-hiv-africa.htm

THE WIDESPREAD IMPACT OF HIV AND AIDS ON AFRICAN SOCIETY

- **On life expectancy:** in many countries of sub-Saharan Africa, AIDS has erased decades of progress made in extending life expectancy. Average life expectancy in sub-Saharan Africa is now 52 years and in the most heavily affected countries in the region is below 51 years. In five of the six sub-Saharan African countries where life expectancy is lower than it was in the 1970s, this decline has been directly linked to HIV and AIDS.

- **On households:** the effect of the AIDS epidemic

on households can be very severe, especially when families lose their income earners. In other cases, people have to provide home-based care for sick relatives, reducing their capacity to earn money for their families. Many of those dying from AIDS have surviving partners who are themselves infected and in need of care. They leave behind orphans who are often cared for by members of the extended family.

- **On healthcare:** in all affected countries, the epidemic is putting strain on the health sector. As the epidemic develops, the demand for care for those living with HIV rises, as does the number of healthcare workers affected.

- **On schools:** schools are heavily affected. This a major concern, because schools can play a vital role in reducing the impact of the epidemic through HIV education and support.

- **On productivity:** the HIV and AIDS epidemic has dramatically affected labour, which in turn slows down economic activity and social progress. The vast majority of people living with HIV and AIDS in Africa are between the ages of 15 and 49 - in the prime of their working lives. Employers, schools, factories and hospitals have to train other staff to replace those who become too ill to work.

- **On economic growth and development:** the HIV and AIDS epidemic has already significantly affected Africa's economic development and, in turn, has affected Africa's ability to cope.

Source: http://www.avert.org/aids-hiv-charity-avert.htm
(based on UNAIDS 2010 figures)

BIO-MEDICAL VULNERABILITY

Women are physically more vulnerable than men to infection through sexual intercourse. This is due to a number of factors including genital physiology, unequal power relations between men and women, sexual behaviour - the nature and pattern of relationships (number of partners and rate of partner change), sexual practices and condom use. HIV infection also depends on the stage of infection, the presence of other Sexually Transmitted Infections (STIs), the availability of STI treatment and, access to condoms.

Male to female HIV transmission is seven times more likely than female to male transmission (often referred to as 'biological sexism'). This is due to the fact that the mucous membranes on the cervix of the uterus are especially vulnerable to sexually transmitted infections, including HIV. The smallest cut (often invisible and unknown) is enough for an STI or HIV virus gain access to a body. As women have a greater mucous surface, they are more at risk of injury and infection through the blood stream. It is for this reason that it is estimated that women are two to four times more susceptible to HIV infection through sexual intercourse. In particular, the vulnerability of teenage girls is further aggravated by how susceptible their immature cervix and genital tract is to tearing, lacerations and infection during intercourse, with this risk doubling during and just after pregnancy.

Although there has been a significant increase in the number of campaigns promoting the use of condoms in order to prevent HIV transmission, research has shown that condoms are more generally used in commercial sex than in the home. The stark reality for some women is that unless they are empowered to have some degree of control in a sexual relationship, the use of a condom will depend on the male, thus highlighting gender inequality.

Since 2005, Anti-Retroviral Drugs (ARVs) have been supplied free from clinics and hospitals across Zambia. Despite this, not all people in need of treatment are receiving it. This is due to the fact that the *'accessibility costs'* can be very high in terms of distance (especially for those unable to afford transport or the associated food and accommodation costs) plus the length of time people have to wait (often for days) because there is no doctor or because the machine is unavailable or broken. ARVs are not always readily available from some clinics especially those in the more rural areas. If a person is bedridden or cannot afford to pay, they regularly end up defaulting on their medication. This is detrimental to the patient's long- term treatment and wellbeing. Although some clinics have support teams to service their clients, this is not true for the majority of clinics.

The biological vulnerability of women is further exacerbated by a range of additional vulnerabilities explored below.

'It thus remains that the vulnerability of women due to biomedical factors is exacerbated by a deep-rooted lack of social capital, income inequality, and social and gender justice, in themselves highly important predictors of HIV."

Helen Rees and Matthew Chersich, 2008

CASE STUDY 1: CHIKU ZULU

Chiku Zulu is a nurse working with HIV programmes in Chikankata. She is fifty years old and separated from her husband. She has four children - two girls and two boys. Her eldest is twenty and works in the laboratory in the local hospital while her second born is studying to be a mechanic. Her first daughter was born in 1990 and is training to be a secretary and her last child was born in 1996 and has just started secondary school.

Chiku argues that the main problem in terms of medication is trying to reach the people who live a long distance from services - this is when the Care and Prevention Team (CPT) helps as they bring the drugs directly to the people. This is not ideal however, because people still need to see the doctor. Often however, it is too far for them to travel, especially if they are bed ridden. Some CPT members have bicycles, but a lot of the time, they have to walk to the communities they serve.

Chiku believes women are more vulnerable to the virus, especially as a result of traditional customs. If you are married, you have to say yes to sex, even if you know your partner has been with other women, or has other wives. Women are also the primary caregivers which also leaves them vulnerable to infection.

'...When a woman gets married, her husband pays a bride price to her relatives and this then gives him the power to demand sex at any time. In my culture, sexual issues are taboo, so you cannot talk about it. If you say no to sex because you think he has been with someone else, then he will go and find someone else to satisfy him. This brings infection into the relationship. If you ask your husband to use a condom, this often results in fights and quarrels...'

ECONOMIC VULNERABILITY

This reality has been very effectively described by Linda Fuller (2008) who argues that although African women produce three-quarters of the continent's food; are described as the *'backbone of the rural African economy'*; farm small plots selling fruit and vegetables and are the main providers for their families, they remain amongst the poorest of the poor. Despite the fact that women do the majority of the informal work within the economy, they are still heavily dependent on men due to the lack of access to capital or credit or control over household resources as well as traditional practices including those that relate to the ownership of land or property and its inheritance. Lacking power or control over household or communal resources makes women subservient to men and relatively powerless in important negotiations.

Only a minority of economically active women earn wages in the formal economy and due to the informal nature of much of their work, women's vulnerability is heightened by the fact that if they (or a family member) become ill, they do not get paid for days missed while caring for themselves or

CASE STUDY 2: FLORENCE HAGILA

Florence Hagila is forty, married with seven children, four of whom are married, two have just finished school, and one is still in school. She herself left school in Grade 10. She also has three dependents, all of whom are orphans because their parents died from HIV related illnesses. She is a farmer growing tomatoes, maize, ground nuts and sweet potatoes. She is the local chairperson of the Society for Women and AIDS in Zambia (SWAAZ – one of Zambia's most active HIV and AIDS focused NGOs) in her village Mulendi, volunteers at the local health centre and visits the local community areas. She instructs people on how to use things like mosquito nets and on personal hygiene and family planning.

Florence feels that as so many people are infected, she is not sure if it is men or women who are most affected. What is important to her is the need to use condoms and to go for VCT. Women are the most significantly affected because they care most for those who are infected, and caring for orphans usually falls on them. Because women stay at home, not many of them have access to adequate information about ARVs and so on. They see their husbands taking drugs, but they do not know what they are. Women are also very affected because they do not have any power within society. Men can infect more women, because they can propose to more than one woman.

'The world is cruel sometimes and that HIV and AIDS are not curable, but that it is not the end of your life. People can still have negative children if they are positive, so long as they find out how to protect their child, and follow the instructions they are given. I feel sad and angry about people who deliberately infect others'

family. Having someone who is infected with HIV or AIDS can have a devastating impact on families with increased costs for medication (though anti-retroviral drugs are free), travel, food, and care. Increased costs in one area means that there is less money to go around and it can mean that children stop going to school, farm work remains unfinished or there is less food. Funeral expenses are yet another heavy burden for families.

It is for reasons such as these that many women are often forced into prostitution or risky sexual relationships with older men (or *sugar daddies*) as a survival strategy in order to provide for their families, despite knowing that this may lead to the transmission of HIV. Additionally, many girls are taken out of school early in order to help at home or to provide care and are subsequently deprived of education, thus reinforcing subordination, vulnerability and disempowerment.

Sociologist Felly Nkweto Simmonds describes the issue in the following terms:

'Economic empowerment for individual girls is first of all internal. Get the girls to think about themselves, to value themselves – that is internal self-confidence, self-esteem, and then the money can work because you are your own self. You can make your own decisions about your life, your children's lives, where they go to school etc. What makes women feel less empowered, even if they have money in their hand, is that they have no control, no power. All of these things undercut women. All women need to be empowered, but there is more work to be done to get there.'

Quoted in Valerie Duffy and Ciara Regan 2010

SOCIAL AND CULTURAL VULNERABILITY

'Many cultures and religions give more freedom to men than to women. For example, in many cultures it is considered normal -- and sometimes encouraged -- for young men to experiment sexually before marriage. Also, in many cultures, it is considered acceptable for men -- even married men -- to have sex with sex workers. These cultural attitudes towards sex are leading to HIV infections in both men and women -- often the men's wives.'

UNAIDS, 2001

A critical factor aggravating the problem that AIDS poses for African women is the definition of the place of women in many societies. Of particular significance in terms of why sub-Saharan Africa has been hardest hit by HIV and AIDS is the subordinate social status of women and the many negative cultural practices and traditions that sustain subordination. The gender role prescribed for women and *femininity*, demands a submissive role, passivity in sexual relations, while the role prescribed for men requires them to be more dominating, knowledgeable and experienced about sex. This also puts many young men at risk as such perceptions prevent them from seeking information and also promotes promiscuity.

The majority of the vulnerabilities women face are not only maintained but are reinforced by damaging *'customary'* practices that potentially increase women's vulnerability to infection during sexual intercourse (e.g. sexual cleansing; wife inheritance; dry sex); harmful practices in society (older men having sex with young girls; multiple concurrent partnerships); those relating to women's health (traditional *'infertility treatments'*); initiation ceremonies (where women often are required to publicly display subordination and the messages communicated at time of initiation and at kitchen parties and regressive child-rearing practices (girls reared to be submissive, non-assertive, subordinate).

Currently the social group with the highest risk of HIV infection are married women where infection routinely occurs through external affairs by husbands and partners and where women do not have sufficient power to negotiate condom use within a relationship. Furthermore, women have insufficient power outside of a relationship to leave it if they are at risk of infection. Other cultural practices and traditions contribute, such as polygamy, levirate (marriage by a man's brother to his widow) or sororate (concurrent marriage with a wife's sister) and sexual cleansing.

Violence towards women compounds the link between gender inequalities and vulnerabilities where some women are continually subjected to abuse and rape. This is particularly of concern in countries with high prevalence rates (such as South Africa and Zambia) as there is a high possibility of HIV transmission if a woman is raped.

CASE STUDY 3: ERIC MUBITA

While Eric was married, he travelled up and down the country for his father collecting orders for his shop. This is when he contracted the virus. He used to travel to Livingstone and the Copperbelt where they were ordering clothes and groceries. On these trips, he would meet lots of different sexual partners but in 2007 he began to feel sick; his dad encouraged him to go for medication and accept his status. Eric felt very ashamed when he found out he was HIV positive, but his wife agreed not to leave him. Initially his wife thought it would be better not to take the drugs but she slowly changed her mind. They went together when he went for VCT, she did not blame him.

His first son is positive, but his other children are negative. The people in the community knew Eric was positive because he was sick. Eric's family would be considered high class in the community. People did not believe that he was positive because he was so fit, due to the drugs (ARVs).

According to Eric, cultural practices contribute greatly, for example, there is a practice where if someone is suspected of being poisoned by witchcraft, they cut the person with a razor blade and syphon out the blood with a horn to get rid of the poison. Circumcision is also an issue; one knife is used for fifteen to twenty children, so the virus can be transmitted this way also.

Eric argues that *'it is better for us to fight this pandemic, especially with initiation ceremonies – behavioural change is the only thing which can make this country a better country. Information dissemination is key. There is too much sexual activity'.*

LEGAL AND POLITICAL VULNERABILITY

'Zambia's constitution prohibits the enactment of any law that is discriminatory on the basis of sex or has such discriminatory effect. But it also recognizes a "dual legal system", which allows local courts to administer customary laws, some of which discriminate against women.'

Human Rights Watch, 2007

Most cultures in sub-Saharan Africa are patrilineal, so when a woman marries through customary law, she will then be a part of her husband's family or tribe and therefore any property will be passed along through the males in the family. Women can often only access land or property through their fathers, brothers, husbands or male relatives and cannot legally own land. If a relationship ends between a woman and her husband, there is a good chance that the woman will lose her home, land, livestock, household goods, money and any other property. These violations thus perpetuate women's dependence on men and undercut their social and economic status.

Although equality, reproductive and sexual rights are supposed to be guaranteed under international and regional human rights treaties, unless they are recognised and enforced by national-level courts, they are of little or no value. This situation is exacerbated by the fact that in much of rural sub-Saharan Africa, there is limited access to legal information or to African national courts in particular when it comes to the rights of women.

In Zambia, there are two legal systems – the *'civil'* court system and the *'traditional'* court system and, depending on location and practice, these systems do not view issues identically with the traditional system being predominant. Women, and especially rural women, are routinely at the mercy of traditional courts and this can greatly affect women especially in terms of finance, and specifically in relation to owning property. Poor educational capacity is often further compounded by lack of access to even basic information on, for example, *'property grabbing'* by the family of a deceased husband or partner and what the law allows.

The link between powerlessness and the risk of HIV infection is key to understanding the sources of women's vulnerability.

'The legal vulnerability of women in Zambia stems from the fact that discrimination against women has been legalised and that, as a result, women are restrained in exercising their right to self-determination, to autonomy and to physical integrity. Zambia has a non-discrimination clause in the Constitution, but that clause does not apply to customary law and yet everybody lives their daily lives governed by customary law because we are born into it and we are socialised by it. By failing, reusing or neglecting to exclude non-discrimination under customary law, we have effectively left the door wide open to discrimination against women.'

Joyce MacMillan - a legal practitioner and analyst with the
Zambian Law Development Commission.

EDUCATIONAL VULNERABILITY

'There is so much excitement when you go to a community and the women have learned to write their name - it brings a lot of pride. Traditionally, when a girl comes of age she is sent to her grandmother to be taught about womanhood and marriage. But if the grandmother is unable to read or write, she will only teach the girl about the more traditional aspects of marriage, and nothing about the danger of HIV and AIDS. If those women are taught how to read and write, and have information about HIV and AIDS that is the time they can teach girls about protecting themselves. Sometimes the children just have to depend on what they learn from television, radio and their friends.'

Edith Ng'oma of Zambian NGO FAWEZA – the Forum for African
Women Educationalists of Zambia

Education can help change key aspects of behaviour (knowledge, attitudes and perceived control) by influencing social networks and by leading to a change in socio-economic status.

For economic, social, family, health and cultural reasons, many young girls are forced to leave school early; there are many factors that may influence whether a girl remains in school or not including the costs of school fees and books, school uniforms, transportation making school attendance an economic impossibility. This reality contributes greatly to lowering female literacy rates and to generally poor educational attainment. In the context of women's health and HIV and AIDS, it puts them at serious risk, not only prior to infection, but also post-infection.

Additionally young people and particularly girls may face a number of school-related difficulties that can put them in danger of exposure to HIV and AIDS. The expense of school fees may lead school children (young girls) into the sale of sexual favours and pressure to succeed in the classroom may lead to sexual relationships with teachers or brighter students. Distance to school may prove a difficulty in terms of time to and from the classroom but also the potential hazards along the way (sexual harassment from school-mates or from strangers).

The practice of *'weekly boarding'* (Tanzania, Kenya, Malawi and Zambia) aggravates the risk – where, according to Michael Kelly SJ, the possible need for food, accommodation, security, recreation, pleasure and exploration or experimentation makes the weekly boarders susceptible to sexual activities with members of the local community or with one another.

There are a number of key issues which must be highlighted and addressed with regards to girls and women's education. The main focus points are the threat HIV and AIDS poses to the progress already made in girls access to, and completion of, basic primary education; how education is a *'critical mitigating force'* for developing life skills and knowledge in terms of supporting themselves and their families; and that by assisting girls in overcoming the effects of HIV and AIDS and supporting them in gaining access to education, they become more empowered to support themselves, their families and communities and also contribute to national development.

Additionally young people and particularly girls may face a number of school-related difficulties that can put them in danger of exposure to HIV and AIDS.

HIV AND AIDS IN INDIA

India is one of the largest and most populated countries in the world, with over one billion inhabitants and more than 300 million people, about one in four, living below the poverty line. There is an estimated 2.4 million currently living with HIV. The virus emerged later in India than it did in many other countries but infection rates soared throughout the 1990s and, now the epidemic affects all sectors of Indian society.

However, HIV prevalence among certain groups (sex workers, injecting drug users, truck drivers, migrant workers, men who have sex with men) remains high and is currently around 6 to 8 times that of the general population.

The primary drivers of the HIV epidemic in India are unprotected paid sex/commercial sex work, unprotected anal sex between men and intravenous drug use (IDU). Given that condom use is not optimal or consistent; men who buy sex are the single most powerful driving force in India's HIV epidemic. For example, recent data on HIV infection patterns in India from the UNAIDS Global Report 2010 reveals that 90% of women in India were infected within long-term relationships This would mean that women are at increased risk from HIV not due to their own sexual behaviour, but because they are partners of men who are within a high risk group - clients of female sex workers, men who have sex with men, drug users. The wider implication of this situation is that in almost 6% of cases in 2008, the route of transmission of infection was from mother to child (Source: UNAIDS 2010 Country Data.

According to the World Health Organisation 2010, Gender inequality, male dominance, stigma, low literacy and barriers to health care services are some of the key issues for higher vulnerability of women to HIV in the South-East Asia Region.

The majority of states within India have a higher population than most African countries and exhibit very high incidences of HIV prevalence (states such as Andhra Pradesh, Goa, Karnataka, Maharashtra and Tamil Nadu). In 2006 UNAIDS estimated that there were 2.4 million people living with HIV in India; the figure was revised in 2009 to between 2 million and 3.1 million with a prevalence of 0.3%. With a population of around a billion, a mere 0.1% increase in HIV prevalence would increase the estimated number of people living with HIV by over half a million.

At the beginning of 1986, despite over 20,000 reported AIDS cases worldwide, India had no reported cases but later that year India's first cases of HIV were diagnosed among sex workers in Tamil Nadu. It was noted that contact with foreign visitors had played a role in initial infections among sex workers, and as HIV screening centres were set up across the country there were calls for visitors to be screened for HIV. Gradually, these calls subsided as more attention was paid to ensuring that HIV screening was carried out in blood banks.

In 1987 a National AIDS Control Programme was launched to co-ordinate the response covering surveillance, blood screening, and health education. By the end of 1987, out of 52,907 who had been tested, around 135 people were found to be HIV positive and 14 had AIDS. Most of these initial cases had occurred through heterosexual sex, but at the end of the 1980s a rapid spread of HIV was observed among injecting drug users in Manipur, Mizoram and Nagaland - three north-eastern states of India bordering Myanmar (Burma).

At the beginning of the 1990s, as infection rates continued to rise, the government set up NACO (the National AIDS Control Organisation), to oversee policy, prevention and control programmes. In the same year, the government launched a

Strategic Plan, the National AIDS Control Programme (NACP) for HIV prevention. By this stage, cases of HIV infection had been reported in every state of the country and it became clear HIV had spread to the general population. Increasingly, cases of infection were observed among people that had previously been seen as 'low-risk', such as housewives and richer members of society.

In 1999, the second phase of the National AIDS Control Programme (NACP II) came into effect with the stated aim of reducing the spread of HIV through promoting behaviour change. During this time, the prevention of mother-to-child transmission (PMTCT) programme and the provision of free antiretroviral treatment were implemented for the first time. In 2001, the government adopted the National AIDS Prevention and Control Policy and former Prime Minister Atal Bihari Vajpayee referred to HIV/AIDS as one of the most serious health challenges facing the country.

The third phase began in 2007, with priority placed on reaching 80% of high-risk groups with targeted interventions, including sex workers, men who have sex with men, and injecting drug users. Targeted interventions are generally carried out by civil society or community organisations in partnership with the State AIDS Control Societies. They include outreach programmes focused on behaviour change through peer education, distribution of condoms and other risk reduction materials, treatment of sexually transmitted diseases, linkages to health services, as well as advocacy and training of local groups.

Source: http://www.avert.org/aids-hiv-charity-avert.htm (based on UNAIDS 2010 figures)

HIV AND AIDS IN SOUTH AFRICA

South Africa has a population of 50 million people – one-sixth the population of the United States of America and yet South Africa has over five times as many people living with HIV and AIDS. In 2009, an estimated 5.6 million people (3.3 million women) were living with HIV and AIDS, more than in any other country and it is estimated that in 2009, some 310,000 South Africans died of AIDS. Prevalence is 17.8% among those aged 15-49, with some age groups being particularly affected. Almost one-in-three women aged 25-29, and over a quarter of men aged 30-34, are living with HIV. HIV prevalence among those aged two and older also varies by province with the Western Cape (3.8%) and Northern Cape (5.9%) being least affected, and Mpumulanga (15.4%) and KwaZulu-Natal (15.8%) at the upper end of the scale.

Recent research suggests a slowing of HIV incidence amid some signs of a shift towards safer sex among young people with the annual HIV incidence among 18-year olds declining sharply from 1.8% in 2005 to 0.8% in 2008, and among women 15–24 years old it dropped from 5.5% in 2003–2005 to 2.2% in 2005–2008. However, South Africa is one of the few countries in the world where child and maternal mortality has risen since the 1990s. AIDS is the largest cause of maternal mortality in South Africa and also accounts for 35% of deaths in children younger than five years.

There are several interlinked issues critical to understanding the scale of the HIV epidemic in South Africa and, by implication, in other countries in Southern Africa including:

• Rising unemployment and social inequalities against a backdrop of collapsing agrarian and waged livelihoods, which are leaving some groups, especially poor women, extremely vulnerable

- Steeply falling marital rates
- Increasing numbers of women migrating to urban and peripheral urban areas, and periodically returning to rural areas. In South Africa, for example, migration has intensified and increasingly involves young women. Female entrants into the labour market rose by two million in 1995-1999, while median wages for women fell sharply – against a backdrop of collapsing agrarian and wage livelihoods generally – with important consequences for household formation, marriage, sexual networking patterns and HIV risk. A KwaZulu-Natal study, for example, found very high levels of HIV infection among migrating young women – 23% among sexually-active 17- to 18-year-olds and 65% among 22- to 24-year-olds.

Source: http://www.nelsonmandela.org

Impact upon children and families

South Africa's HIV and AIDS epidemic has had a devastating effect on children in a number of ways. There were an estimated 330,000 under-15s living with HIV in 2009, a figure that has almost doubled since 2001. HIV in South Africa is transmitted predominantly through heterosexual sex, with mother-to-child transmission being the other main infection route. The national transmission rate of HIV from mother to child is approximately 11%. Because the virus is transmitted from the child's mother in cases of mother-to-child transmission, the HIV-infected child is born into a family where the virus may have already had a severe impact on health, income, productivity and the ability to care for each other.

The age bracket that AIDS most heavily targets – younger adults – means it is not uncommon for one or more parents to die from AIDS while their offspring are young. The number of premature deaths due to HIV/AIDS has risen significantly over the last decade from 39% to 75% in 2010.

AIDS treatment in South Africa

The scaling up of treatment in South Africa in recent years has succeeded in vastly increasing the numbers of people receiving HIV treatment. At the end of 2007, it was estimated that only 27% of people in need of treatment were receiving it, below the average at the time for low and middle-income countries. By the end of 2009, this had increased to 56% above the average for low and middle-income countries. However, these estimates are based on earlier WHO guidelines and may underestimate the problem. The latest guidelines (2010) recommend starting treatment earlier, and have therefore increased the number of people estimated to be in need of HIV treatment, so that now treatment coverage is only 37%.

Table 10:1 - Estimated HIV prevalence among South Africans, by age and sex, 2008

AGE GROUP	MALE PREVALENCE %	FEMALE PREVALENCE %
2-14	3	2
15-19	2.5	6.7
20-24	5.1	21.1
25-29	15.7	32.7
30-34	25.8	29.1
35-39	18.5	24.8
40-44	19.2	16.3
45-49	6.4	14.1
50-54	10.4	10.2
55-59	6.2	1.8
60+	3.5	1.8
Total	**7.9%**	**13.6%**

Sources: http://www.avert.org/aids-hiv-charity-avert. htm (based on UNAIDS 2010 figures and the South African National HIV Survey, 2008) and UNAIDS Global Report: UNAIDS report on the global AIDS epidemic 2010, New York

READING

Linda Fuller (2008) African Women's Unique Vulnerabilities to HIV/AIDS: Communication Perspectives and Promises, New York, Palgrave Macmillan

UNAIDS Global Report: UNAIDS report on the global AIDS epidemic 2010, New York

Valerie Duffy and Ciara Regan (2010) This is what has happened... HIV and AIDS, women and vulnerability in Zambia, Bray, 80:20 Educating and Acting for a Better World

Michael J Kelly (2006) HIV and AIDS: A Justice Perspective, Lusaka, Jesuit Centre for Theological Reflection

Alan Whiteside (2008) HIV/AIDS: A Very Short Introduction, Oxford University Press

Stephen Lewis (2006, 2nd ed.) Race Against Time: Searching for Hope in AIDS Ravaged Africa, Toronto, Anansi Press

MORE INFORMATION & DEBATE

http://www.avert.org/aids-hiv-charity-avert.htm - UK based international AIDS charity

http://www.unaids.org/globalreport/Global_report.htm - the major international report on the epidemic

http://www.heard.org.za – Health Economics and HIV/AIDS Research Division is based at the University of KwaZulu-Natal in South Africa

http://www.worldaidscampaign.org – major international campaign on HIV and AIDS

Photo: Garry Walsh

[*'Human security privileges people over states, reconciliation over revenge, diplomacy over deterrence and multilateral engagement over coercive unilateralism.'*]

Archbishop Desmond Tutu, 2005

Chapter 11

'FREEDOM FROM AND FREEDOM TO...' HUMAN SECURITY AND DEVELOPMENT

Knut Elvatun

This chapter reviews the debates and evidence around the concept of human security and how it relates to the associated concepts of human rights and human development. Having reviewed some of the key philosophical roots of human security and the tension between state security and individual security, the chapter discusses genocide, democide and politicide. Recent trends in conflicts worldwide are reviewed in addition to evidence on conflicts involving non-state actors.

KEYWORDS

Human rights, **human development**, human security, Cold War, 'broad and narrow' definitions, **genocide**, democide, **politicide**, deaths from conflict

INTRODUCTION

Stressing the *'human'* in human security, establishes a counter-balance or tension with the traditional concept of state security narrowly defined as *'military defence of state interests and territory'*. The idea of human security extends security beyond the *'national'* and seeks to convince states that human security is, in reality, state security. The argument goes that as a matter of self-interest, governments should actively participate in the protection of citizens in other states against immediate and ongoing threats to their security. A corresponding implication accompanying the idea of human security is that the international community has a duty to protect the rights and interests of people and not simply the interests of states. This debate becomes all the more important when we consider the evidence and arguments of US political scientist RJ Rummel who has argued that more people were killed in the 20th century by their own governments than by all wars combined (see discussion below).

This chapter explores the development of the concept of human security, the various approaches adopted and how this concept relates to other terms such as human rights and human development. The primary focus is on patterns and trends as regards conflict, genocide and other forms of organised and systematic killing as a key element in human security, as many other chapters have focused on the other dimensions. At the outset, it is necessary to note that *'human security'* is not a term set in stone but is rather a concept in development, one that is contested not only in terms of its own logic but also in terms of its international implications.

DURING THE LAST 100 YEARS FAR MORE PEOPLE HAVE BEEN KILLED BY THEIR OWN GOVERNMENTS THAN BY FOREIGN ARMIES

HUMAN SECURITY - PHILOSOPHICAL ROOTS

Many of the basic principles of human security are reflections of the ideas and writings of Montesquieu, Rousseau and Condorcet, as are principles of state security rooted in the work of Kant, Hobbes and Grotius. Historically, this latter *'state-centred world view'* prevailed over the broader, more pluralistic perspective. The pluralism of the eighteenth century focused on the protection of the individual as philosophers such as Montesquieu focused on freedom and the perceived rights of the individual over the dictated security provided by the state. Adam Smith defined security as protection of the individual from *'sudden or violent attack on one's person or property'* – this security being the most important prerequisite for a successful society. Similarly Condorcet described a social contract in which the security of the individual was the central principle.

This liberal perspective was, however, not unanimous; there was some agreement over the vital role of individual safety, but some believed that this could be best achieved as a consequence of the security of the state. Hobbes argued that there was little difference if an individual's insecurity was at the hands of a thief or an army and that protection from either was the absolute responsibility of the state. Kant also looked to the state to provide security, and he aimed even higher by envisaging a universal international order, or a global society based on the moral standards of its members. Between these two, Grotius argued for a more moderate international dynamic under *'supranational'* law but with a balance of power between states and a social contract between the state and its citizens.

Placing the primary responsibility of protecting individual security in the hands of the state became the dominant world view until the end of the Cold War.

The traditional state-centred security thinking reached a high-point during the Cold War. The idea was straightforward enough - if the state was secure, then so would its citizens be; security amounted to protection from invading armies and was provided primarily by technical and military capabilities. Wars were never to be fought at home, but as the Cold War ended it became clear that within the dominant world order citizens were not necessarily safe. People were not necessarily killed by outright military confrontation, but rather by hunger, violence, human rights abuses, poverty and disease (this reality has been examined in detail recently by historian Timothy Snyder (2010) in *Bloodlands: Europe between Hitler and Stalin*, London, Vintage Books). The protection of the individual had been steadfastly neglected or ignored by an over-focus on the state and, post-1945 this prompted a shift away from inter-state relations towards the interests and needs of the individual which, in turn, meant that research and policy shifted more directly on to issues that actually threatened peoples' lives.

There have been many attempts to define the complex concept of human security but there is, as yet no universally agreed definition. At its simplest security is the absence of insecurity – to be free from fear (abuse, violence, persecution, death) and want (gainful employment, food, health) – the emphasis is on identifying threats, avoiding them (if possible) and mitigating their effects should they be realised. Most definitions can, however, be located in two main schools of thought crudely described as *'broad'* and *'narrow'*. The concept in its broad sense incorporates a detailed list of potential threats ranging from traditional threats such as war to more development-oriented threats such as ill-health, poverty and environmental vulnerability. In the more narrow conception, human security is still focused on the individual but incorporates a much shortened list, limited to violent threats (e.g. landmines, small arms and intra-state conflict). However, it is important to remember that both conceptions are focused on the individual and are complementary rather than contradictory. An excellent introduction to the concept and the debates surrounding it is contained in the UNDP Human Development Report for 1994 entitled *New Dimensions of Human Security*.

BROAD OR NARROW – TWO CONCEPTIONS OF HUMAN SECURITY

As already noted, the broad definition does not limit itself to situations of conflict but also focuses on issues as broad as fair trade, access to healthcare, patent rights, access to education and basic freedoms. The more focused and narrow approach looks more towards removing the use of, or threat of, force and violence from people's everyday lives as illustrated by initiatives such as the campaign to ban anti-personnel landmines, the agenda of the International Criminal Court and the broad area of security sector governance.

Most human security definitions come from the *'broad'* school of thought. Analyst and researcher Taylor Owen argues that clarity can emerge from the broad conception if three key attributes are taken into account; its scope of coverage, its systems-based approach to human security and its focus on the vital core of the individual. These critical aspects can be illustrated through the approach and advocacy of the UNDP, Jorge Nef and the Independent Commission on Human Security (2003).

The approach of the 1994 UNDP Human Development Report best illustrates the broader approach to human security and was, arguably the first significant attempt to establish an *'operational'* approach to human security. The report describes two key dimensions: freedom from chronic threats such as hunger, disease and repression and protection from sudden calamities. The Report itemises seven components to human security – *economic; food; health; environmental, personal, community, and political security.*

- **Economic security threatened by poverty** – economic security generally requires an 'assured basic income' or a publicly financed safety net. In 1994 only one quarter of the world's population were secure in this sense.

- **Food security threatened by hunger and famine** – food security means that all people at all times have physical and economic access to basic food.

- **Health security threatened by injury and disease.**

- **Environmental security threatened by pollution, environmental degradation and resource depletion** – people rely on a healthy physical environment.

- **Personal security threatened by all forms of violence** – this can take several forms: threats from the state, other groups, individuals or gangs, threats directed towards women (rape, domestic abuse), threats directed towards children, threats to self (suicide, drug use).

- **Political security threatened by political repression** – people should be able to live in a society that honours their basic human rights.

- **Community security threatened by social unrest and instability** – though most people derive security from their membership in a group, they should be secure from oppressive practices ranging from ethnic clashes to female genital mutilation.

Human security is concerned with reducing and – when possible – removing the insecurities that plague so many human lives.

The work of geographer Jorge Nef illustrates the importance of the systems-based approach; he describes five interconnected sub-systems of human security - *ecosystem, economy, society, polity and culture,* comprising five major elements; context, culture, structures, processes and effects (see table on page opposite).

For Nef, the five sub-systems exist in a complex set of relationships and therefore acknowledging the interconnectedness of all human security components is critical for understanding causality and properly addressing policy. For Nef:

'The very idea of national security, under the umbrella of collective defence and centred on a single actor, the nation state, in an eminently bipolar world, has been replaced by the need to cope with a much different kind of collective and cooperative security. Nor it is possible to assume a resurgence of pre-World War II multi-polarity. The new type of security is a much more complex varied and nuanced concept. Ethnic conflict, cultural diversity, national disintegration, civil war, systemic and sub-systemic restructuring have become paramount. Issues of poverty, trade, finance, health, environment, gender, communications, resource depletion, population, migration, technology, drugs, human rights, and refugees are also part of the equation; and the list could go on.'

Jorge Nef (1995) Human Security and Mutual Vulnerability: An Exploration into the Global Political Economy of Development and Underdevelopment, Ottawa, IDRC)

VARIABLES	ECOLOGY (life)	ECONOMY (wealth)	SOCIETY (support: well-being, affection, respect, rectitude)	POLITY (power)	CULTURE (knowledge, skill)
CONTEXT	Natural setting: the biophysical surroundings of societal action	Styles of development: economic models	Societal expectations and traditions	Internal and external conflicts: capabilities/ expectations elite/mass sovereignty/ dependence	Images of the physical and social world and collective experiences
CULTURE	Ecoculture: place of environment in cosmovision	Economic doctrines: ways of understanding the economy	Social doctrines: values, norms and attitudes; identity and personality	Ideologies: the function of the state and its relation to the citizen	Philosophy (axiologies, teleologies, and dentologies), moral and ethical codes
STRUCTURES	Resource endowment and spatial distribution: relation between environment and resources	Economic units: consumers/ producers; labour/ capital	Status and roles: social structures, groups, classes, fractions	Brokers and institutions: interest groups, parties, cliques, governments, bureaucracies	Educational structures, formal and informal: schools, universities, learning institutions
PROCESSES	Depletion/ regeneration of air, water, land, flora and fauna	Production and distribution of goods and services	Interactions: cooperation, conflict, mobilisation and demobilisation	Conflict-resolution: consensus, repression, rebellion, stalemate	Learning: building of consciousness, cognitions, basic values, procedures and teleologies
EFFECTS	Sustainability/ entropy	Prosperity/ poverty	Equity/ inequity	Governance/ violence	Enlightenment/ ignorance

Another key aspect in human security is the focus on the vital core of the individual as outlined by the Independent Commission on Human Security in 2003:

'Human security in its broadest sense embraces far more than the absence of violent conflict. It encompasses human rights, good governance, access to education and health care, and ensuring that each individual has opportunities and choices to fulfil his or her own potential…Freedom from want, freedom from fear and the freedom of the future generations to inherit a healthy natural environment – these are the interrelated building blocks of human, and therefore national security'

In contrast, the narrower view focuses on crisis situations that require international remedies. Sometimes, the human security agenda can go beyond professional distinctions such as those between *'humanitarian relief, development assistance, human rights advocacy and conflict resolution'*, requiring new, coordinated mechanisms of international co-operation or intervention to replace the slow, institutional approach that characterised international attempts in the past. The narrow approach primarily focuses on violent threats, and clearly separates itself from the more expansive field of human development. The definition limits the parameters of human security to violent threats against the individual.

Many, and some may argue most, international policy initiatives have been undertaken under the narrow approach to human security. Examples of these are the Mine Ban Convention (see chapter 16), the International Criminal Court (see chapter 4) as well as international focus on child soldiers and small arms. Advocates of the narrow approach argue that as the list of included harms increases, so does the difficulty in articulating and measuring the concept – it becomes increasingly difficult to operationalise. Canada has incorporated the narrow approach into its foreign policy, as is illustrated by former Canadian Minister for Foreign Affairs Lloyd Axworthy who argues that:

'...in essence, an effort to construct a global society where the safety of the individual is at the centre of the international priorities and a motivating force for international action; where international human rights standards and the rule of law are advanced and woven into a coherent web protecting the individual; where those who violate these standards are held fully accountable; and where our global, regional and bilateral institutions – present and future – are built and equipped to enhance and enforce these standards.'

GENOCIDE, POLITICIDE, DEMOCIDE AND INTERNATIONAL CONFLICT

The shocking reality is that more people were killed in the 20th century by their own governments than by all wars combined:

- Under Lenin and Stalin, the **Soviet government** became one of the world's greatest killers. Lenin's policies in 1921-1922 caused an estimated 4 million deaths and, in 1932, Stalin ordered that the **Ukraine** be starved to enforce collectivisation policies and to crush Ukrainian nationalism resulting in the murder of at least 8 million Ukrainians

- Between 1917 and 1953 (the year of Stalin's death), the **Soviet Union** executed, in one way or another, some 40 million people and many Russian and international historians estimate the figure to be even higher

- In **China**, under Mao Tse Tung, 2 million dissidents or *'class enemies'* were shot and another one million **Tibetans** and **Turkestani Muslims** were *'liquidated'*. Between 1950 and 1975, an estimated 30 million people starved to death with another two million dying during the Cultural Revolution. This amounts to a total of some 35 million people

- **Hitler and the Nazis** were responsible for the deaths of 12 million civilians, half of them **Jews**

- An estimated two million **German** civilians were killed in 1945, and at least 200,000 died in concentration camps between 1945 and 1953. The Allied forces *'handed back'* an estimated 2 million **Soviet citizens** to Stalin in 1945: half of them were shot and the rest sent to Arctic camps where many of them died

- During World War I, the **Ottoman Empire** murdered or starved up to 2 million Armenians, the first great genocide of the new century

- In the early 1960's, 600,000 **ethnic Chinese** were massacred in **Indonesia** by government-encouraged mobs and soldiers

- During the Marcos era in the **Philippines**, 75,000 **Muslims** were massacred by government paramilitary gangs

- In 1971, **Pakistani troops** killed tens of thousands of **Bengalis** in former **East Pakistan**. Indian security forces and police have massacred great numbers of tribes people in border regions and many civilians in **Kashmir** and **Punjab**

- Between 1975 and 1979, an estimated 2 million people were killed in **Cambodia** during the regime of **Pol Pot**

- In the 1980's, **Ethiopia's then Marxist regime** through a series of policies including forced relocation caused the deaths of an estimated million people, often through starvation

- 1994 witnessed the slaughter of an estimated 800,000 **Tutsis** and **Hutu** *'dissidents'* at the hands of **Rwanda's** Hutu government

- **Serbia's nationalist regime** orchestrated the massacre of 200,000 **Muslim civilians** in **Bosnia** between 1992 and 1995

The scale of the figures above (even if their complete accuracy is contested) has led many researchers to develop a series of concepts designed to capture and define the nature of such systematic killing. 'Democide' is the term developed by US political scientist RJ Rummel who defines democide as:

'…the murder of any person or people by a government, including genocide, politicide, and mass murder'.

He created the term as a concept which included forms of government murder not covered by the legal definition of genocide. 'Politicide' is the term used to describe the killing of groups of people not because of shared ethnic or communal traits (the types of groups covered by the United Nations Convention on the Prevention and Punishment of the Crime of Genocide) but because of their position in the power structure of a state or their political opposition to a regime and its dominant groups.

Rummel's research concludes that the death toll from democide is far greater than the death toll from war. Having studied over 8,000 reports of government-caused deaths, he suggests that there have been 262 million victims of democide in the last century. According to his figures, six times as many people have died from the actions of people working for governments than have died in battle. Another of his major findings is that liberal democracies have much less democide than authoritarian regimes and he argues that there is a relationship between political power and democide. Political mass murder grows increasingly common as political power becomes unconstrained and, conversely, where power is *'diffuse, checked, and balanced'*, political violence is a rarity.

Genocides have declined dramatically since the end of the Cold War. Yet, determining exactly what genocide is has proven problematic. The UN Genocide Convention defines genocide as *'acts committed with intent to destroy, in whole or part, a national, ethnical, racial or religious group.'* If this definition was applied strictly to Cambodia (1975-79), only the ethnic Chinese, Vietnamese and Chams, and not the one million Khmers that were killed, would be considered victims of genocide. For these reasons, the terms *'politicide'* and *'democide'* were developed.

The number of wars waged internationally declined by 78% between 1988 and 2008 with a corresponding decrease in the number of battlefield deaths until 2003 (due largely to the war in Iraq). The more recent increase in battle deaths needs to be seen in the context of the overall decline in estimated war-death tolls since 1946:

- In 1950 (the first year of the Korean War) there were an estimated 600,000 battle deaths worldwide
- in 1972 (the deadliest year of the American war in Vietnam) the toll was more than 300,000
- in 1982 (the height of the Iran-Iraq War) it was 270,000
- in 1999 (when wars were being fought between Ethiopia and Eritrea and in East Africa's Great Lakes region) it was 130,000.

While the accuracy of figures for such deaths is the subject of intense debate and disagreement, the estimated number of deaths in 2008 was 27,000, so the overall trend downward in not universally accepted. The patterns are clearly highlighted in the diagrams opposite.

Between 1992 and 2003, the number of *state-based armed conflicts* - those involving a government as one of the warring parties - dropped by 40% but since 2003, the international incidence of armed conflicts increased by 25%. Meanwhile, *non-state conflicts* - violent confrontations between communal groups, rebels, or warlords that do not involve a state as a warring party - increased by a significant 119% with a quarter of the conflicts that began or were reignited between 2003 and 2008 being associated with Islamist political violence and the so-called *'War on Terror'*.

The Human Security Report for 2009 – 2010 identified a number of issues with significant impact today or implications for the future:

- In 2008 four of the five most deadly conflicts in the world - Iraq, Afghanistan, Pakistan and Somalia highlight the struggle between Islamist insurgents and national governments backed by the United States. In 2010 these conflicts remained unresolved with little sign of progress and with very significant human consequences
- The current world economic crisis threatens to propel millions of people below the poverty line in the developing world, heightening the possibility of new conflicts
- Those conflicts that continue are becoming more and more difficult to resolve with significant potential for them to have consequences elsewhere internationally.

The Report concludes:

'The real worry is that we may again be witnessing a long-term trend of steadily rising political violence around the world reminiscent of the Cold War years when conflict numbers tripled over some four decades.'

 MORE INFO

RJ Rummel came up with the term 'Democide' in his 1994 book Death by Government: Genocide and Mass Murder since 1900, New Brunswick, Transaction). You can read an excerpt of the book here - http://www.hawaii.edu/powerkills/DBG.CHAP2.HTM

For a detailed analysis and discussion of Genocide, see Samantha Power's excellent 2002 book A Problem from Hell: America and the Age of Genocide (New York, Basic Books).

Genocide Watch have put together a table listing all the various genocides, politicides and other mass murders since 1945. Besides calculating the civilian death toll of each, the table also mentions the main perpetrators for each genocide or politicide and ranks them according to a genocide measure developed by analyst Gregory Stanton.

http://www.genocidewatch.org/aboutgenocide/genocidespoliticides.html

http://www.genocidewatch.org/aboutgenocide/8stagesofgenocide.html

ARMED CONFLICTS BY TYPE, 1946–2008

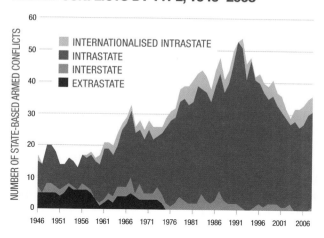

Extra-state conflicts or anticolonial struggles largely ceased by the mid-1970s. Interstate conflicts (conflicts between states that do not involve full scale war) were never very numerous and have become even rarer in the past two decades. Intrastate conflicts (conflicts within a state) declined dramatically from 1992 but increased modestly since 2003.

The graph opposite (taken from the Human Security Report 2009 – 2010) is a 'stacked graph', meaning that the number of conflicts in each category is indicated by the depth of the band of colour. The top line shows the total number of conflicts of all types in each year.

AVERAGE NUMBER OF INTERNATIONAL CONFLICTS PER YEAR, 1950–2008

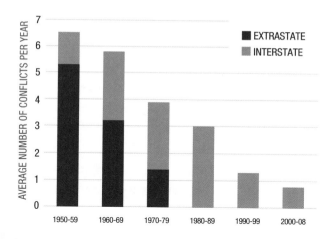

There has been a steady decline in the number of international conflicts (including interstate and extrastate conflicts) over the last five decades. In contrast, non-state armed conflicts continue to affect regions, paricularly Sub-Saharan Africa, which has experienced more such conflicts than all other regions combined.

THE RISING TIDE OF DEMOCRACY, 1946-2008

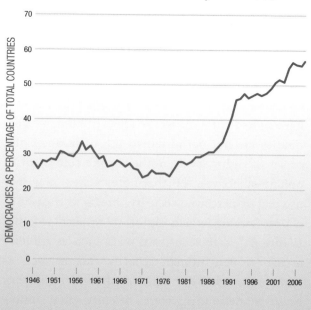

NON-STATE ARMED CONFLICTS 2002–2008

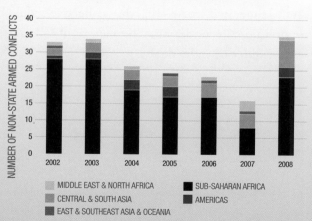

Source: Human Security Report 2009 – 2010

HUMAN SECURITY, HUMAN RIGHTS AND HUMAN DEVELOPMENT

Human rights were originally designed to protect the individual against the state. However, human rights have gradually evolved to also protect individuals against non-state actors. Protection from gross violations of human rights is a necessary but not sufficient condition of human security. Moreover, human rights should be seen as one of many components of human security - a necessary but not sufficient condition - human security and human rights are intertwined. Just like some, but not all, environmental disasters cross the threshold of severity to become human security threats, so too do some, but not all, human rights abuses. Some analysts argue that human security is a human right. Article 3 of the Universal Declaration of Human Rights affirms that that *everyone has the right to life, liberty and the security of person*. At least one can argue that when human rights are respected, then human security is well advanced.

Looking at the differences between the international human rights legal regime and the literature on human security, a main point is that respect for and protection and fulfilment of human rights are not policy choices. States may not pick and choose among which rights to protect, whose rights to protect, or when to protect. States that have signed and ratified the relevant human rights treaties are not permitted to prioritise one right, or set of rights, over another in the fulfilment of policy objectives. Furthermore, states may not use real or perceived security threats as excuses to choose among which rights to respect, whether traditional state security threats such as military attack or new human security threats such as climate change. Although some human rights may be suspended during states of emergency, some - such as the protection against torture - may not be derogated from regardless of the situation.

On the other hand, an individual can rarely appeal to a state to protect their human rights if not a citizen of that state. This leaves stateless individuals unprotected, while migrants, whether legal or illegal, frequently have no recourse against violation of their human rights even if they are formally citizens of a state where they no longer reside. Human security's broadening of states' responsibilities to include non-citizens, even if in principle rather than practice, is thus a significant change from the international human rights regime, with its insistence primarily on states' responsibilities to their own citizens. Some critics even argue that although the human rights legal regime is extensive, it has not had much real positive effect since 1945. And some scholars argue that there is no evidence that the actual human rights performance improves when a state signs a human rights treaty.

In discussions of human security, human rights appear to be merely a subset of human security concerns, and as such less worthy of attention than they so far have been. This is also seen in practice as some states could be tempted to focus more strongly on human security issues in order to deflect attention away from human rights abuses. A human security perspective may also unintentionally lead to a more collective approach which prioritises the collective over the individual. An exaggerated focus on the collective could unintentionally undermine human rights claims within states by individuals.

HUMAN SECURITY AND DEVELOPMENT

Protection from human rights violations is only one component of ensuring human security. Individuals also need protection from poverty, disasters, conflict, and disease. Human security emphasises the complex relationships and often-ignored linkages between disarmament, human rights and development. Today all security discussions demand incorporation of the human dimension. Human security and human development are often described as twin concepts, being simply *'freedom from fear'* and *'freedom from want'*, respectively. Human security could be said to be a necessary but not sufficient precondition for human development. If human security could cover the most urgent threats, development would then address social well-being. This interconnectedness is illustrated in the work of Frances Stewart who argues that:

'...human security forms an important part of people's well-being, and is therefore an objective of development ... lack of human security has adverse consequences on economic growth, and therefore development. Imbalanced development that involves horizontal inequalities is an important source of conflict.'

Human security provides the underlying basis for development of policies and strategies to improve crisis prevention, mitigation and recovery in Sub-Saharan Africa. Human security clarifies the critical link between disaster and development and provides a basis on which relief and development interventions can be more effectively planned and coordinated. Transition from crisis to development is too rarely achieved in Sub-Saharan Africa. Instead, many societies are stuck in a cycle of underdevelopment and crisis. Human security must be achieved before societies can effectively achieve broad-based sustainable development.

Human security is the ability to enjoy the fruits of human development in a safe environment. Human development is one important means to create human security. The two initiatives are complementary and mutually reinforcing. Without one, the other becomes difficult, if not impossible. Human security is achieved when population groups have minimum standards of human rights as well as livelihood and environmental security. Again, Frances Stewart summarises the issues effectively:

'Some development costs are obvious. People who join the fighting forces, who are killed or flee, can no longer work productively; schools, power stations and/or roads that are destroyed reduce the productive capacity of the economy. There are also more complex interactions between events associated directly with war (fighting, movement of people, deaths, physical destruction, international embargoes, military expenditures) and developments in the macro, meso and micro-economy which mostly lead to adverse changes in individual entitlements, both economic and social. To take one example, movement of manpower may reduce the production of exports, thereby reducing foreign exchange earnings, import potential and consequently further constraining output, leading to a decline in employment and earnings. However, we should note that there can be positive indirect effects as resources are used more fully and efficiently. These positive effects seem to have occurred in some countries in the Second World War.'

Frances Stewart (2004), Development and Security, Working Paper 3, CRISE, University of Oxford

Based on the discussion above it seems clear that neither the broad or narrow approach to human security can exclude the other nor it is possible to argue that any one of them is more valuable than the other. To some extent it can be argued that the narrow approach to human security moves closer to the concept of human rights while the broad approach moves more towards development. And furthermore, to achieve a high level of human security, human rights standards and a certain level of development must be present.

'Human security means protecting vital freedoms. It means protecting people from critical and pervasive threats and situations, building on their strengths and aspirations. It also means creating systems that give people the building blocks of survival, dignity and livelihood. To do this, it offers two general strategies: protection and empowerment. Protection shields people from dangers. Empowerment enables people to develop their potential and become full participants in decision-making.'

Report of the UN Commission on Human Security (2003)

COUNTRIES WITH THE GREATEST NUMBER OF CONFLICT YEARS - 1946-2008

BURMA 246
INDIA 180

BURMA HAS, ON AVERAGE, EXPERIENCED FOUR CONFLICTS IN EACH OF THE 61 YEARS SINCE ITS INDEPENDENCE IN 1948 UNTIL 2008

ETHIOPIA 113
PHILIPPINES 100
UNITED KINGDOM 91
FRANCE 89
ISRAEL 79
VIETNAM 71

Sources: figures from the Uppsala Conflict Data Program and the International Peace Research Institute, Oslo

READING

Human Security Report Project (2009) Human Security Report 2009/2010: The Causes of Peace and the Shrinking Costs of War, Vancouver

Jorge Nef (1999) Human Security and Mutual Vulnerability, Ottawa, International Development Research Centre

Samantha Power (2002) A Problem from Hell: America and the Age of Genocide (New York, Basic Books)

Report of the UN Commission on Human Security (2003) Human Security Now, New York, UN

Timothy Snyder (2010) Bloodlands: Europe between Hitler and Stalin, London, Vintage Books

Frances Stewart (2004), Development and Security, Working Paper 3, CRISE, University of Oxford

UNDP (1994) Human Development Report: New Dimensions of Human Security, New York, UNDP

UN Secretary General (2010) Report of the Secretary General on Human Security, New York, General Assembly

MORE INFORMATION & DEBATE

http://www.hsrgroup.org/human-security-reports/human-security-report.aspx - human security report project site

http://www.eassi.org - the site of the Eastern African Sub-Regional Support Initiative for the Advancement of Women (EASSI)

http://www.iss.co.za/default.php - Institute for Security Studies, South Africa

http://www.ehs.unu.edu - United nations University Institute for Environment and Human Security site

http://www.humansecurityindex.org - site dedicated to developing a human security index

CIVIL & POLITICAL RIGHTS WORLDWIDE

FREEDOM IN TODAY'S WORLD

" The past decade began at a high point for freedom and concluded with freedom under duress. The next decade could witness a new wave of democratic development if democracy's champions remember that freedom is more powerful—both as an idea and as the basis for practical governance—than anything its adversaries have to offer. " Arch Puddington, Freedom in the World, 2011: The Authoritarian Challenge to Democracy

THE 'WORST OF THE WORST'

Of the 47 countries designated **'Not Free'**, 9 have the lowest possible rating of 7 for political rights and civil liberties.

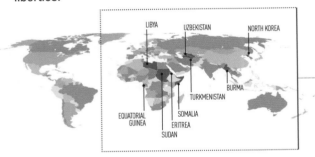

They represent a narrow range of systems and cultures - **North Korea** is a one-party, Marxist-Leninist regime; **Turkmenistan** and **Uzbekistan** are ruled by dictators; Until 2011, **Libya** was ruled by a secular dictatorship and **Sudan** is ruled by a military junta with elements of radical Islamism

Burma is ruled by a military dictatorship; **Equatorial Guinea** by a highly corrupt and brutal regime; **Eritrea** by an increasingly repressive police state and **Somalia** is a failed state

The one worst-rated territory **Tibet**, is ruled by **China**

An additional 10 countries had scores slightly above those of the worst-ranked countries - **Belarus**, **Chad**, **China**, **Côte d'Ivoire**, **Cuba**, **Laos**, **Saudi Arabia**, **South Ossetia**, **Syria**, and **Western Sahara**.

FREEDOM IN THE WORLD 2010:
GLOBAL DATA

COUNTRY BREAKDOWN BY STATUS

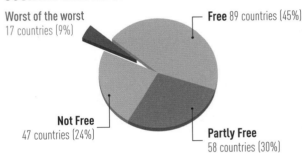

Worst of the worst
17 countries (9%)

Free 89 countries (45%)

Not Free
47 countries (24%)

Partly Free
58 countries (30%)

POPULATION BREAKDOWN BY STATUS

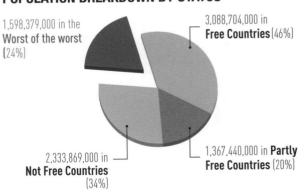

1,598,379,000 in the
Worst of the worst
(24%)

3,088,704,000 in
Free Countries (46%)

2,333,869,000 in
Not Free Countries
(34%)

1,367,440,000 in Partly
Free Countries (20%)

Source: Freedom House (2010) Freedom in the World, Washington

THE IBRAHIM INDEX

The Ibrahim Index assesses governance against 88 criteria, making it the most comprehensive collection of qualitative and quantitative measures of governance in Africa.

REGIONAL AVERAGES

NORTH AFRICA 55
WEST AFRICA 50
EAST AFRICA 45
CENTRAL AFRICA 38
SOUTHERN AFRICA 57

TOP TEN SCORES 2008/09

1. Mauritius (83)
2. Seychelles (79)
3. Botswana (76)
4. Cape Verde (75)
5. South Africa (71)
6. Namibia (67)
7. Ghana (65)
8. Tunisia (62)
9. Egypt (60)
10. Lesotho (60)

BOTTOM TEN SCORES 2008/09

44. Côte d'Ivoire (37)
45. Guinea (36)
46. Equatorial Guinea (35)
47. Sudan (33)
48. Central African Republic (33)
49. Zimbabwe (33)
50. Eritrea (32)
51. Congo, Democratic Rep. (31)
52. Chad (29)
53. Somalia (8)

FREEDOM ON THE NET

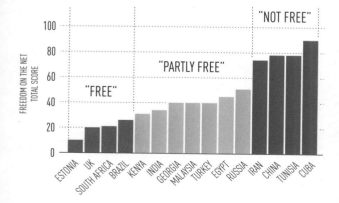

"FREE" "PARTLY FREE" "NOT FREE"

ESTONIA, UK, SOUTH AFRICA, BRAZIL, KENYA, INDIA, GEORGIA, MALAYSIA, TURKEY, EGYPT, RUSSIA, IRAN, CHINA, TUNISIA, CUBA

FREEDOM ON THE NET TOTAL SCORE

Source: Freedom House (2010) Freedom in the World, Washington

IN 2009, AMNESTY INTERNATIONAL RECORDED:

Human rights abusers enjoyed impunity in at least 61 countries

In at least 111 countries, people were tortured or ill-treated

There were unfair trials in at least 51 countries

Freedom of expression was restricted in at least 96 countries

Prisoners of conscience were being held in at least 48 countries

Photo: Gareth Bentley

[*'Starvation is death by deprivation; the absence of one of the essential elements of life. It's not the result of an accident or a spasm of violence, the ravages of diseases or the inevitable decay of old age. It occurs because people are forced to live in the hollow of plenty. For decades, the world has grown enough food to nourish everyone adequately.'*]

R. Thurow and S. Kilman (2009) Enough: Why the Poorest Starve in an Age of Plenty

Chapter 12

'LIVING IN THE HOLLOW OF PLENTY' WORLD HUNGER TODAY

Michael Doorly

Initially, this chapter describes the scale and character of hunger today; it then defines hunger and related terms and discusses some of the challenges in measuring hunger and identifying who the hungry are. The causes of hunger are then reviewed as is the evolution of the concept of the right to food and what is termed the 'food paradox'. Finally the chapter briefly identifies four key challenges in respect to hunger — recent food price increases, population growth, bio-fuels and gender inequality.

KEYWORDS

Definition and measurement of hunger, who are the hungry, the right to food, undernutrition, malnutrition, food security, causes of hunger, the hunger paradox, food prices, biofuels, gender inequality and hunger, population and hunger

INTRODUCTION

The world has witnessed remarkable changes over the last 50 years. There has been a surge in population, significant levels of economic growth (despite the recent global financial crises), technological advancement and increased global economic integration. Hundreds of millions of people have increased their standard of living and have escaped from absolute poverty. However, with nearly one billion people suffering from under-nutrition, tackling hunger remains one of the most immediate and fundamental challenges we face. While the proportion of the world's population that is suffering from hunger has fallen from 35% in 1969 to 16% in 2010, the absolute number has been rising since the mid-1990s.

Hunger, like so many other global issues, is a stark reminder of our unequal and unjust world. At its most immediate level, hunger is about the lack of appropriate food but, ultimately, it is about poverty and power (or the lack of it) and, as such is a political issue, it is a human creation by and large and therefore can only be ended by political will and concerted action. Sadly, the words of John F Kennedy remain as true today as they were when he spoke them in 1963 - *'We have the means; we have the capacity to eliminate hunger from the face of the earth in our lifetime, we need only the will.'*

But of course we know that ending hunger is not only a political issue, it is much more fundamentally an issue of justice and human rights; a right included in the Universal Declaration of Human Rights, expanded in the International Convention on Economic, Social and Cultural Rights (ICESCR) and included in over 200 additional UN instruments, declarations, treaties and conventions.

'The right to food is a human right. It protects the right of all human beings to live in dignity, free from hunger, food insecurity and malnutrition. The right to food is not about charity, but about ensuring that all people have the capacity to feed themselves in dignity.'

Jean Ziegler, the first UN Special Rapporteur on Food

This chapter reviews what the right to food means in practice for the hundreds of millions of people who simply do not have enough to eat; it sketches the current realities and explores a number of interventions to combat hunger by people and communities *'on the ground'*, as well as steps being taken by those in positions of power to overcome some of the many issues that frustrate its realisation.

HUNGER IN THE WORLD TODAY

Whereas significant progress was made in reducing chronic hunger in the 1980s and the first half of the 1990s, hunger has been slowly but steadily on the rise for the past decade. The number of hungry people increased between 1995-97 and 2004-06 in all regions except Latin America and the Caribbean (and even in this region, gains in hunger reduction have been reversed as a result of high food prices and the current global economic crisis). Almost all of the world's undernourished live in developing countries and account for some 16% of their total population and, according to the Food and Agriculture Organisation (2009), the regional breakdown is as follows:

- An estimated 642 million in Asia and the Pacific
- 265 million in Sub-Saharan Africa
- 53 million in Latin America and the Caribbean
- 42 million in the Near East and North Africa
- An additional 15 million in developed countries

UNDERNOURISHMENT BY REGION, 2010

ASIA & THE PACIFIC 578 MILLION

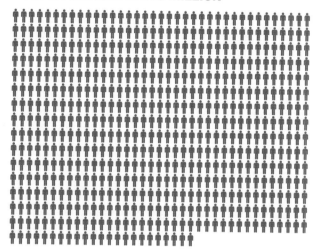

SUB-SAHARAN AFRICA 239 MILLION

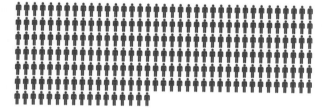

The majority of the world's undernourished people live in developing countries. Two-thirds live in just seven countries (Bangladesh, China, the Democratic Republic of the Congo, Ethiopia, India, Indonesia and Pakistan) and over 40% live in China and India alone.

Projections for 2010 indicate that the number of undernourished people will decline in all developing regions, although with a different pace. The region with most undernourished people continues to be Asia and the Pacific but with a 12% decline from 658 million in 2009 to 578 million, this region also accounts for most of the expected global improvement.

Source: FAO 2010 The State of Food Insecurity in the World

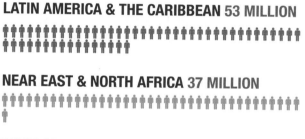

LATIN AMERICA & THE CARIBBEAN 53 MILLION

NEAR EAST & NORTH AFRICA 37 MILLION

DEVELOPED COUNTRIES 19 MILLION

WHO ARE THE WORLD'S HUNGRY?

In 1983 the United Nations appointed the first Special Rapporteur on the Right to Food with the specific mandate to receive information and highlight violations of the right to food, to cooperate with UN agencies and other international organisations and NGOs to put the right of food into practice around the world, and to ensure that a gender perspective is incorporated as part of any strategy in addressing the right. The current Special Rapporteur, Professor Olivier De Shutter, describes four broad categories of people who make up the world's hungry:

- **Small landholders** who depend on less than 1.5 hectares of land for their livelihood, a group

that NGO Concern Worldwide has termed the *'the farming yet hungry'*. This group is made up of nearly 500 million people, the majority of whom are women, who produce food for themselves and their families on small plots of land in remote rural areas. When their harvests run out, they often have little to eat and lack the financial means to buy sufficient or adequate food which could see them through the *'hunger gap'*. Frequently they are located far from markets where they could sell their produce or buy goods and services that could increase their productivity.

- **Agricultural labourers** - of the 450 million such labourers in the world nearly 200 million are hungry. Many of them work on large plantations and are employed on a seasonal basis, often without a contract or any legal or social protection.

- **Artisan farmers** comprised mostly of indigenous peoples, who do not own land or are in waged employment. They depend on artisan fishing, raising livestock and the products of the forest. This group comprises approximately 10% to 15% of the world's hungry.

- **The Urban Poor** - a rapidly expanding group with the percentage of urban residents in sub-Saharan Africa alone expected to rise from 30% to 47% by 2020. Access to food in urban areas depends almost entirely on cash purchase, posing serious problems for those who lack a fixed income. Although a wider variety of food is available, the food consumed in urban areas is not necessarily of superior nutritional quality.

Source: Olivier De Schutter, Facing up to the Scandal of World Hunger, Trócaire Development Review, 2010

DEFINING HUNGER

The public image of hunger is the starving African child, whether witnessed as a victim of the Ethiopian famine of 1984, Niger in 2005 or the Horn of Africa drought and famine in 2011 – shocking and immensely tragic events. Yet less than 10% of the world's hungry are involved in such crises or emergency situations – the scandal of our times is that millions more people suffer from daily and persistent undernourishment, the less visible form of hunger, where its victims are unable to access the nutrients and energy their bodies need, where they become trapped in a cycle of hunger, their bodies stunted or underdeveloped, suffering sickness and weakness with little or no energy to earn a living, access services or ever reach their potential.

The terminology used to refer to different concepts of hunger can be confusing, but nonetheless it is important to make a distinction since their underlying causes and solutions differ. Similarly although all three are food related, solutions are not uniform: there is no blanket, one-size-fits all approach to eliminating world hunger in all its forms.

- **Hunger:** the Food and Agriculture Organisation (FAO) defines it specifically as consumption of fewer than 1800 kilocalories a day; the minimum that most people require to live a healthy and productive life. The vast majority of the nearly 1 billion people deemed as hungry in our world suffer from **Chronic Hunger** caused by a lack of both sufficient quality and quantity of food.

- **Undernutrition** signifies deficiencies in energy, protein, essential vitamins and minerals, (or any or all of these). It is the result of inadequate intake of food, in terms of either quantity or quality, or poor utilisation of nutrients due to infections or other illnesses or a combination of these two factors. Severe undernutrition can be referred to as **Acute Hunger** and is associated with 35% of all under five mortalities, accounting for over 3 million deaths per year.

- **Malnutrition** refers more broadly to both undernutrition (problems of deficiencies) and overnutrition (problems of unbalanced diets, such as consumption of too many calories in relation to requirements with low intake of micronutrient-rich foods, e.g. green vegetables). This type of hunger can also be referred to as **Hidden Hunger** and includes almost two billion people worldwide, including many people in the 'developed' world with improper diets.

- **Food security** covers availability, access, utilisation and stability and exists when all people, at all times, have physical, social and economic access to sufficient safe and nutritious food which meets their dietary needs and food preferences for an active and healthy life.

MORE INFO

For more information, see John B Mason (2002) 'Measuring hunger and malnutrition' available at http://www.fao.org/docrep/005/Y4249E/y4249e0d.htm

SOME CAUSES OF HUNGER

Low agricultural productivity: 400 million people live on small farms that do not produce enough food to feed their families. Aid to agriculture decreased from a high of 17% of global funds in 1982 to 3.7% in 2006. This reduction in aid, along with changes in agricultural trade, has meant that poor farmers have not had sufficient access to seeds, tools, training, fertilisers, extension to credit or secure access to land.

Poverty is the root of human underdevelopment and also a result of it. Many of the world's poor are unable to afford sufficient food or food of the required quality. As a result, their diets tend to be based on starchy foods, such as rice and bread, with few fruits and vegetables or animal products leading to low calorie intake and micronutrient deficiencies.

Gender Inequality: More than 60% of the chronically hungry people in the world are women. Former UN Secretary General Kofi Anan argued that in the developing world; *'Women are on the front lines. They grow, process and prepare; they gather water and wood and they care for those suffering from HIV and AIDS. Yet women lack access to credit, technology, training and services and are denied their legal rights such as the right to land.'*

Conflict and Poor Governance: It is a fact that no country with a strong democracy has ever suffered from famine and it is also true that many nations with the highest levels of hunger are also gripped by violent conflicts or civil wars. As well as restricting access to fields to sow or harvest crops, armed groups often use hunger as a weapon by cutting off food supplies, destroying crops and hijacking relief aid.

Unjust Trade Regimes: The terms of international trade are for the most part stacked against poor and developing countries. Currently, markets work against the poor in both directions. The poor cannot afford the price of food and, at the same time, they are unable to sell their produce on the market to earn a living. Reform of Europe's Common Agricultural Policy and of World Trade Organisation rules is a priority and must allow for fair access to international markets.

Climate Change and Natural Disasters: The worst affected hunger hotspots are to be found in regions of climatic extremes; areas which experience drought, flooding, unpredictable precipitation patterns, soil erosion and desertification (increasingly as a result of Climate Change patterns).

Falling Remittances: Worldwide, nearly $300 billion was sent in remittances to developing countries in 2007 from relatives living outside their native country, surpassing global official development assistance (ODA) by 60%. With the global financial crises and consequent increase in unemployment, many of these remittances have now dried up leaving family *'back home'* even more vulnerable to poverty and hunger

Sources: Olivier De Schutter (2010), Facing Up to the Scandal of World Hunger, Trócaire Development Review, Kofi Annan (2006) UN Aids Report, New York

THE EVOLUTION OF THE RIGHT TO FOOD

The Right to Food (as with the majority of other rights) did not arrive *'fully formed'*, but has *'evolved'* in substance, understanding and detail over the past six decades. It is an evolution that has moved from a broad and somewhat theoretical concept to the detailing of concrete and practical steps needed to be taken by states and other parties to ensure the fulfilment of the right.

It is worth noting three significant milestones in this evolutionary journey;

1. The first is the 1966 International Covenant on Economic, Social and Cultural Rights (ICESCR) which built comprehensively upon the right to food 'as part of an adequate standard of living', articulated in article 25 of the 1948 Universal Declaration of Human Rights (UDHR).

The ICESCR acknowledges that hunger is not caused simply by a lack of food but is 'multi-dimensional' in nature in that it is caused above all by poverty, income disparities and lack of access to clean water, education and health care. It recognises too the important difference between the 'right to adequate food' as distinct from the 'right to be free from hunger'. The concept of freedom from hunger requires the state to provide food to those who are unable to meet their food needs for reasons beyond their control such as age, disability, economic downturn, famine, disaster or discrimination, whereas the right to food requires a progressive improvement of living conditions that will result in regular and equal access to resources and opportunities so that every individual is enabled to provide for his/her own needs.

It further noted that the Right to Food (as part of the indivisible nature of all rights) is to be included in the legal, administrative, political and social culture of all nations and such is legally binding on all its signatories (158 countries have ratified the Covenant to date). Article 11 of the ICESCR (see below) clearly lays out the obligations of state parties in regard to food and freedom from hunger.

ICESCR Article 11:

1. The States Parties to the present Covenant recognize the right of everyone to an adequate standard of living for himself and his family, including adequate food.... The States Parties will take appropriate steps to ensure the realisation of this right, recognising to this effect the essential importance of international cooperation based on free consent.

2. The States Parties to the present Covenant, recognising the fundamental right of everyone to be free from hunger, shall take, individually and through international cooperation, the measures, including specific programmes, which are needed:

(a) To improve methods of production, conservation and distribution of food by making full use of technical and scientific knowledge, by disseminating knowledge of the principles of nutrition and by developing or reforming agrarian systems in such a way as to achieve the most efficient development and utilisation of natural resources;

(b) Taking into account the problems of both food-importing and food-exporting countries, to ensure an equitable distribution of world food supplies in relation to need.

Some 30 years later, in 1999, a further detail was added to the Convention by adopting General Comment 12 on the Right to adequate food which highlighted three obligations incumbent on every State; to respect, protect and fulfil (see below). It further stated the right to adequate food is realised *'when every man, woman and child, alone or in community with others, has physical and economic access at all times to adequate food or means for its procurement.'* And that the right must not be interpreted *'in a narrow or restrictive sense which equates it with a minimum package of calories, proteins or other specific nutrients.'* In other words, delivering food as part of emergency relief, while in the short-term vital and necessary, is not compatible with human dignity in the long term.

Three State Obligations: Respect, Protect and Fulfil

The obligation to respect requires governments not to take any measures that arbitrarily deprive people of their right to food, for example, by measures preventing people from having access to food.

The obligation to protect means that states should enforce appropriate laws and take other relevant measures to prevent third parties, including individuals and corporations, from violating the right to food of others.

The obligation to fulfil (facilitate and provide) requires governments to pro-actively engage in activities intended to strengthen people's access to, and utilisation of, resources so as to facilitate their ability to feed themselves. As a last resort, whenever an individual or group is unable to enjoy the right to adequate food for reasons beyond their control, states have the obligation to fulfil that right directly.

The right to food means that governments must not take actions that result in increasing levels of hunger, food insecurity and malnutrition. It also means that governments must protect people from the actions of powerful others that might violate the right to food. States must also, to the maximum of available resources, invest in the eradication of hunger.

---

2. The second milestone is the 1996 World Food Summit in Rome set against a backdrop of increased global hunger which had risen to over 800 million people. At a time too when many States continued to ignore or misunderstand what the right to food meant in practice, this Summit came to be seen as a milestone in the efforts to bring increased attention and renewed political will to implementing the right. The Summit adopted a Plan of Action expressed in seven commitments, one of which challenged the international community, governments, civil society and international institutions to take concrete steps toward a right-to-food based approach by stipulating the need to; *'...clarify the content of the right to food and the fundamental right of everyone to be free from hunger...and to give particular attention to the implementation and full and progressive realisation of this right as a means of achieving food security for all'.* (Commitment 7.4).

In addition, the Plan of Action invited; *'... the UN High Commissioner for Human Rights in consultation with others to better define the rights related to food in Article 11 of the Covenant (ICESR) and propose ways to implement and realise these rights. The call for a further clarification on the right to food brought a sharper focus to the task of 'giving concrete content to this right and promoting its adaptation to real life situations in different countries'.*

3. The third significant milestone occurred in 2004 after two years of *'intense and constructive negotiations'* between governments, civil society organisations and others, a set of voluntary guidelines to support Member States' efforts to achieve the progressive realisation of the right to food, was formally adopted by the FAO. Their objective is to *'provide an additional instrument to combat hunger and poverty and to accelerate*

attainment of the Millennium Development Goals.' The Guidelines cover the full range of actions to be considered by governments at the national level in order to build an *'enabling environment for people to feed themselves with dignity'*.

Perhaps the two most significant achievements of the Guidelines are first, that they put *'the entitlements of people more firmly at the centre of development'* and, second, that they dispel any lingering notion of the right to adequate food as a theoretical concept by creating a practical, clear and detailed *'map'* for assisting Governments to design appropriate policies, strategies and legislation for implementing the right in real and tangible ways. For example, by assisting people in accessing legal assistance to better assert the progressive realisation of the right to adequate food; protecting land tenure rights, especially with regard to women; and appropriate market regulation and the creation of social safety nets.

Despite the huge advances represented by each of these (and many other) milestones, the fact remains that nearly 1 billion people do not have access to adequate food and that key instruments, such as the Voluntary Guidelines referred to above, are just that – voluntary, and not legally binding on member states. The challenge, therefore, remains not only to build a supportive environment at national level but to clarify the international obligations that go beyond those of a state's responsibility to its own people. Just as genocide, torture and crimes against humanity are recognised as gross violations of human rights which demand an *'extra-judicial'* or *'extra-territorial'* response on the part of the international community, so too should the denial of the right to food demand similar action.

Only 22 countries have enshrined the right to food in their constitutions either for all citizens or specifically for children. Unfortunately, no country has yet instituted specific legislative measures to implement this right. Laws still need to address common areas, such as land tenure, access to water, minimum wage levels, social safety nets, rural markets and food production and quality.

THE HUNGER PARADOX: INCREASED FOOD PRODUCTION = INCREASED HUNGER

Perhaps the best known efforts in ending world hunger took place in the decades from the 1960's to 1990's in what came to be known as the Green Revolution, spearheaded by American scientist Norman Borlaug. From the 1940s, Borlaug had worked on developing, disease resistant wheat varieties capable of producing hugely increased yields. The improved seeds were instrumental in boosting Mexican wheat production and averting famine in India and Pakistan. By combining wheat varieties with new mechanised agricultural technologies and increased use of fertilisers and irrigation, Mexico was able to produce more wheat than was needed by its own people, leading to it becoming an exporter of wheat in the 1960s. Even in the United States where Green Revolution technologies were widely used, the country moved from importing half of its wheat in the 1940s to becoming self-sufficient in the 1950s to being an exporter in the 1960s. In India, Borlaug developed a new variety of rice named IR8 and today the country continues to be one of the world's leading rice producers.

The Green Revolution also saw dramatic improvements in corn yields throughout the developing world and Borlaug's work inspired the creation of centres in India, the Philippines, several countries in Africa and, of course, in Mexico itself where he established The International Maize and

Continued on page 220 →

MEASURING HUNGER

Measuring hunger like measuring literacy, poverty or child mortality is an attempt to capture the *multi-dimensional* aspects of human development. Measurement tools, however, are not simply about counting numbers or noting trends but, when used effectively, are about raising awareness, drawing attention to regional disparities, providing insights into barriers to development and, perhaps most importantly, highlighting successful interventions. In the introduction to its 2010 Global Hunger Index report the authors remind us that even though *abundant technological tools exist to collect and assess data almost instantaneously, there are still enormous time lags when it comes to reporting vital statistics on hunger.*

International Food Policy Research Institute - the Global Hunger Index (GHI)

The Global Hunger Index provides important information on how individual countries are dealing with their hunger problem. It measures progress being made, or not, in working towards the achievement of the first Millennium Development Goal (MDG), to halve, by 2015 the proportion of the world's population who are hungry. The Index was developed by the International Food Policy Research Institute (IFPRI) and it incorporates three interlinked hunger-related indicators:

- The proportion of undernourished as a percentage of the population (reflecting the share of the population with insufficient dietary energy intake)

- The prevalence of underweight children under the age of five

- The mortality rate of children under the age of five (reflecting the fatal linkages between inadequate dietary intake and unhealthy environments)

The GHI ranks countries on a 100 point scale, with 0 being the best score (no hunger) and 100 being the worst, although neither of these extremes is ever reached in practice. Values between 0 - 5.0 reflect low hunger; 5.0 - 9.9 are considered to have moderate hunger, 10 – 19.9 indicate serious hunger with 20 – 29.9 indicating alarming levels and over 30 reflecting *extremely alarming* levels of hunger.

According to the GHI 2010 Report, the index for hunger in the world remains at a level characterised as *serious* - a result that is unsurprising given that the overall number of hungry people surpassed 1 billion people in 2009.

Countries that have seen the largest progress in improving their GHI between 1990 and 2010 include Angola, Ethiopia, Ghana, Mozambique, Nicaragua and Vietnam, by scaling up investment in small farms and introducing social protection schemes such as public works employment, cash transfers, food rations and free school meals. However, 29 countries still have levels of hunger rated as *extremely alarming* or *alarming*. The majority of these countries are in sub-Saharan Africa and South Asia. The largest deterioration in GHI scores was seen in the Democratic Republic of Congo, largely because of conflict and political instability whereas:

'...In contrast, poor economic growth, strong agricultural performance and increasing gender equity can reduce hunger below what would be expected based on income.'

2010 Global Hunger Index 2010

Wheat Improvement Centre. During the decades of major Green Revolution advances (between the 1970s and 1990s) the total food available per person in the world rose by 11% while the estimated number of hungry people fell from 942 million to 786 million, a 16% fall.

Frances Moore Lappe et al (1998, 2nd edition) World Hunger: 12 Myths, New York, Grove Press

Despite the obvious successes and production advances of the Green Revolution it was by no means without its critics. Its high dependency on fertilisers, pesticides, irrigation and other ecologically harmful practices, as well the loss in biodiversity (in India there were about 30,000 rice varieties prior to the Green Revolution, today there are around 10) and long-term soil degradation have all been well documented. Critics further point out that the benefits of the Green Revolution were skewed toward wealthier and bigger farmers who not only could afford the agro-chemical inputs but because of their political clout were the recipient of generous government subsidies.

Perhaps one of the most important lessons from the Green Revolution is that increased productivity on its own cannot, and does not, alleviate hunger. Recent experiences of hunger crises in many parts of the developing world have shown us time and again that hunger (paradoxically) is not caused by a shortage of food, but by a lack of purchasing power and stable environments as well as the inadequate distribution of resources to the poor. The International Food Policy Research Institute claims that '...simply adding to the pile of food will not be enough' and that policymakers will need to target the poor '...more precisely to ensure they receive greater benefits from new technologies.'

Simply put, if the poor don't have money to buy food, no amount of increased production is going to help them.

HUNGER: FOUR KEY CHALLENGES

FOOD PRICE CRISIS

Many of us may still remember the images in our papers and on our TV screens of people rioting in the streets of over 30 developing world countries in 2008; riots that were not due to any of the 'usual' causes (political coups etc.) but were as a result of the high price of food (not the lack of food). Having remained stable for the best part of 15 years, the increase in global food prices caught many people by surprise with the doubling in price of staple crops in early 2008. Analysts claimed that it was a 'perfect storm' of factors that gave rise to the increases, including rising energy prices, demand for biofuels, depreciation of the U.S. dollar, and a variety of trade shocks including panic purchases, export restrictions and unfavourable weather. In March of 2008 global food prices were 75% higher than they had been in 2005. It is estimated that the food price crises, heightened by the global recession, was responsible for adding a further 100 million people to the ranks of the world's hungry, bringing the global figure in 2009 to over 1 billion people for the first time in history (2010 figures show a slight but unsteady decrease to 925 million).

With many poor families spending between 60% - 70% of their income on food, even the slightest increase in food prices can have a devastating impact as people are forced to make tough choices about how to spend the little income they have. Not only is there a reduction in the quantity and quality of food consumed but reductions in spending in areas such as health-care and education follow suit. As long as food prices remain high, the poor (and hungry) begin to sell the very things they need to help them in the long-term; important assets such as land, tools, livestock and household goods on which they depend to make a living. In many instances seeds set aside for next year's crop are consumed to meet the food needs of today.

In 2009 a broad-based coalition of non-governmental organisations, advocacy groups and religious organisations produced a Roadmap to End Global Hunger, that called on President Obama, Congress and other leaders in the United States to provide *'global leadership'* in generating the needed political will to effectively address the problem. The Roadmap reminded us that the hunger crises did not begin with the Global Financial problems of 2008 but even before then over 850 million people lacked sufficient food. In effect the Roadmap stated, yet again, that what is needed is a *'comprehensive strategy that ensures support is balanced between short-term emergency and long-term development needs.'*

At roughly the same time other governments through the G8 and G20 as well as the World Bank met to discuss the global food crises. A food security initiative was created at the L'Aquila G8 summit in Italy where over US$20 billion was pledged over three years for sustainable agricultural development. In 2010 the World Bank's Board of Executive Directors approved the Global Agricultural and Food Security Program (GAFSP) to support country-led efforts to improve food security, especially among small holder farmers. Regrettably, as of July 2011 only US$400 million of the US$1 billion pledged has been realised. It appears that food price volatility like hunger itself will be with us for some time to come.

BIOFUELS AND HUNGER

'Taking food from the poor to burn in the cars of the rich - biofuels is a crime against humanity.'

George Monbiot, Guardian Newspaper, 2008

Following a number of initiatives to reduce global carbon emissions, northern industrialised nations embarked on an ambitious plan to invest and grow biofuels as an alternative to rapidly depleting fossil fuels worldwide. The EU adopted a biofuels directive in 2003 setting targets for use of biofuels in transportation at 5.75% by 2010 increasing to 8% by 2015. Similarly the United States embarked on an ambitious biofuel strategy which aims to provide 30% of US fuel needs by 2050.

With tax incentives, subsidies and relief schemes in place for biofuel producers, there are fears that the price of staple foods could increase by as much as 15% by 2020. According to a 2010 report by NGO Action Aid, *Who's Really Fighting Hunger*, 30 to 40 million hectares of land will be needed by 2020 to meet the EU's demand for biofuels, half of which will be in the Developing World. EU companies have already acquired or requested 5 million hectares of land for industrial biofuels in developing countries. All of the biofuel produced on this land is for export, a familiar cash crop problem for poor nations. Biofuels are conservatively estimated to have been responsible for at least 30% of the global food price spike in 2008 that pushed 100 million people into poverty and drove some 30 million more into hunger.

The most immediate effect of the push for industrial biofuels is to compete with food for feedstock, thereby inflating food prices. With the exception of Jatropha (a group of shrubs and plants particularly suited to biofuels) the main agricultural crops used for industrial biofuels (palm, soy, sunflower, rapeseed used in making biodiesel, and maize, wheat and sugars used in ethanol), are all food crops.

The UN Food and Agriculture Organisation estimated that in 2008 – 2009, 125 million tonnes of cereals were diverted into biofuel production. In 2010, more cereal (1,107 million tonnes) was diverted into animal feed and industrial uses than for feeding people (1,013 million tonnes). Independent analysts have concluded that industrial biofuels have been responsible for 30% to 75% of the global food price increase in 2008.

For many critics of biofuel policies, it is clear that biofuels are socially unsustainable in competing for land that '...*should be growing food, increasing food prices and landlessness, causing widespread hunger, and depriving millions of the poorest of their livelihood.*' According to Action Aid, evidence from real production data bears out what many scientists have been saying: most if not all biofuels offer no savings in energy or carbon emissions, especially when indirect emissions from deforestation and other land use changes are taken into account.

In concluding its 2010 analysis, Action Aid suggested a number of actions that needed to be taken regarding biofuels and their potential impact on hunger:

- A moratorium on further expansion of industrial biofuel production and investment
- Ensure EU member states do not lock into industrial biofuels in their 2010 national action plans
- Reduce transport and energy consumption
- End targets and financial incentives for industrial biofuels
- Support small-scale sustainable biofuels in the EU and abroad

Action Aid, 2010 Who's Really Fighting Hunger

POPULATION GROWTH AND HUNGER

The population of the planet is projected to reach over 10 billion by 2050 which has given rise to a heated debate about the carrying capacity of the earth across a whole range of issues, not least of which is the world's ability to feed its ever-growing population. This issue has already been addressed in Chapter 3, and we have summarised many of the arguments in the current debate on the page opposite.

GENDER INEQUALITY

Estimates show that women produce up to 80% of the basic food in many developing countries; they undertake the majority of work associated with farming, such as collecting water, sowing, weeding, fertilising and harvesting staple crops. Yet women are the most vulnerable to chronic poverty and hunger, they have fewer benefits under legal systems than men (such as in owning land) do not receive adequate training and support, lack access to credit and are denied decision-making power.

While the role of women in any given society is rooted in historical and cultural norms, a lesson learned from international development efforts over the past four decades is that we must '*look with fresh eyes at women's role in the agricultural economy and see women, not merely as subsistence farmers and caretakers of their own families, as often comes to mind, but also as vital actors in the agricultural economy and the expanding world of commercial agriculture.*' Women in Africa constitute the majority of farmers, yet they receive less than 10% of small farm credit and own just 1% of the land (on this topic, see UNDP 1995 Human Development Report and Linda K. Fuller 2008 *African Women's Unique Vulnerabilities to HIV/AIDS*, New York: Palgrave Macmillan).

In addressing the issue of equality for women in agriculture, the UN Food and Agriculture Organisation believes that giving women the same tools and resources as men, including financial services, education and access to markets, could increase agricultural production in developing countries by up to 4% which, in turn, could reduce the number of hungry people in the world by up to 150 million.

DEBATE IT: Population & World Hunger

Reduce Population to feed the world

First of all, there are about 1 billion people who are chronically hungry. Given the tremendous burden of food production on the earth's ecosystems, given the fact that we have not properly fed all the people on the planet to date and given the fact that pressures are going to increase, population growth remains a significant problem.

Norman Borlaug, founder of the Green Revolution, said that the success of the past century is only a 'breathing space' to deal with the need to slow population growth. In order for the food supply to meet the demand, we must slow down population growth, and at the same time, invest in agricultural research.

About 70% of consumption of freshwater is for irrigation for agriculture. Our heavy uses of resources, along with intensive methods of farming degrade the environment over time. These practices aren't sustainable into the future.

We have over 7 billion people, and they are on the search for enough food, water and energy to meet their needs, to make economic progress. But when you add it all up, we are already an unsustainable world society. Climate change, water stress, environmental degradation, species extinction are all increasing in importance and the challenges they pose are immense without adding larger numbers.

The FAO estimates that food production must rise by 50 percent by 2030 to meet growing demand...an unsustainable prospect

Population is not a cause of hunger

Rapid population growth is not the root cause of hunger. Like hunger itself, it results from underlying inequities that deprive people, especially poor women, of economic opportunity and security.

Rapid population growth and hunger are endemic to societies where land ownership, jobs, education, health care, and old age security are beyond the reach of most people.

Nowhere does population density explain hunger. For every Bangladesh, a densely populated and hungry country, we find a Nigeria, Brazil or Bolivia, where abundant food resources coexist with hunger. Or we find a country like the Netherlands, where very little land per person has not prevented it from eliminating hunger and becoming a net exporter of food.

If the so called developed world is serious about protecting the planet its people simply need to start consuming less; it is the developed world which is more of a threat to the long term sustainability of the planet.

The carbon footprint of the average Tanzanian (and indeed the average African) is a fraction of that of their counterpart in the West.

Keeping girls in education, upholding women's rights, and improving their access to effective reproductive health services is the ultimate answer to not only hunger but many of the challenges the world faces

Part of the problem has been that policy interventions, designed to help communities become more food secure and better nourished, have not addressed the specific needs, constraints and multiple roles of poor women. Women farmers must be recognised as 'farmers' in their own right and donors, policy-makers and other experts must create opportunities for them to participate in policy-making processes through capacity building initiatives and investing in long term strategy of empowerment and gender equality within the wider community. Any intervention must address the situation of women like Sandra Kabbabu in Zambia;

'The kind of life I am leading – we're suffering and there's little we can do. I just have to be strong and cultivate my land. I have to rely on farming because this is how I can feed my children. I want to continue farming – I want to improve my situation by continuing to farm so I can produce enough to at least feed my family properly the whole year around.'

John Madeley and Robin Willoughby, Unheard Voices, Women Marginal Farmers – working hard to fight hunger, Concern Worldwide, 2010

READING

Asbjørn Eide (2008): The human right to adequate food and freedom from hunger, FAO Corporate Document Repository

FAO (2011) State of World Food Insecurity: how does international price volatility affect domestic economies and food security? Rome, FAO (annual)

B. Halweil and D. Nierenberg, (2011) State of the World, Innovations that Nourish the Planet, The Worldwatch Institute

International Food Policy Research Institute (2011) Global Hunger Index, Washington (annual)

Wayne Roberts (2008) The No-Nonsense Guide to World Food, London, New Internationalist

R. Thurow and S. Kilman (2009) Enough: Why the Poorest Starve in an Age of Plenty, New York, PublicAffairs

Jo Walker (2010) Who's Really Fighting Hunger, London, Action Aid

MORE INFORMATION & DEBATE

http://www.fao.org - UN Food and Agriculture Organisation

http://www.ifpri.org - International Food Policy Research Institute

http://www.worldhunger.org – solid information and education site on world hunger

http://www.globalissues.org/issue/6/world-hunger-and-poverty - extensive website on a range of development issues including world hunger

http://www.bread.org/hunger/global - US, faith based site on world hunger and related issues

http://www.wfp.org/hunger - World Food Programme site

A WORLD OF 7 BILLION

"Today 95 percent of population growth is occurring in the world's developing countries, already home to 82 percent of the world's people yet producing just 34 percent of gross world product in absolute dollar terms"

Robert Engelman, 2010

WORLD POPULATION
(IN MILLIONS)

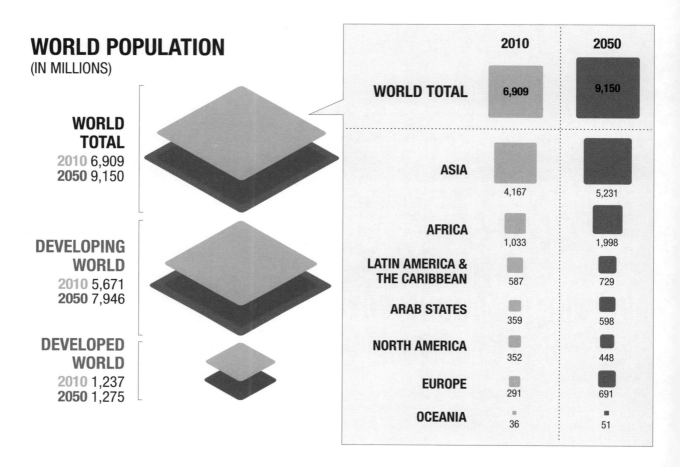

WORLD TOTAL
2010 6,909
2050 9,150

DEVELOPING WORLD
2010 5,671
2050 7,946

DEVELOPED WORLD
2010 1,237
2050 1,275

	2010	2050
WORLD TOTAL	6,909	9,150
ASIA	4,167	5,231
AFRICA	1,033	1,998
LATIN AMERICA & THE CARIBBEAN	587	729
ARAB STATES	359	598
NORTH AMERICA	352	448
EUROPE	291	691
OCEANIA	36	51

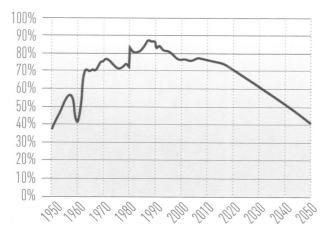

◀ **Annual world population change 1950-2050**

In addition to growth rates, another way to look at population growth is to consider annual changes in the total population. The annual increase in population peaked at about 87 million in the late 1980s. The peak occured then, even though annual growth rates were past their peak in the late 1960s, because the world population was higher in the 1980s than in the 1960s.

Source: U.S. Census Bureau, International Data Base. December 2010 update.

10 LARGEST CITIES
(2010)

9 **DHAKA** 14.64 MILLION

8 **KOLKATA** 15.55 MILLION

7 **SHANGHAI** 16.57 MILLION

6 **NEW YORK** 19.42 MILLION

10 **KARACHI** 13.2 MILLION

5 **MEXICO CITY** 19.46 MILLION

4 **MUMBAI** 20.04 MILLION

1 TOKYO
36.7 MILLION

3 SAO PAULO
20.26 MILLION

2 DELHI
22.5 MILLION

> *He let his mind drift as he stared at the city, half slum, half paradise. How could a place be so ugly and violent, yet beautiful at the same time.*
>
> Nigerian author Chris Abani 2004

A WORLD OF SLUM DWELLERS

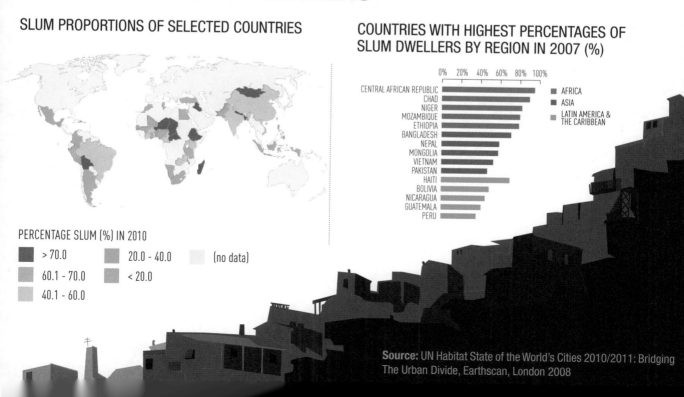

SLUM PROPORTIONS OF SELECTED COUNTRIES

PERCENTAGE SLUM (%) IN 2010

- \> 70.0
- 60.1 - 70.0
- 40.1 - 60.0
- 20.0 - 40.0
- < 20.0
- (no data)

COUNTRIES WITH HIGHEST PERCENTAGES OF SLUM DWELLERS BY REGION IN 2007 (%)

0%   20%   40%   60%   80%   100%

CENTRAL AFRICAN REPUBLIC
CHAD
NIGER
MOZAMBIQUE
ETHIOPIA
BANGLADESH
NEPAL
MONGOLIA
VIETNAM
PAKISTAN
HAITI
BOLIVIA
NICARAGUA
GUATEMALA
PERU

- AFRICA
- ASIA
- LATIN AMERICA & THE CARIBBEAN

Source: UN Habitat State of the World's Cities 2010/2011: Bridging The Urban Divide, Earthscan, London 2008

Photo: Gareth Bentley

[*'I can't tell you what art does and how it does it, but I know that art has often judged the judges, pleaded revenge to the innocent and shown to the future what the past has suffered, so that it has never been forgotten.*

'I know too that the powerful fear art, whatever its form … and that amongst the people such art sometimes runs like a rumour and a legend because it makes sense of what life's brutalities cannot, a sense that unites us, for it is inseparable from a justice at last. Art, when it functions like this, becomes a meeting-place of the invisible, the irreducible, the enduring, guts and honour. ']

John Berger

Chapter 13

READ, LISTEN, SEE
THE ARTS, DEVELOPMENT AND HUMAN RIGHTS

Bertrand Borg

Having addressed the question of what art and literature can contribute to our understanding of human rights and development, this chapter briefly explores common themes and understandings within literature, art, music and film. It discusses the importance of context, geography and history in approaching literature, popular music, art and film, citing examples and case studies from Africa, the United States, Asia and Latin America. Issues such as post-colonialism; women's rights, the impact of the American war in Vietnam, and post-Apartheid South Africa are also introduced.

KEYWORDS

Development and human rights education, emotional understanding, **idea and experience,** American protest literature, **post-colonialism,** reflecting and shaping consciousness, **music as protest,** visual arts, **film**

INTRODUCTION

WHY BOTHER WITH ART AND LITERATURE?

Consider this - in 2010, 288,355 books were published, 1,419 Hollywood movies opened in theatres and 75,000 music albums were released – *in America alone!* There is more choice in the art, culture and entertainment industries than ever before. A simple internet connection provides decades of music, film and literature at one's fingertips. While it would be foolish to deny the undoubted benefits this dizzying variety provides, it also runs the risk of devaluing these works of art. When you can simply download an album, film or book, it becomes much easier to treat it as a disposable commodity.

At this juncture, the question arises: why bother with literature, music, art or film? They entertain us and serve as a form of escapism from everyday reality – is this not enough? How can – *why should*, some would ask – the arts enhance our understanding of the world around us? And, *how does this happen?*

The human rights and development fields are laden with technical terms, indices, economic and political theories and even complex court judgements. Ask about human rights and many *'experts'* will point towards international conventions and tribunals, theoretical frameworks, the various *'generations of rights'* and growing volumes of case law. Learning about development is no better - HDI, GDP, IMF; colonisation, stages of growth, conditionality - a jungle of terms, acronyms and figures.

Yet human rights and development are primarily about people; people who together make up families, communities, towns, cities and nations, with jobs, ambitions, concerns and prejudices.

It is people who must respect human rights and who are, in turn, protected by them; development policies and initiatives begin as abstract figures and concepts, but they manifest themselves in the lives of people across the globe.

This visceral, instinctive aspect to human rights and development, however, is often hard to grasp and even harder to teach. It is relatively straightforward, for instance, to memorise a timeline concerning the abolition of slavery: it is somewhat harder to fully come to terms with what slavery meant for those forced to live with its yoke. Rattling off poverty statistics is all well and good, but facts, devoid of context, understanding and empathy, only go so far. We were all born under different stars, and grew up in different circumstances - boys or girls, rich or poor, healthy or not. Some will be dark-skinned, others fair; many start working at puberty, others will pursue their education until well into their 20s. Our lives shape the way in which we see the world and others in it.

The arts, in whatever form, help bridge this gap between theoretical concept

Yet human rights and development are primarily about people; people who together make up families, communities, towns, cities and nations, with jobs, ambitions, concerns and prejudices. It is people who must respect human rights and who are, in turn, protected by them; development policies and initiatives begin as abstract figures and concepts, but they manifest themselves in the lives of people across the globe.

and emotional understanding. They allow us to access scenarios which would otherwise be alien to us: through Achebe's descriptions, Dylan's lyrics or Toulouse-Lautrec's brush strokes, we have the opportunity to see the world anew and this is especially valuable when that world is alien to our own experience and understanding. Great literature, music, art or film (or any other art form) has the power to transform an abstract idea, such as the equality of women, into an intuitive experience. It brings humanity closer together by spanning the chasms of understanding between people with empathy and shared experiences. Mario Vargas Llosa, the Peruvian writer and Nobel laureate, expressed as much in his essay *'Why Literature'*:

'Nothing better protects a human being against the stupidity of prejudice, racism, religious or political sectarianism, and exclusivist nationalism than this truth that invariably appears in great literature: that men and women of all nations and places are essentially equal, and that only injustice sows among them discrimination, fear, and exploitation.

'Nothing teaches us better than literature to see, in ethnic and cultural differences, the richness of the human patrimony, and to prize those differences as a manifestation of humanity's multi-faceted creativity.'

Mario Vargas Llosa (2001) 'The premature obituary of the book: Why Literature?'
http://www.uwec.edu/pnotesbd/Llosa_article.htm

Art can illuminate and educate. Art, with its bridging between our intellect and emotions, can make us think and feel about issues on a level that more straightforward, fact-based knowledge cannot. For example, reading Tahar Ben Jelloun's classic *This Blinding Absence of Light*, or Arthur Koestler's *Darkness at Noon*, makes us understand what it must be like to be arbitrarily arrested and imprisoned better than a dossier or report ever could. Dave Egger's *What is What* allows us to *'know better'* what life can be like for 'lost boys' in the Sudan beyond the statistics.

Why, then, have the arts been so relatively neglected when it comes to development and human rights education? Is it possible that the entire literary, musical and motion picture canon has so little to teach us on issues of human rights, women's rights, war and conflict or development? This chapter seeks to begin to redress this imbalance. It does not claim to provide an exhaustive guide to the potential of the arts in development and human rights education, nor is it especially comprehensive. Its aim is simply to stimulate readers into finding their own way of appreciating literature, music or film from a development and human rights perspective. An important caveat: although non-Anglophone books, music and films are touched upon, the vast majority of the works discussed in this chapter come from the English language, for obvious reasons.

> *Art can illuminate and educate. Art, with its bridging between our intellect and emotions, can make us think and feel about issues on a level that more straightforward, fact-based knowledge cannot.*

WHEN DID YOU SAY THAT? HISTORICAL CONTEXT

Art, literature and music cannot be divorced from the societies in which they are conceived and created. While a few works manage to transcend their era and/or geographical space, these remain the exception rather than the rule. There are times when a creative vision captures the zeitgeist and becomes *the* emblematic vision of its time: witness, for example, how we associate urban Victorian England with Charles Dickens; fin-de-siècle Paris with the paintings and illustrations of Toulouse-Lautrec or the trenches of the Great War with the vivid imagery of Wilfred Owen's poems. In most cases, however, a thorough understanding of a work of art, music or literature necessitates a modicum of comprehension of the context in which it was created. Without a basic understanding of apartheid, its victims and its effect on South African society, a book such as J.M. Coetzee's *Disgrace* loses much of its power; the folk protest songs that emerged out of 1960s America gain resonance when placed in the context of the American war in Vietnam and the various uprisings on US university campuses while the music of Victor Jarra needs to be understood in the context of Chile following the murder of Salvador Allende in 1973.

There is an inherent contradiction in our established canon of literary classics - many of the books we today consider to be literary classics would most likely not be passed for publication had they been written today. Racism; sexism; the subjugation of the poor; the association of poverty and evil; all these themes (and many more) can be found in a number of classics, from Conrad's *Heart of Darkness* to the children's classic, *Babar the Elephant*. In effect, we learn about these issues through works which capture the zeitgeist of their time.

This is perhaps one of literature's greatest strengths in its role as enlightener. Reports, factual texts and testimonials must, by their very nature, be literal. They spell out inequality and highlight

> *There is an inherent contradiction in our established canon of literary classics - many of the books we today consider to be literary classics would most likely not be passed for publication had they been written today.*

injustices. Fiction, on the other hand, has no such metaphorical chains. It is left up to the reader to understand, interpret and react to the fictional universe laid out before them.

Reading books written in another age, when a different set of social norms and values prevailed, can be beneficial to the student of development and human rights in two ways. First, it helps the reader make sense of the world as it was in a different era. It may at times make the reader feel distinctly uncomfortable - for instance, the use of the word 'nigger' in Mark Twain's *Huckleberry Finn* – but it provides the reader with a window into the past. Second, reading such texts today can help frame progress. Conrad's depiction of Africans in *Heart of Darkness* may shock in 2011 but that in itself is a sign of progress.

These points are most relevant to literature – not because song or film are, as mediums, naturally impervious to such provincial mind frames - but simply because it is only in the 20th century that recorded music and film came into existence. There are, nevertheless, exceptions. Take, for instance, D.W. Griffith's film *The Birth of a Nation*. Its openly racist imagery shocked people even in 1915 when it was first released, but Griffith's portrayal of the American South was honest to reality at the time. Even closer to the present day, as recently as 1974 Gilbert O'Sullivan felt justified in writing and singing a song called 'A Woman's Place' (sample lyric: *I'm all for a woman who can make it on her own/ But I believe a woman's place is in the home*).

PHLEGMATIC PROSE: PROTEST IN LITERATURE

A tradition of American protest in print goes back to the angry plain talk of pamphleteer Thomas Paine and the orderly defiance of Thomas Jefferson's Declaration of Independence. From there, protest literature opens for the reader an alternative window onto the American history of slavery, the forced migration of American Indians, women's suffrage, labour unrest, lynching, the Great Depression, Civil Rights, *'second-wave'* feminism, Vietnam, and gay rights. All reformist writers are *'picture-makers'*, said Trodd, quoting ex-slave and abolitionist Frederick Douglass. They see the world as it is, but they also *'see the world as it could be'*, she said.

Rarely are the alternative visions of protest literature part of the mainstream - though radical heroes of the past sometimes become the mild staples of future popular cultures. In her novel *Beloved*, US author Toni Morrison enters the world of black slavery and the myriad ways in which it dehumanises. In the novel, Sethe, an enslaved woman, is driven to kill her own two-year old daughter rather than see her re-captured and enslaved.

Development and decolonisation have captured the imagination of many a writer. In fact, a cursory glance at the greatest novels of the past 40 years reveals a startling slant towards 'New World' novels and novelists: from Salman Rushdie and Amitav Ghosh in India, to Wole Solyinka in Nigeria, Chinua Achebe in Kenya, VS Naipaul in Trinidad, Garcia Marquez in Colombia and Vargas Llosa in Chile.

'This is worse than when Paul D came to 124 and she cried helplessly into the stove. This is worse. Then it was for herself. Now she is crying because she has no self.'

Ralph Ellison's classic *Invisible Man* (not to be confused with the HG Wells science fiction novel *'The Invisible Man'*) uses some extremely powerful imagery to hammer home the reality of living in a racist society as a member of its underclass. The (unnamed) narrator sets the tone on the very first page:

'I am an invisible man…I am invisible, understand, simply because people refuse to see me.'

As the story progresses, we are gradually exposed to the narrator's growing disillusionment with society. The manager at a paints manufacturer tells the narrator that:

'Our white is so white you can paint a chunka coal and you'd have to crack it open with a sledge hammer to prove it wasn't white clear through.'

He is talking about their *'Optic White'* paint, but Ellison's symbolism is clear: American society, like Optic White paint, covers up blackness and hides it from view. In order for a black person to prosper in society, Ellison is implying, they must subjugate their black identity and make themselves *'whiter'*. *Invisible Man* was published in 1953, just as popular notions of Negritude, Black Consciousness and pan-Africanism were coming into vogue. The book, partly the fruit of this movement, and partly its cause, continues to be widely read and discussed to this day.

Development and decolonisation have captured the imagination of many a writer. In fact, a cursory glance at the greatest novels of the past 40 years reveals a startling slant towards 'New World' novels and novelists: from Salman Rushdie and Amitav Ghosh in India, to Wole Solyinka in Nigeria, Chinua Achebe in Kenya, VS Naipaul in Trinidad, Garcia Marquez in Colombia and Vargas Llosa in Chile.

THINGS FALL APART – BIG IDEAS, LITTLE WOMEN?

A work of literature/fiction is as important for what it doesn't say or describe as for what it does. In *Things Fall Apart*, Achebe opened many eyes to the systematic oppression inherent within African colonialism but, counter-wise, he failed to tackle the gender imbalances within both that system and the traditional Ibo way of life. Women are portrayed as objects, equal to yams or barley; even in ceremonies supposedly dedicated to them (the payment of a bridal price/gifting of kola nuts to future in-laws) they play a peripheral role, seen but not heard, cooking but not eating.

Writing about contemporary African fiction and the politics of gender in 1994, Florence Stratton argues that *Things Fall Apart* is *'a story whose concern is wholly for men and their dilemmas, one in which what happens to women is of no consequence....Achebe does not tell African women 'where the rain began to beat them'. Nor does he attempt to restore 'dignity and self-respect' to African women.'*

See Florence Stratton (1994) Contemporary African Literature and the Politics of Gender, London, Routledge

POST-COLONIAL LITERATURE: INDEPENDENCE AND IDENTITY

As the sun set on the age of empire and colonisation, it rose for a new generation of nation states. Entire peoples who for generations had been subjects of a foreign power suddenly found themselves in the limelight. As flags and anthems were established, patriotism and nationalism flourished. In literary fields, meanwhile, this post-colonial fever adopted a more introverted form. As they sifted through the remains of their country's colonial past and tried to understand their compatriots, a new generation of writers – Indian, Pakistani, Nigerian, Ghanaian, Sri Lankan and many more – discovered their literary voice. Chinua Achebe's classic novel, *Things Fall Apart*, is one of the most regularly cited examples of post-colonial literature. The book tells the story of Okonkwo, a hard-working and ambitious man who is one of his clan's leaders. His fear of being considered weak and 'womanly', however, leads him to join in the murder of a boy he had adopted as his son. Okonkwo is exiled for seven years in order to appease the gods: when he returns, he finds the village irrevocably changed by the arrival of white missionaries.

Things Fall Apart can be read as a direct response to the literature about Africa that existed at the time – fiction written by white males who saw the 'dark continent' (to borrow one favourite phrase) as a sort of adventure playground for adult boys, replete with exotic animals and peoples to subjugate and shoot at. *Things Fall Apart* was certainly not the first book written about Africa, or about life in Africa; what distinguished it from previous novels – for example, *Conrad's Heart of Darkness* or Blixen's *Heart of Africa* – was that it was written from an African perspective, and very consciously at that.

Achebe had experienced colonialism first-hand, and like most other post-colonial writers, he saw relationships of power everywhere he looked. He also understood the power of the written word, and that literature and fiction were not power-neutral tools. *'Stories are not innocent'* he said. *'They can put you in the wrong crowd, in the party of the man who has come to dispossess you'.*

To Achebe's mind, much existing literature about Africa fell into this category of story. His views on *Heart of Darkness* – a literary classic whose title has subsequently become a lazy euphemism for the African continent – warrant quoting:

'Conrad was a bloody racist. That this simple truth is glossed over in criticism of his work is due to the fact that white racism against Africa is such a normal way of thinking that its manifestations go completely undetected. Students of 'Heart of Darkness' will often tell you that Conrad is concerned not so much with Africa as with the deterioration of one European mind caused by solitude and sickness...Which is partly the point: Africa as setting and backdrop which eliminates the African as human factor. Africa as a metaphysical battlefield devoid of all recognisable humanity, into which the wandering European enters at his peril.....The real question is the dehumanisation of Africa and Africans which this age-long attitude has fostered and continues to foster in the world. And the question is whether a novel which celebrates this dehumanisation, which depersonalises a portion of the human race, can be called a great work of art. My answer is: No, it cannot.'

Chinua Achebe 1975

Other African post-colonial writers have also tackled such issues – of power and powerlessness, the struggle between traditional culture and modernity,

> Other African post-colonial writers have also tackled such issues – of power and powerlessness, the struggle between traditional culture and modernity, and the need to break free of the past.

and the need to break free of the past. Books such as Ngugi Wa Thiong'o's *Petals of Blood* and *The Beautiful Ones Are Not Yet Born* by Ayi Kwei Armah deal with similar issues – the former in post-independence Kenya, the latter post-independence Ghana.

Fast forward to the present day, and you can still find African literature trying to reconcile past and present. Other concerns, however, also emerge. The magical realism of Ben Okri's *The Famished Road* couches a social commentary on Nigerian poverty and ghetto life; in *Half of a Yellow Sun*, Chimamanda Ngozi Adichie weaves her narrative around the events of the Biafran war in the 1960s, painting a lexical picture of the complexity surrounding identity and communal belonging in her native Nigeria.

Look to the southernmost tip of the African continent and a significantly different literary pedigree emerges. The shadow of Apartheid and the gross injustices it engendered inevitably hangs low in South African writers' minds. Nadine Gordimer, JM Coetzee, Mongane Serote or Alex La Guma: all these leading contemporary South African writers have inevitably grappled with Apartheid – explicitly, as in Serote's poem *'No Baby Must Weep'*, or in a more oblique manner, as with Coetzee's novels *Disgrace* or *Life and Times of Michael K*. An excerpt from the former:

> *'...these streets stink like apartheid*
> *these streets are traps*
> *hold my hand my mother*
> *let's look at them*
> *these the squeaking blood-stained hungry-rat battlegrounds*
> *where children pick the idea of making children here*
> *only whores know how to breathe in the dust*
> *and only the murderers live long*
> *and the cops shoot first and think after*
> *these streets of this town*
> *are dirty*
> *these streets go nowhere'*

In 'Disgrace', JM Coetzee seeks to dissect the aftermath of Apartheid, and explores (through the eyes of a disgraced (white) university professor who moves to the country to live with his daughter) how a society which for decades was centred upon hatred can heal:

'It was so personal,' she says. 'It was done with such personal hatred. That was what stunned me more than anything. The rest was…expected. But why did they hate me so? I had never set eyes on them.'
He waits for more, but there is no more, for the moment. 'It was history speaking through them' he offers at last. 'A history of wrong. Think of it that way, if it helps. It may have seemed personal, but it wasn't. It came down from the ancestors.'

When analysing the world's literary plains for correlations with the twin planets of human rights and development, certain common traits begin to emerge. Novels, like all forms of artistic expression, are generally a reflection of the societies in which they are conceived. In the west, novels have often touched upon issues of civil and political rights; African contemporary literature has been overwhelmingly concerned with the aftermaths of decolonisation and the friction between traditional ways of life and modernity; Asian literature – Indian writing in particular – often questions its own identity and place in history; Latin American literature, finally, has regularly returned to issues concerning authoritarian rule, family ties and the societal effects of war.

> ### 🔗 MORE INFO
> - Pearl Buck (1931) *The Good Earth*
> - John Le Carré (2001) *The Constant Gardener*
> - Mario Vargas Llosa (1974) *Conversation in the Cathedral*
> - Ngugi Wa Thiong'o (1977) *Petals of Blood*
> - Harriet Beecher Stowe (1852) *Uncle Tom's Cabin*

OF BRAS AND STRAPLINES – WOMEN'S RIGHTS IN LITERATURE

The panorama of women's rights-related literature can sometimes appear to be a somewhat barren one. The reputation, however, is undeserved. For although it is difficult to find fiction which deals *directly* with the rights of women, there is a plethora of literature that speaks of and to women about gender inequality. That this literature is so diverse and so rarely grouped together is simply evidence of the diversity of women's rights issues across the world.

Canadian author Margaret Atwood, for instance, concerns herself with women's reproductive rights. Her classic novel *The Handmaid's Tale* (1985) approaches female identity and women's rights in a somewhat oblique manner. In the speculative world Atwood creates, women are categorised and segregated: from the reproductive Handmaids – child-bearing concubines by another name – to the domestic servant Marthas or chaste Wives. In this dystopian society, women cannot own property, have jobs or freely voice their opinions. Doctors who are caught performing abortions are publically hanged. The *Handmaid's Tale* is Atwood's warning of what would come to pass if western society came to be governed by religious fanaticism.

Conversely, African female novelists tend to write of women eager to assert their independence outside of the social roles that straightjacket them. Flora Nwapa's *Efuru* (1966) for instance, plants the reader firmly within the world women faced in post-colonial Nigeria. A beautiful, headstrong young woman, Efuru wants it all: a career, a husband and children. Her optimism is soon shredded, however, as two failed marriages and the death of a child put paid to her aspirations. Efuru, dispirited and disillusioned, maintains a wilful independence throughout. Through *Efuru*, Flora

Nwapa challenges the widespread perception that a woman is only as valuable as her family and children.

Conversely, African female novelists tend to write of women eager to assert their independence outside of the social roles that straightjacket them.

Buchi Emecheta's novel *Second Class Citizen* (1974) rails against the discrimination African women face in getting a good education. Like *Efuru*, Adah soliders on, disregarding the put-downs and sexism she encounters. When she finally makes it to the UK, she soon realises that discrimination occurs under cloudy skies too.

In South America, writers such as Cristina Peri Rossi and Isabel Allende grapple with women's distinctly subservient role within patrilineal societies. In Peri Rossi's short story *The Fallen Angel* (1983) an angel that falls from heaven is kept company by a woman Peri Rossi describes as a *'rather intelligent woman with a great sense of independence.'* The woman casually speaks to the angel as she smokes a cigarette. Eventually though, the army appear, the woman stands up, prepares her papers, and resigns herself to her fate. The angel eventually gets up and brushes itself off. Peri Rossi writes: *'Afterwards, it wondered if anyone would miss the woman who had fallen before being violently forced into the armoured car'*. Peri Rossi's feminism is of the softer variety – defiant, stoical women who refuse to compromise on their dignity, even as the societies they live in crush them.

Women writers in the Asian world, meanwhile, often focus on the physical and cultural oppression their gender face. In the short story *'A Real Durwan'*, Jhumpa Lahiri (1999) tells the tale of an impoverished stair sweeper who, in exchange for sleeping quarters on the roof, entertains residents of the building she sweeps with her improbable stories. The moment the stairwell sink goes missing, however, the residents turn on her and accuse her of theft. The sweeper, a victim of the 'new India' in which women from lower castes are destined to lead lives of destitute poverty, finds herself jobless, homeless, and with nowhere to go.

Taslima Nasrin, a Bangladeshi writer whose feminist stances have often landed her in trouble in her native country, uses her poetry to encourage women to stand up to oppression. In *'You Go Girl!'*, for instance, she writes:

They will say— 'You have a loose character!'
When you hear that, just laugh louder…

They will say—'You are rotten!'
So just laugh, laugh even louder…

Hearing you laugh, they will shout,
'You are a whore!'

When they say that,
just put your hands on your hips,
stand firm and say,
"Yes, yes, I am a whore!"

They will be shocked.
They will stare in disbelief.
They will wait for you to say more, much more…

The men amongst them will turn red and sweat.
The women amongst them will dream to be a whore like you.

📎 MORE INFO

- Bessie Head (1969) *When Rain Clouds Gather*
- Doris Lessing (1962) *The Golden Notebook*
- Grace Ogot (1980) *The Graduate*
- Alifa Rifaat (1983) *Distant View of a Minaret and other stories*

Photo: Garry Walsh

Photo: Gareth Bentley

'IT IS PEOPLE WHO MUST RESPECT HUMAN RIGHTS AND WHO ARE, IN TURN, PROTECTED BY THEM; DEVELOPMENT POLICIES AND INITIATIVES BEGIN AS ABSTRACT FIGURES AND CONCEPTS, BUT THEY MANIFEST THEMSELVES IN THE LIVES OF PEOPLE ACROSS THE GLOBE.'

Photo: Gareth Bentley

Photo: Gareth Bentley

"NOTHING BETTER PROTECTS A HUMAN BEING AGAINST THE STUPIDITY OF PREJUDICE, RACISM, RELIGIOUS OR POLITICAL SECTARIANISM, AND EXCLUSIVIST NATIONALISM THAN THIS TRUTH THAT INVARIABLY APPEARS IN GREAT LITERATURE: THAT MEN AND WOMEN OF ALL NATIONS AND PLACES ARE ESSENTIALLY EQUAL, AND THAT ONLY INJUSTICE SOWS AMONG THEM DISCRIMINATION, FEAR, AND EXPLOITATION."

A SHORT HISTORY OF MUSIC AS PROTEST: FROM ABOLITION TO FELA TO RAGE

Nowadays, the arts' capacity to serve as a vehicle for protest and dissent is taken for granted – to the extent that the arts (music and visual arts especially) are frequently associated with anti-establishment values and norms. Yet while literature and the written word have been associated with protest and resistance for centuries, it has taken art and music somewhat longer to enter this politically-charged fold. The first true protest songs were by and about slaves. These songs often had religious overtones, and echoes of these pioneer tracks can be heard in the gospel-driven songs that are now associated with the 1960s civil rights movement – from 'We Shall Overcome' to 'Oh Freedom'. By the mid-1800s, American protest music had been so firmly established that groups such as the Hutchinson Family Singers were singing politically-charged songs such as 'Get Off the Track' at the White House to the President. A sample verse:

'Politicians gazed, astounded,
When, at first our bell resounded:
Freight trains are coming, tell these foxes,
With our votes and ballot boxes.

Roll it along! Roll it along!
Roll it along! Thro' the nation
Freedom's car, Emancipation'

It was not until the invention of Thomas Edison's phonograph towards the end of the 19th century, however, that the protest song grew in stature. The phonograph did to music what the printing press had done to literature, allowing it to break free of its medium's constraints. Songs could now be recorded and played hundreds of miles away.

The Great Depression of the 1930s brought with it misery and ruin, but also a flourishing of artistic talent: in literature (Steinbeck's masterpiece The Grapes of Wrath), art (most notably in the rise of photography as an art form, but also in the works of Thomas Hart Benton and growing popularity of wall murals) and also music. As America grew more cosmopolitan and urbanised, its music developed too. 'Strange Fruit', which Billie Holliday would subsequently turn into a worldwide hit, condemned racism and the lynching of blacks in the American South.

At the same time, many Americans, out of work and with little hope of improving their lot, identified with Marc Blitzstein's musical 'Nickel Under The Foot':

'Go stand on someone's neck while you take him
Cut into somebody's throat as you put
For every dream and scheme, depending on whether
All through the storm
You've kept it warm
That nickel under your foot'

The post-war period was to prove to be the heyday of western protest music. Music records were more widely available, and musicians drew influences from a wider pool of talent. American society continued to be a deeply divided one and songs such as Leadbelly's 'Bourgeois Blues' reflected this division:

'Well, me and my wife we were standing upstairs
We heard the white man say "I don't want no niggers up there"
Lord, in a bourgeois town
Uhm, bourgeois town
I got the bourgeois blues
Gonna spread the news all around

Nowadays, the arts' capacity to serve as a vehicle for protest and dissent is taken for granted – to the extent that the arts (music and visual arts especially) are frequently associated with anti-establishment values and norms.

Well, them white folks in Washington they know how
To call a colored man a nigger just to see him bow
Lord, it's a bourgeois town
Uhm, the bourgeois town
I got the bourgeois blues
Gonna spread the news all around'

The post-war world was, in many ways, a much broader place than the world that was. Countries became far more involved in each others' affairs, and radio and television brought international disputes and wars straight into domestic living rooms. As interest in more specific issues grew, music addressing these concerns developed alongside it. The 60s and 70s were the heyday of the protest song. Although still closely associated with the folk tradition (Bob Dylan, Pete Seeger and Phil Ochs being particular examples) musicians and singers of all genres got in on the act. The civil rights movement brought soul and funk music into the protest fold, while the rock scene adopted the American war in Vietnam as its cause.

But this period of time is significant for more than the sheer number of protest songs which began to appear as such songs stopped simply reflecting society's zeitgeist, and began to shape it. The music served as an informational tool, increasing awareness and drumming up support for civil rights/pacifism/anti-government sentiment. Folk singers such as Phil Ochs and Pete Seeger carved out well-defined niches as protest or resistance singers. Broadly left-wing, anti-war and with a disdain for the American middle-class (see Och's track *'Love me I'm a Liberal'*) they captured the mood among many American young people. It was also around this time in the 60s that protest music went truly global expanding its reach and impact.

What of other parts of the world? In Chile, Salvador Allende's assassination and Augusto Pinochet's brutal ascent to power in 1973 led to a crackdown on the *Nueva Cancion* movement which had developed alongside Allende. At the forefront of this was Victor Jara, a softly-spoken folk singer whose songs concerning peace, justice and equality were poorly looked upon by the Pinochet administration. Jara was arrested alongside 5,000 other dissenters, locked in Chile's national stadium in Santiago, and shot. His final poem, written while in captivity, was called *Somos cinco mil* [We are five thousand]. Here's the first verse, in the original Spanish and English translation.

Somos cinco mil
En esta pequeña parte de la ciudad.
Somos cinco mil
¨Cuántos seremos en total?

*Cuánta humanidad
Con hambre, frío, pánico, dolor
Seis de los nuestros se perdieron
En el espacio de las estrellas.

Un muerto, un golpeado como jamás creí
Se podría golpear a un ser humano.
Los otros cuatro quisieron
Quitarse todos los temores.

We are five thousand
Confined in this little part of town
We are five thousand
How many of us are there throughout the country?

Such a large portion of humanity
With hunger, cold, horror and pain
Six among us have already been lost
And have joined the stars in the sky.

One killed, another beaten
As I never imagined a human being
could be beaten
The other four just wanted to put an end
To their fears and pain
Six among us have already been lost
And have joined the stars in the sky.

In Jamaica, Bob Marley was making a name for himself across the globe with his politically-aware reggae. Songs such as *Babylon System* and *Blackman Redemption* spoke about black oppression; hits such as *Get Up, Stand Up* and *War* were more universal in their concern. *War* is an especially good example of the crossover between music and politics. Its lyrics are almost directly lifted from a speech given by Emperor Haile Selassie of Ethiopia to the United Nations General Assembly in 1963. An extract from the speech:

'*.... until the philosophy which holds one race superior and another inferior is finally and permanently discredited and abandoned; That until there are no longer first-class and second-class citizens of any nation; That until the colour of a man's skin is of no more significance than the colour of his eyes; That until the basic human rights are equally guaranteed to all without regard to race; That until that day, the dream of lasting peace and world citizenship and the rule of international morality will remain but a fleeting illusion, to be pursued but never attained.*'

http://www.hermosarecords.com/marley/selassiespch2.html

While Bob Marley and the Wailers were achieving superstardom, another equally fiery musician was shaking up Africa; Fela Kuti, with his flamboyance, frantic energy and classical music training blended jazz, the highlife sound that was so popular in his native Nigeria and more traditional sub-Saharan rhythms into a new form of music he called Afrobeat. What most distinguished Fela from his peers, however, was the overtly political nature of his music. He sang of economics ('*International Thief Thief*') of water shortages ('*Original Sufferhead*') of mindless violence ('*Zombie*'), Apartheid ('*Beasts of No Nation*'), and corruption ('*Army Arrangement*'). From the folk and singer-songwriter of the 60s and 70s, the protest song grew angrier, faster and louder from the 80s onwards. Not that the softer-spoken song disappeared: singers like Tracey Chapman and Suzanne Vega brought issues of domestic violence into the limelight with songs like '*Behind the Wall*' and '*Luka*' respectively, while Nena's hit, '*Ninety-Nine Red Balloons*' protested the Cold War and threat of nuclear war.

It was the rockier bands, however, which became the vehicle for protest music. The Dead Kennedys aimed their ire at the spoilt middle classes who saw the Third World simply as an exotic holiday destination ('*Holiday in Cambodia*') while REM protested the violent repression in Guatemala ('*Flowers of Guatemala*'). The Manic Street Preachers got in on the act with songs like '*A Design for Life*' (about class divisions in the UK) or '*Ifwhiteamericatoldthetruthforonedayit'struthwouldf allapart*' [sic], an indictment of the US specifically but western attitudes more generally, while U2 filled stadiums with '*Sunday Bloody Sunday*'.

Rage Against the Machine, like Phil Ochs and Pete Seeger had done in the 60s, made a career out of protest music, with songs like '*Killing In the Name Of*', '*People of the Sun*' (about the Zapatista movement in southern Mexico), '*Year of the Boomerang*' (a phrase borrowed from a speech by Frantz Fanon) or '*Bulls on Parade*' ('Weapons not food, not homes, not shoes/not need, just feed the war cannibal animal' they sing).

Rap and hip-hop also got in on the act. '*The Message*' by Grandmaster Flash sang about inner-

While Bob Marley and the Wailers were achieving superstardom, another equally fiery musician was shaking up Africa; Fela Kuti, with his flamboyance, frantic energy and classical music training blended jazz, the highlife sound that was so popular in his native Nigeria and more traditional sub-Saharan rhythms into a new form of music he called Afrobeat.

city poverty ('Got a bum education, double-digit inflation/can't take a train to the job, there's a strike at the station') while Public Enemy encouraged their fans to *'Fight the Power'*. This protest tradition is continued today by rappers such as Immortal Technique, who raps about 9/11 (*'Bin Laden'*) and the war on drugs (*'Peruvian Cocaine'*).

THE HUMAN RIGHTS EASEL: VISUAL ARTS

'Art, like morality,' wrote G.K. Chesterton, *'consists in drawing a line somewhere.'*

The line may seem a droll one, but beneath its humour lie layers of truth. Art does not operate within a moral vacuum – it influences and is influenced by all that surrounds it – and it is this moral strand that makes art so relevant to human rights and development. Unlike music or literature, art has not, as a rule of thumb, taken to wearing its political sentiments on its sleeve. It is a more oblique medium, requiring patience in order to fully grasp.

Art, then, is often ambiguous. It can be political, overtly or in more subtle ways; it often reflects its surroundings. Art is itself a form of storytelling – at times, the storytelling becomes a collective process, as with the Ottoman miniaturist tradition. More than music, literature or film, art lends itself to the nuance of contradiction. The best art is both ambiguous and complex, and it is these qualities that make it sometimes difficult to deconstruct. Its ambiguity, however, can also be its strength: Picasso's *Guernica* continues to resonate so strongly with all those who see it because it is about more than just the bombing of the little Basque village of Guernica – it could just as easily be a portrayal of Iraq, Somalia, East Timor, Vietnam, Sierra Leone or any other conflict.

Art is an inherently political medium. For centuries, overtly political art (such as Delacroix's *Liberty Leading the People*) sought to drum up nationalistic fervour and pride in the nation-state. In the post-war period, however, such nationalism became synonymous with fascism, and openly political art assumed a diametrically opposite tack. Gone was the bombast and glorification of conquest. In its place came art that was deeply suspicious of government, power and society. Pop art parodied the society it mirrored, while artists such as the Mexican José Hernández Delgadillo dedicated their careers to protesting oppressive governments – in Delgadillo's case through his explicitly political wall murals.

The anti-Apartheid movement had a strong grounding in art too, from posters to murals and political cartoons. As far back as the late 1940s black South African artists such as George Pemba and John Koenakeefe Mohl were making bold anti-apartheid statements through their art and, in the 1960s, Dumile Feni came to prominence for his angry depictions of black oppression in South African townships.

The Ghanaian sculptor El Anatsui and Benin's Romuald Hazoumé use debris and discarded materials to create their art – art which itself comments on their surroundings.

The Ghanaian sculptor El Anatsui and Benin's Romuald Hazoumé use debris and discarded materials to create their art – art which itself comments on their surroundings. El Anatsui, based in Nigeria and part of the so-called *Nsukka* group of artists, established his reputation for his use of wood and clay, but he has increasingly turned to using detritus in his work. Visa Queue, one of his most moving works, is made entirely of small shards of wood. The work represents West African migration to Europe. *'Migration, especially for economic* reasons,' El Anatsui has said, *'produces these desperate situations in which people are ready to be subjected to all forms of dehumanisation.'* El Anatsui is also renowned for his weaving of recycled bottle tops into large tapestries; the words on the bottle tops, often western brands, serve to remind Africans of their colonial history.

Romuald Hazoumé's work treads a similar path: using everyday objects in an unfamiliar way, thereby altering their meaning. In his work *La Bouche du Roi* (which focuses on the history of the slave trade in Africa) jerry cans become masks, and the masks mould together to form a slave ship. There is also a deeper meaning to be taken: jerry cans carry fuel, and to Hazoumé, Africa continues to be a commodity teat from which the rest of the world suckles – once plundered for slaves, now for resources. But Hazoumé is not an accusatory artist. One of the masks in La Bouche du Roi represents the King of Benin, who sold the slaves. As Hazoumé says:

'The gun has been brought to him by the white man, so that he can sell his brother. We cannot say that it was only the white man who was responsible for slavery. We in Africa have to take responsibility too.'

FILM: ANALYSING THE DEVELOPMENT REEL

Of the various artistic mediums discussed in this chapter, perhaps none has the direct potential to educate and inform than film. It is popular, accessible, blends a variety of aesthetic qualities (visual, aural, narrative) and extremely varied. A comprehensive breakdown of the various films that deal with development and human rights issues would take up an entire book. What follows, therefore, is a brief overview of some key films which touch upon a variety of issues. As with the sections concerning music or literature,

the films mentioned are simply a sample of those available.

Although war is an extremely popular film theme, it is not all that easy to identify war films which can be applied to an increased understanding of human rights issues. I have limited myself to two examples: *Apocalypse Now*, the Francis Ford Coppola masterpiece which is both a harrowing look at the war in Vietnam as well as a reimagining of Joseph Conrad's *Heart of Darkness* and *The Battle of Algiers*, which tells the story of the Algerian War against French colonisation and is often considered to be one of the most politically influential films of all time.

Apocalypse Now needs little introduction, infamous and renowned as it is. What distinguishes Francis Ford Coppola's masterpiece from other excellent war films is that *Apocalypse Now*, although ostensibly about the American war in Vietnam, goes beyond this. It is ultimately a film about human barbarity and the depths to which mankind plunges in order to wage war. Kurtz, played by Marlon Brando, gasps 'The horror! The horror!' (echoing Conrad's Kurtz in Heart of Darkness) before dying, and that is precisely what it all is. *The Battle of Algiers* pits the Algerian guerrilla movement during their war of independence against the French colonisers between 1954 and 1957. Neither side is glamorised: the cold brutality of civil war is simply presented for the viewer to assess. Gillo Pontecorvo purposely cast non-professional Algerians for most of the film's parts, and what emerges is one

Although war is an extremely popular film theme, it is not all that easy to identify war films which can be applied to an increased understanding of human rights issues.

of the most layered and realistic portrayals of guerrilla warfare to have ever been committed to film.

The Pianist and *Sometimes in April* both deal with genocide. They are, however, significantly different films, not least because of their differing subject matter. *The Pianist* tells the story of a Polish Jewish pianist in Warsaw at the outbreak of war in 1939. As the Nazis close in and the Holocaust grinds into motion, we see the gradual erosion of the Jewish population's rights. The desolation genocide engenders is beautifully encapsulated by the film's cinematography. Watching *The Pianist*, as with *Apocalypse Now*, gives one the sense of what the Manic Street Preachers titled an *'intense humming of evil'*. *Sometimes in April*, on the other hand, leaves behind a bitter aftertaste simply because it is hard to shrug off the sense that more could have been done to prevent the Rwandan genocide of 1994. The film illustrates the various stages of the genocide – its preparation, the gradual dehumanisation of the Tutsi, rapid escalation of violence, right through to the attempts at justice and reconciliation. A child asks Augustine, the central character, whether their ID card will carry their ethnic background, *'Yes'* replies Augustine *'but one day I hope it will just say Rwandan'*.

It might seem hard to believe that just 120 years ago, no country in the world afforded women the right to vote when nowadays, only Saudi Arabia continues to continues to resist. *Iron*

Art, in all its forms, remains a particularly powerful medium through which to undertake analysis of our obscenely unequal world today and is especially adept at suggesting other ways of imagining another world.

Jawed Angels tells the story of the American suffragette movement, from the establishment of the National Women's Party to lobbying Washington, arrest and imprisonment and finally suffrage at last. Although not to be taken literally – a number of characters are fictional – *Iron Jawed Angels* succeeds in bringing across the spirit of the suffragette movement. In similar fashion, the female factory workers in *Made in Dagenham* faced an uphill battle in their quest to achieve pay parity with their male counterparts. Ken Loach's film manages to bring across two perspectives which are often ignored: that of women, naturally, but also that of the industrial working class. Not that gender discrimination was a class-based issue. The prevailing attitude is perhaps best encapsulated in the words one of the women campaigners speaks:

'I'm Lisa Burnett, I'm 31 years old and I have a first class honours degree from one of the finest universities in the world, and my husband treats me like I'm a fool.'

Slavery, racism, the civil rights movement in the USA and Apartheid in South Africa: all have attracted their fair share of motion picture attention. The vast majority of movies dealing with these sensitive topics, however, have been mediocre at best. Naturally, there are exceptions. *Glory* tells the story of the US Civil War's first all-black volunteer company, and in doing so manages to be both a fantastic telling of the American civil war, as well as an excellent dissection of race relations in America right at the cusp of abolition. *The Great Debaters* is based on the true story of the all-black Wiley College debating team which won the national debating championship in 1935. Setting aside the various Hollywood clichés and its occasional heavy-handedness, *The Great Debaters* is an excellent film to reflect upon the struggles faced by the black minority in the American south. As a young James Farmer Jr. says, *'We do what we have to do in order to do what we want to do'*. From a historical perspective, *The Birth of a Nation*, D.W. Griffith's 1915 silent film is an eye-opener. The Ku Klux Klan are depicted as heroic patriots, with African Americans as sexual predators. The film provoked significant outcries when it was released – but still became the highest-grossing film of that decade.

Cry Freedom is one of the better Apartheid-themed films available. Through the story of Donald Woods, a (white) South African journalist forced to flee his country, we learn about the Apartheid regime, its violent and unequal nature, and Steve Biko, the founder of South Africa's Black Consciousness Movement who was brutally tortured and murdered by security police forces. As a vehicle for thinking and discussing the social structures behind Apartheid, however, it is hard to beat *District 9*. Ostensibly a science-fiction film about aliens which are interned in Johannesburg, the film is in fact an oblique reference to Apartheid-era South Africa.

Art, in all its forms, remains a particularly powerful medium through which to undertake analysis of our obscenely unequal world today and is especially adept at suggesting other ways of imagining another world. Art allows us to more easily understand issues that underpin the heart of our 80:20 world – the immense contradictions, complexities, connections, ambiguities and confusions that cloud and yet shape our understanding of that world and, more importantly, our engagement with it. In an unsettling way, art is simultaneously tentative about many of the issues we seek to address and yet is also strikingly definitive.

Art facilitates enquiry, exploration, investigation and experimentation – in matters of development and human rights it is often less prescriptive and yet compelling and unforgiving.

READING

Bertrand Borg and Colm Regan (2010) 'Seeing comes before words; Human rights, art and education', Commonwealth: Youth and Development 8:2

Margaret Busby, ed. (1992) Daughters of Africa: An International Anthology of Words and Writings by Women of African Descent from the Ancient Egyptian to the Present, New York Random House

John Johnston, Stary Mwaba and Colm Regan (2009) Thinking art: Making art in development and human rights education. Bray: 80:20 Educating and Acting for a Better World

David Lewis, Dennis Rodgers and Michael Woolcock (2008) 'The Fiction of Development: Literary Representation as a Source of Authoritative Knowledge', Journal of Development Studies, Vol. 44:2, pp 198-216.

Edward Said (1993), Culture and Imperialism, London: Chatto and Windus

UNESCO (2006) Road Map for Arts Education, The World Conference on Arts Education: Building Creative Capacities for the 21st Century, Lisbon

MORE INFORMATION & DEBATE

http://www.pen-international.org - London based international organisation to celebrate literature and promote freedom of speech

http://www.bwpi.manchester.ac.uk/research/fictionofdevelopment/welcome.html - site maintained by Brooks World Poverty Institute in Manchester with extensive list of literature

http://www.elanatsui.com - view the work of artist El Anatsui

http://www.tannerlectures.utah.edu/lectures/atoz.html - online library of Tanner lectures on a vast range of topics including art, literature and human values

http://www.amnesty.org.uk/content.asp?CategoryID=10836 – Amnesty International UK site exploring human rights issues through film

Photo: Gareth Bentley

[*'In Bolivia, we consider water to be a common good – a human right, not a commodity. It is central to life and all that it embraces. It is collective property, yet in another sense it belongs to no one.'*]

Oscar Olivera, Secretary of the Bolivian Federation of Factory Workers

Chapter 14

'PARADIGM WARS'
RESOURCES, CONFLICT AND DEVELOPMENT

Colm Regan

This chapter explores the debates surrounding the link between natural resources and development through three case studies, those of water, minerals and oil. The debates are placed in the framework of what has been described as 'paradigm wars' – conflicting ways of viewing the world and the role of resources within development. The chapter focuses on Bolivia, India, the Democratic Republic of the Congo (DRC) and China to highlight some of the issues. Two key issues emerge – are resources a blessing or a curse in development and is a resource, such as water, a 'human right' or an 'economic commodity'

KEYWORDS

Water wars, **privatisation**, water footprinting, **virtual water**, India, **slum populations and water**, conflict minerals, **the DRC**, oil and oil as a liability, **the 'resource curse'**, China in Africa

A DECADE OF 'WATER RIOTS'

2000	COCHABAMBA IN BOLIVIA
2001	IRAN, PAKISTAN, ARGENTINA
2002	PERU, BOLIVIA
2003	MAPUTO IN MOZAMBIQUE, CEBU IN THE PHILIPPINES
2004	SOUTH AFRICA, PERU
2005	IRAQ, INDIA, PAKISTAN
2006	ETHIOPIA AND KENYA
2007	INDIA, BURKINA FASO, GHANA AND COTE D'IVOIRE
2008	NIGERIA, ZIMBABWE
2009	SOMALIA, SOUTH AFRICA
2010	PHILIPPINES

INTRODUCTION

In the New York Times in 1995, World Bank vice-president Ismail Serageldin argued that '...*many of the wars of this century were about oil, but wars of the next century will be about water*'. Similarly, in 2000, UN Secretary General Kofi Annan stated that '... *fierce competition for freshwater may well become a source of conflict and war in the future*'.

The assumption frequently associated with these assertions is that such conflict is most likely to occur between states, especially those which experience advanced water stress. However, there are dissenting voices, most notably that of Indian environmental activist Vandana Shiva who argues:

'*When water wars are referred to, people usually imagine militaristic attacks between countries, but the water wars that are spreading around the world are, at one level, paradigm wars. They are about two ways of looking at the world.*

'*In one view, water is nature's gift, and we need to maintain its flow as a gift. Even now, if you come to India on a hot day, you will see people put out water in street corners. It's called the gift of water or the temple of water. Anyone who's thirsty can go there and drink. Instead of accumulating wealth, these people are accumulating the good act of giving and meeting other people's needs for basic survival... The other view has it that water can be appropriated and sold to make huge profits or wasted.*'

Oscar Olivera, Secretary of the Bolivian Federation of Factory Workers, agrees:

'*In Bolivia, we consider water to be a common good – a human right, not a commodity. It is central to life and all that it embraces. It is collective property, yet in another sense it belongs to no one. These ideas, which have their roots in indigenous people's thinking, are what mobilised working people, both in the countryside and in the cities. The struggle to take control of our water supplies in 2000 became known as the 'water war' of Cochabamba.'*

Oscar and thousands of other Bolivians took to the streets to protest at the privatisation of water supplies in the city of Cochabamba (and later elsewhere throughout Bolivia) in what has become the most well known of such water wars internationally.

For decades, the World Bank and International Monetary Fund have not only supported but actively promoted privatisation of water services through clauses in trade agreements and through conditions for loans to developing countries. The strategy was to force countries to provide key services such as water through private rather than state companies on the assumption that they would be more efficient at providing such services to the benefit of all citizens and for economic growth. One such country was Bolivia; under World Bank guidance and direction (and faced with the threat of withholding debt relief), the water systems of some of Bolivia's poorest regions were put up for sale to private interests and in the area of Cochabamba, a US-owned company, Bechtel, was awarded a long-term contract to manage and deliver water in that region.

Bechtel promised to extend the availability of water access to previously deprived areas and, to achieve this took over local wells, water pumps, and the public system infrastructure that was already in place and added to the existing infrastructure to reach new communities. The costs for these improvements and additions were passed on to customers, doubling the cost many people had been paying; in many cases the monthly water bill reached US$20 in a city where the minimum wage is less than $US100 a month. Many residents were simply unable to pay and even though water was now available, they couldn't access it because they couldn't afford it.

In response to these price increases, an alliance of the citizens of Cochabamba called La Coordinadora de Defensa del Agua y de la Vida (The Coalition in Defence of Water and Life) was formed in January 2000. Through mass mobilisation, the alliance shut down the city for four days and within a month millions of Bolivians marched to Cochabamba and held a general strike, stopping all transportation. The protesters then issued the Cochabamba Declaration, which called for the protection of universal water rights for all citizens. In 2000, riots broke out in Cochabamba as protestors (mainly students, sweatshop employees, and street vendors) became increasingly upset and angry, hundreds filled the streets in protest and the violence impacted considerably on the local government. Other cities followed and, in response, the Bolivian government promised to reverse the price hike.

In this sense, water wars are global wars, with diverse cultures and ecosystems, sharing the universal ethic of water as an ecological necessity, pitted against a corporate culture of privatisation, greed, and enclosures of the water commons.

This did not occur and in February 2000, La Coordinadora organised a march demanding the repeal of the legislation permitting the privatisation and the ending of Bechtel's contract. Slogans such as *'Water Is God's Gift and Not A Merchandise'* and *'Water Is Life'* were frequently used. Their demands were strongly rejected by the government which in the following April declared martial law to try to end the protests. Many were arrested, protesters were killed and the media was censored. The protests continued and in late April Bechtel left Bolivia and the government was forced to repeal the legislation. Bechtel filed a lawsuit against Bolivia in a World Bank Trade Court, demanding US$50 million in compensation; the protests continued for four years and in 2006, Bechtel settled for a token 30 cent payment although the case had cost the Bolivian Government over US$1 million in legal fees.

According to Vandana Shiva:

'Paradigm wars over water are taking place in every society, East and West, North and South. In this sense, water wars are global wars, with diverse cultures and ecosystems, sharing the universal ethic of water as an ecological necessity, pitted against a corporate culture of privatisation, greed, and enclosures of the water commons. On the one side of these ecological contests and paradigm wars are millions of species and billions of people seeking enough water for sustenance. On the other side are a handful of global corporations... assisted by global institutions like the World Bank, the World Trade Organisation (WTO), the International Monetary Fund (IMF), and G-7 governments.'

Source: Water Wars: Privatisation, Pollution and Profit, 2002

This chapter examines three case studies of different levels of conflict around resources and resource use in the context of development – water, minerals and oil and relates these to arguments about *'paradigm wars'*, democracy and human security.

WATER WORLDWIDE – 'POWER, POVERTY AND INEQUALITY'

In the introduction to the 2006 Human Development Report (Beyond scarcity: power, poverty and the global water crisis), the authors argue that:

'The scarcity at the heart of the global water crisis is rooted in power, poverty and inequality, not in physical availability'

Worldwide, over 1.1 billion people do not have access to even the most basic and safe water supply while more than 2.6 billion lack access to a basic

toilet and an estimated 10,000 people die each day as a result of diseases linked to lack of clean water and sanitation. People are left with no choice but to consume water obtained from unsafe sources, such as unprotected wells, ditches, rivers or lakes; they are forced to use buckets, plastic bags, fields and public places for sanitation purposes. Household water requirements represent but a tiny fraction of demand (usually less than 5%) but, as with so many other dimensions of basic needs, there is massive inequality in access to both water and sanitation. For the UNDP:

'There is more than enough water in the world for domestic purposes, for agriculture and for industry. The problem is that some people – notably the poor – are systematically excluded from access by their poverty, by their limited legal rights or by public policies that limit access to the infrastructures that provide water for life and for livelihoods. In short, scarcity is manufactured through political processes and institutions that disadvantage the poor.'

On Thursday, 30 September, 2010, the UN Human Rights Council adopted a resolution affirming that water and sanitation are human rights; the resolution adopted by the Council took an important step in asserting:

'...the human right to safe drinking water and sanitation is derived from the right to an adequate standard of living and inextricably related to the right to the highest attainable standard of physical and mental health, as well as the right to life and human dignity.'

Catarina de Albuquerque, the UN Independent Expert on human rights obligations in relation to access to safe drinking water and sanitation commented that:

'This means that for the UN, the right to water and sanitation is contained in existing human rights treaties and is therefore legally binding... The right to water and sanitation is a human right, equal to all

other human rights, which implies that it is justiciable and enforceable.'

Currently, 178 countries worldwide have now recognised the right to water and sanitation at least once in an international resolution or declaration and, as a result, governments can no longer deny their legal obligations to ensure the rights to water and sanitation. While recognising the right to water, 10 countries did not recognise the right to sanitation (in international or regional declarations) - Albania, Austria, Belize, the Czech Republic, Malta, Sweden, Trinidad and Tobago, Turkey, Turkmenistan and the UK.

Figure 14:1 - *Water prices: the poor pay more, the rich pay less*

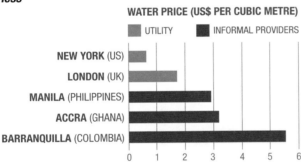

Source: 2006 UNDP Human Development Report

WATER FOOTPRINTS WORLDWIDE

There are about 326 million trillion gallons of water in the world; 97.5% of it is salt water, leaving a mere 8.15 million trillion gallons (2.5%) as fresh water, 70% of which is frozen in the icecaps of Antarctica and Greenland leaving less than 1% available for direct human use.

The water footprint concept has been developed in order to have an indicator of water use in relation to consumption by people. The water footprint of a country is defined as the volume of water needed for the production of the goods and services consumed by the inhabitants of the country. Closely linked to the water footprint concept is the virtual water concept. Virtual water is defined as the volume of

water required to produce a commodity or service. International trade of commodities implies flows of virtual water over large distances. The water footprint of a nation can be assessed by taking the use of domestic water resources, subtract the virtual water flow that leaves the country and add the virtual water flow that enters the country.

The internal water footprint of a nation is the volume of water used from domestic water resources to produce the goods and services consumed by the inhabitants of the country. The external water footprint of a country is the volume of water used in other countries to produce goods and services imported and consumed by the inhabitants of the country. The water footprint of a product (good or service) is the volume of fresh water used to produce the product, summed over the various steps of the production chain. *'Water use'* is measured in terms of water volumes consumed (evaporated) and/or polluted. The water footprint is a geographically explicit indicator, not only showing volumes of water use and pollution, but also the locations and timing of water use.

Differences in global water footprints are significant – the USA has an average footprint of 2,480m³ per capita per year whereas China's is currently 700 (but growing rapidly). There are four key factors explaining this pattern:

- the total volume of consumption related to the gross national income of a country (e.g. USA, Italy and Switzerland)
- high water intensive consumption patterns, especially relating to the consumption of meat (e.g. USA, Canada, France, Spain, Portugal, Italy and Greece)
- climate, especially related to high levels of evaporation (e.g. Senegal, Mali, Sudan, Chad, Nigeria and Syria)
- inefficiencies in agricultural practices (e.g. Thailand, Cambodia, Turkmenistan, Sudan, Mali and Nigeria)

Source: Water Footprint Network – http://www.waterfootprint.org

WATER &
WATER FOOTPRINTS

" Human activities consume and pollute a lot of water. At a global scale, most of the water use occurs in agricultural production, but there are also substantial water volumes consumed and polluted in the industrial and domestic sectors. "

World Water Assessment Programme, 2009

THE GLOBAL WATER GAP

▼ **AVERAGE WATER USE**
PER PERSON PER DAY, 1998-2002 (LITRES)

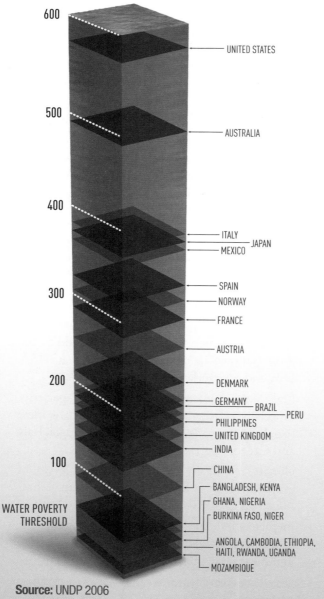

Source: UNDP 2006

Less than 3% of the world's water is fresh and drinkable – the rest is seawater.

Of this 3% over 2.5% is frozen – in Antarctica, the Arctic and glaciers and is not available for use.

We therefore rely on 0.5% for all of our own and the ecosystem's fresh water needs.

▼ **WATER USE BY REGION AND SECTOR (%)**

■ Municipal
■ Industrial
■ Agricultural

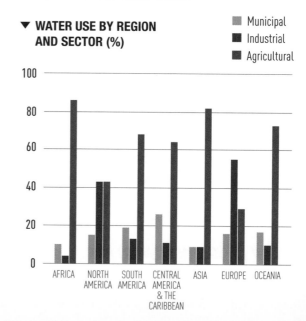

▼ **WATER IMPORT DEPENDENCY (%)**

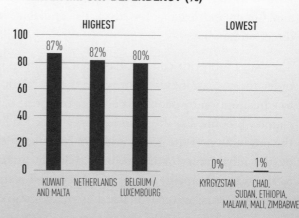

WATER FOOTPRINTS

The total water footprint of a product breaks down into three components: the **blue**, **green** and **gray** water footprint.

👣 BLUE WATER SURFACE WATER AND GROUND WATER

The volume of freshwater that evaporated from the global blue water resources (surface water and ground water) to produce the goods and services consumed by the individual or community.

👣 GREEN WATER RAINWATER STORED IN THE SOIL AS SOIL MOISTURE

The volume of water evaporated from the global green water resources (rainwater stored in the soil as soil moisture).

👣 GRAY WATER POLLUTED WATER

The volume of polluted water associated with the production of all goods and services for the individual or community; it is calculated as the volume of water that is required to dilute pollutants to such an extent that the quality of the water remains above agreed water quality standards.

TOP 5 WATER FOOTPRINTS
PER CAPITA IN M3 (2004)

Australia	3,483
Brazil	3,455
Cyprus	3,219
Belgium/Luxemburg	3,145
Bolivia	3,096

BOTTOM 5

Togo	619
Eritrea	684
Yemen	722
Haiti	738
Ethiopia	740

OTHERS

South Africa	1,384
India	1,062
United States	2,895
United Kingdom	1,695
Ireland	2,948

SOURCE: Ashok Chapagain and Stuart Orr (2004) UK Water Footprint: the impact of the UK's food and fibre consumption on global water resources: Volume Two: appendices, World Wildlife Fund, London

WATER FOOTPRINT BY PRODUCT

T-SHIRT 2,700 litres for one cotton T-shirt - 11,000 litres are needed to make 1 kg of final cotton textile – 45% irrigation water, 41% id evaporated water from the cotton field and 14% to dilute wastewater that result from fertiliser and chemicals.

HAMBURGER 2400 litres for one 150 grams of beef hamburger, most of the water is needed to produce the beef contained in the hamburger.

CHOCOLATE 2400 litres for one 100 gram bar (world average) = combined water footprint of cocoa paste, cocoa butter and milk powder (higher for milk chocolate than for dark chocolate).

LEATHER 16,600 litres for 1 kg of leather, the footprint depends on the type animal used.

RICE 3400 litres for 1 kg of rice (world average) - rice fields consume about 1350 billion cubic meters annually (21 % of global water use for crops).

Source: Water Footprint Network http://www.waterfootprint.org

'...DELHI: DRINKING THE CITY DRY'

'You should come here early in the morning to see the long queues and the fights. Some women bring sticks to fight for the water ... for us poor, each year seems harder than ever. Nowadays we can only wash once every four days for lack of water'

Delhi slum dweller Santi Singh

Indian government maps are colour coded to show the intensity of slum populations in certain cities. The intensity of slum concentration is represented by yellow turning to orange and then red for the highest density. For Bombay, New Delhi and Kolkata, most of these municipal maps are dominated by bright red and orange.

In the 2001 census *'... the definition (of a slum) used was a 'compact area of at least 300 people or about 60 to 70 households of poorly built, congested tenements in unhygienic environments, usually with inadequate infrastructure and lacking proper sanitary and drinking water facilities...'*

One in six people on earth is Indian and it is expected that by 2020, about 50% of India's population will live in cities with estimates over 50% of them living in slums (in Mumbai, the 2001 figure was already 54%).

WATER WORLDWIDE: 10 SELECTED FACTS

1. The international norm for a family of 5 people is 100 litres per day
2. Average daily water use ranges from 200 – 300 litres in most European countries to 575 litres (and up to 1,000 litres in parts) in the US to less than 10 litres per day in countries such as Mozambique, Rwanda and Haiti
3. For the over 1.1 billion who live more than one kilometre from a water source, water use is often less than 5 litres per day (the basic requirement for a lactating mother engaged in even moderate work is 7.5 litres)
4. There are also huge internal differences within countries – in rural Uganda, average consumption ranges from 12 to 14 litres and in arid areas, affected by the dry season, consumption can fall to 5 litres per day (in, for example, western India, the countries of the Sahel etc.)
5. Most of those without clean water live in south Asia while the lowest rates of coverage exist in sub-Saharan Africa; 40 developing countries provide clean water for fewer than 70% of citizens and 54 provide safe sanitation for fewer than half
6. It is agreed that figures generally underestimate the scale of the problem; the poor are routinely undercounted as they live in 'illegal' areas often not recognised by government. For example, while Mumbai officially has a safe water rate of 90%, whole areas with significant population are not counted – one slum Dharavi, home to almost 1 million people is estimated to have one toilet for 1,440 inhabitants
7. Pollution of water sources poses a massive challenge as the separation of water and excrement remains unrealised. For example, in China, 16 cities with populations of more than half a million have no wastewater treatment facilities and in Karachi in Pakistan (with a population of 10 million plus), 40% of water supplies are unfiltered and 60% of effluent remains untreated and, as a result, water-borne diseases are common. The poisoning of water with minerals also poses a major threat
8. One estimate suggests that 40 billion hours are spent each year in sub-Saharan Africa in fetching water, the vast bulk of it undertaken by women, further re-enforcing gender inequalities
9. Realising the right to universal basic water and sanitation would have measurable economic impact in terms of reducing poverty, disadvantage and overall health costs, improving girls' education, freeing up time expended by girls and women in collecting water etc.
10. Realising the right to water and sanitation would also have very significant health benefits. 1.8 million children die annually of diarrhoea linked directly to unclean water and unsafe sanitation.

Sources: UNDP 2006, World Health Organisation 2011, www.water.org and the Blue Planet Network

While official figures suggest a slum population of 1.8 million in Delhi, unofficial estimates place the number living in 'authorised' and unauthorised slums at between 3 and 7 million. Only those living in authorised slums qualify for government services, including water and sanitation.

For slum dwellers such as Santo Singh (a 40 year old mother and long-term resident of Kalianpuri with 10,000 huts on the south-eastern edge of Delhi), the water story is bleak '... *When we fight we use our hands and legs and the police have to come and split us up. It's all about the water...*'

Her story is testament to India's dwindling water resources as well as the government's struggle to meet the fast-growing needs of the urban poor while meeting the expectations of the equally fast-growing middle class.

As Kalianpuri grew, people began installing their own hand pumps in the 1980s, most of them 12m to 15m in depth and by the 1990s, all houses had access to pumps but by 2000, most began to dry up and as Santi observes, '...*no one wants to build new pumps because the water table is still falling and the water is not clean, and putting in wells is too expensive*'. Now, most houses depend on tap-less water pipes which flow twice a day, for two hours at a time.

Santi's stories of endless fights at the standpipes are echoed throughout India as women struggle to get their buckets under the pipes before the weak flow trickles to nothing. A 2005 World Bank study warned of increasing water scarcity due to groundwater depletion, environmental degradation and climate change and argued that unless dramatic changes were introduced by 2025, 3 out of 5 aquifers in India will be critically low and that by 2050 demand for water will exceed all available supply.

The government claims that 90% of Indians enjoy safe drinking water but the World Bank puts the figure at about 40%, leaving some 650 million people without clean water.

'*The cruel irony of being poor in Indian cities is that without a municipal water supply, the poor spend disproportionately more on water. Commercial suppliers make a killing as they freely extract what they want from private wells and sell it to those not served, or served poorly, by the city.*'

The unequal distribution of water in Delhi can be very extreme, even in planned residential areas – in the central area (dominated by the government and the army), the average daily available quantity is 500 litres per capita – 18 times that for northern and southern areas at 30 litres – a household of 5 people is deemed to need at least 120 litres for basic usage. Delhi faces a water shortage of 750 million litres per day and is a city with a perpetual water crisis.

There are alternatives as illustrated by the slum communities of Mongolpuri and Sultanpuri where local residents, especially women, joined together and with the support of a local NGO, Saahas-ee, agitated for basic services with considerable success with the result that now every 24 houses have 3 hand pumps (dug 12 to 15 metres deep) and each individual house has the right to apply for a supply for a nominal fee. The project is managed by local women's groups.

'*The story of water in India is not about government negligence towards the poor, nor the urban poor being singled out for marginalisation. In fact, many praise India's efforts to provide potable water to its fast-growing population, and it may be one of the few countries to come close to meeting the MDGs in this sector. Instead, the story is more universal and entering a critical moment, a collision of supply and demand.*'

Extracted from: tomorrow's crises today: the humanitarian impact of urbanisation OCHA/Irin and UN Habitat, 2007

IS THERE BLOOD ON YOUR MOBILE PHONE?' - DEBATING 'CONFLICT MINERALS'

In the period since the ending of the Second World War in 1945, an estimated 51 million people have been killed in wars with almost a third of these deaths occurring in China. Along with China, Vietnam, Sudan and the Democratic Republic of Congo (DRC) have suffered the highest number of war-related deaths. In the case of the DRC, the country has been devastated for over a century by regional conflict fuelled, in part, by a vicious and deadly scramble for its significant reserves of natural resources. For many commentators, the effort to control those resources has been a principal motivation behind ongoing conflict and its associated atrocities including the use of widespread rape as a weapon of war (estimates vary between 15,000 and 17,500 rapes in 2009 alone). In eastern Congo today, these mineral resources help finance a large number of local armed groups and contribute directly to fuelling the ongoing conflict.

The role of these *conflict minerals* has become the focus of intense debate, political lobbying and popular campaigns aimed at people in Europe and the United States.

Wealth generated from the minerals trade runs into hundreds of millions of dollars each year through four key minerals – the ores that produce tin, tantalum, tungsten, and gold. Many of these mineral resources end up in electronic devices such as cell phones, portable music players, and computers. Anti conflict mineral activists argue that due to lack of transparency around the supply of minerals consumers have no way of ensuring that their products are conflict free with all that this implies.

'Is there blood on your mobile phone?' was the question posed by non-governmental organisation DanChurchAid in 2006, a question repeated many times since. Private companies in Europe and the US and beyond have been accused by campaigners of ignoring agreed principles and international standards of good practice as laid down, for example, in the Guidelines for Multinational Enterprises from the Organisation for Economic Co-operation and Development, which outline how companies should operate and argue that western governments have failed to investigate their behaviour. Neighbouring governments - Uganda and Rwanda – have been accused of supporting Congolese rebel groups and of benefitting from the wealth of the DRC's resources and of undermining attempts at building a peace.

Campaigners scored a major victory when in July 2010 President Obama signed into law a financial reform bill which included a clause requiring U.S. companies importing products containing certain minerals to complete an annual report declaring whether they source their minerals from Congo or from one of nine surrounding countries. Should a company indicate that its supply chain passes through the region, it will be required to explain what steps it is taking to ensure that its trade is not contributing to the conflict or even actually funding it. While the law does not ban companies from importing the minerals it does require them to publish relevant information on their websites in order to allow consumers choice.

Obviously the campaign around conflict minerals and the law signed by President Obama has been the source of vigorous debate and disagreement with many of those involved in the minerals and telecommunication sectors disagreeing fundamentally with the arguments of the activists. Below is a summary of the main arguments put forward on both sides.

Arguments in support of the minerals trade:

- The issues that actually drive the violence in the DRC are land rights, citizenship disputes, the failure of the state, ethnic hostility and violence and the 'story' created by the conflict minerals campaigners makes far too much of the contribution of minerals to the conflict. Mineral campaigners thus highlight one issue allowing international attention to ignore those issues that actually fuel the conflict and the violence.

- If legislation such as that passed in the US became effective, it could significantly reduce formal trade and companies might decide to pull out of the region leaving thousands of Congolese jobless and poorer and this could increase membership of the armed factions. This view is expressed inside and outside the DRC.

- Many leaders in the region argue that the mineral trade is not the cause of violence and that ending the trade is very unlikely to end most of the violence especially in circumstances where there is an absence of effective and real political and security structures. Banning the conflict minerals trade is the wrong way to try to end violence. The argument that armed militias will stop raping, looting, and burning down villages if one of their sources of revenue is cut defies logic and evidence.

- One of the key minerals highlighted in the campaign is Coltan (an African slang word for ore that contains tantalum, a metal prized for electronics use because of its resistance to corrosion and heat) which is used in mobile phones, MP3 players, gaming consoles and aircraft engines (a typical Nokia handset has a tantalum 'capacitor' that temporarily stores electrical charges). The DRC is not a major source of tantalum as most of it comes from Australia, Canada and African countries such as Ethiopia and Mozambique. For example, the US Geological Survey groups the Congo under 'other' sources of tantalum that together account for only 2% of world production. Recycled tantalum is also available and even tantalum from the Congo isn't necessarily 'tainted' – domestic and foreign companies also mine it legally. Because of earlier debates and controversies, some companies no longer obtain tantalum from the Congo.

- Even when they try, it is nearly impossible for companies to say with absolute certainty that no tantalum of 'dubious origin' makes it into the supply chain. Black market operators sell the ore and sell it cheaper because they can avoid customs and clearance costs etc. and while most developed countries have controls, some Chinese ports wave shipments through and once the ore has been refined to non-radioactive tantalum powder, it becomes impossible to trace. The story of conflict minerals demonstrates in reality how difficult it is for companies to be socially responsible.

- African, and especially Congolese voices, have not been heard or listened to because the conflict minerals campaigners use lobbyists and media personalities to promote their cause at the expense of locals. The campaign did not develop in response to a Congolese need or request and is essentially driven by 'outsiders'.

'The consequence...will be that thousands of Congolese will be jobless and might most probably (be) joining the armed groups...'

John Kanyoni, Head of the Association of Mineral Exporters in Congo's Eastern North Kivu province, 22.07.10

Arguments against the minerals trade:

- The DRC is, in the present situation, cursed with some of the world's largest diamond reserves, rich gold fields, as well as huge reserves of cobalt and tantalum and all the warring parties (local and international) have exploited these reserves to finance their military operations and to buy weapons, often committing serious human rights abuses in the process. In this way, the minerals trade contributes directly to the violence and to the ideologies that sustain it. The mining of Tantalum delivers huge wealth to warring sides, causes farmers to be pushed off their land and contributes significantly to environmental degradation.

- The campaign against conflict minerals has brought the issue out into the open; has put pressure on companies to be transparent; has encouraged consumers to ask important questions about the ethics of consumption; has highlighted the plight of people in the region (something those supporting or justifying the trade have never done) and has also highlighted the many other dimensions of the conflict and the need to act pro-actively in support of peace.

- Arguing that companies involved in the minerals trade should develop and exercise 'due diligence' in their activities is, in and of itself, a good and necessary thing for local people and for consumers who do not want their purchases produced in abusive contexts or circumstances. If the campaign succeeds in having legislation introduced that produces greater accountability in the business sector in the eastern Congo, this is a good thing in its own right. Even if the volume of minerals contained in any item is tiny, this is not an argument in support of continuing unjust practices or policies. Promoting meaningful change in the minerals trade is something that should be done anyway.

- By focusing on two key issues – those of conflict minerals and sexual violence, campaigners have 'humanised' the issues and made them relevant and have succeeded in forcing legislators and the media to explore the issues in a way that did not happen before. Campaigners recognise the complexity of the issues in the region and do not believe that the conflict is simply fuelled by economics and greed but these two issues are effective as 'entry points' in the much-needed debate.

- Effective control of the trade is not as complex as the companies would argue – there are only six key stages between the mine and the mobile and each is open to effective regulation if seriously addressed – the mine itself, the trading houses where the ore is sold, the exporting companies, the transit countries, the refining companies and the electronic companies themselves. Tracing, auditing and certifying mineral inputs is both feasible and effective.

- Without sustained campaigning pressure from civil society groups, leadership on the issue is unlikely to come from companies involved or, if it is, it is likely to be 'lowest common denominator' leadership which denies the importance, gravity and urgency of the issue to be addressed. While leadership does come from some companies, others will always need to be pressurised to act and to act responsibly.

'Eastern Congo's hell is an instance of how globalisation generates ungovernable spaces. Where there is a collision of desperate poverty, plentiful guns and a world greedy for natural resources, a brutal chaos results. To combat that, it takes a very tenacious sort of global campaigning – bringing to attention each element of the system and the part it can play in leveraging change – and mercifully, that is what is now finally starting to happen.'

Madeleine Bunting, the Guardian, December 12th, 2010

✎ MORE INFO

- http://www.enoughproject.org
- http://www.globalwitness.org
- http://www.conflictminerals.org
- http://wrongingrights.blogspot.com/2010/07/pointcounterpoint-conflict-minerals-law.html
- http://texasinafrica.blogspot.com/2009/12/show-me-data.html

OIL, RESOURCES, DEMOCRACY AND DEVELOPMENT

'Oil is the one strategic commodity of the world that governments, from superpowers to minor states, will never allow to be free of political control'

Journalist Youssef Ibrahim, January 2004

'Oil vividly illustrates the tendency for resources wealth to support corruption and conflict rather than growth and development'

Worldwatch Institute 2005

Ever since the beginning of the *'industrial age'* worldwide but especially in Europe and the United States, oil has been a strategic economic, political, social and cultural imperative. It has, literally, fuelled development; it has been the foundation of massive wealth and power; it has generated widespread conflict and war and has placed key regions around the world at the centre of international politics – usually at the expense of local populations. Over the past half-century, oil has consistently occupied centre-stage in many of the major crises worldwide from Suez in the 1950s, to the oil price crises of the 1970s, the Iranian revolution of 1979, on to the Gulf crises of 1990 and 2003 and more recently.

In the Worldwatch Institute's State of the World Report 2005, authors Prugh, Flavin and Sawin argue that oil has been transformed from being an asset to becoming a liability and outline the risks and costs of using oil at three levels:

- Oil threatens global economic security as it is a finite resource (without a clear successor to date) and because the gap between supply and demand appears to continue to grow, it makes the world vulnerable to recurring economic shocks

- The value of oil as a commodity undermines civil security by compromising efforts to achieve peace, civil order, human rights and democracy

- Oil threatens climate stability because its continuing use accounts for a major element of global greenhouse gas emissions.

They conclude '...*where oil once helped ensure human security, it now makes us more vulnerable.*'

Figure 14:2 - *Risk from natural resources*

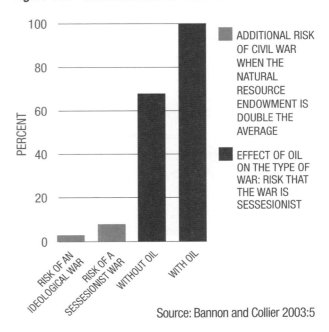

ADDITIONAL RISK OF CIVIL WAR WHEN THE NATURAL RESOURCE ENDOWMENT IS DOUBLE THE AVERAGE

EFFECT OF OIL ON THE TYPE OF WAR: RISK THAT THE WAR IS SESSESIONIST

Source: Bannon and Collier 2003:5

A key moment in the history of oil and its role in international relations was the setting up of the Organisation of Petroleum Exporting Countries (OPEC) in 1960 which transformed the manner in which oil business was conducted. A decade later, OPEC succeeded in forcing oil companies to negotiate directly with the organisation rather than with individual countries; oil reserves had been significantly nationalised in most producer countries and the prices paid to producing countries increased significantly (prior to this date, Gulf producer countries were paid US$1 per barrel for oil that was sold by oil companies for US$12-$14 per barrel).

The net result was a significant loss of western control over oil reserves and the beginning of the era when oil producing countries rather than oil consuming countries came to the fore; a pattern that has remained significantly unchanged to this day. In order to achieve one of its key objectives internationally – ensuring effective access to affordable oil – the United States became extensively involved politically, diplomatically and militarily in the Gulf and the Middle East. In order to improve its energy security, the US has also embroiled itself in the Caucasus region and in Central Asia – regions increasingly important in the geo-politics of oil, especially in the context of the rise of China and its hugely increased demand for oil.

As part of the broader debate on the *'resource curse'*, political scientist Michael Ross, in an important 2001 article, reviewed the relationship between resource wealth, democracy and development and whether resource wealth *'in itself'* might harm or help a country's prospects for development. On the basis of the available evidence, Ross argued that *'oil does hurt democracy'* and that oil does more damage to democracy in poorer states than in richer ones even when exports are relatively small. He argued that the harmful influence of oil is not restricted to the Middle East but has made democratisation harder in states such as Indonesia, Malaysia, Mexico, Nigeria and possibly the oil-rich states of Central Asia. Importantly, Ross also argued that *'nonfuel mineral wealth also impedes democratisation'* noting that while major oil exporters are concentrated in the Middle East, many mineral exporting countries are located in Africa, Asia, and the Americas and include many states where democracy remains weak or elusive e.g. in Angola, Chile, the Democratic Republic of Congo, Cambodia, and Peru.

Ross identified three causal links between oil and authoritarianism:

- *'the rentier effect'* (where governments use low tax rates and high spending to dampen pressures for democracy)
- *'a repression effect'* (where governments build up their internal security forces to ward off democratic pressures)
- *'a modernisation effect'* (where jobs in the industrial and service sectors remain weak and this undermines the drive to democracy).

For Ross, the links between mineral wealth in general and authoritarianism are less immediately clear but nonetheless important. Research has indicated that resource wealth tends to reduce economic growth and increases the likelihood of civil war and Ross adds a third component to the *'resource curse'* – that of authoritarian rule and that these three effects create a *'resource trap'* with significant consequences not just for democracy but also for development. Authoritarian governments may be less able to resolve domestic conflicts and hence more likely to suffer from civil war while slow growth can make domestic unrest difficult to resolve while civil wars *'wreak economic havoc'*. Importantly, Ross notes that there is nothing inevitable about the resource curse as states such as Malaysia, Chile, and Botswana have done relatively well despite their oil and mineral wealth.

CHINA AND AFRICA'S NATURAL RESOURCES

The bulk of the world's oil reserves are located in the politically volatile Middle East with Europe and Eurasia (especially Russia) a distant second; close behind comes Africa with estimated petroleum reserves of over 120 billion barrels. Today, some 20% of US oil imports come from Africa with Nigeria being the 5th largest exporter to the US.

Africa has always been strategically important to western powers in terms of its mineral wealth

MEASURING OIL'S PIVOTAL PLACE IN THE WORLD ECONOMY

- 10 of the world's top 20 companies in 2010 were either oil and gas companies or automobile companies

- 3 of the top 10 US companies by revenue in 2010 were oil companies and 2 were automobile companies

- Most oil is used in transportation and the world's automobile fleet grew from 53 million in 1950 to 539 million in 2003 to 850 million in 2007

- Annual automobile production grew from 8 million in 1950 to over 41 million in 2003 to an estimated 52 million in 2009

- The estimated growth in the sales of cars in China was 24% in 2008 and China became the world's 2nd largest car market in 2010

- China is now the world's second largest importer of oil and consumption continues to rise by an average of 13% per year

- While air travel accounts for a smaller amount of oil consumption, it is increasing dramatically from 28 billion passenger- kilometres in 1950 to 2,942 billion in 2002 to 4.28 trillion in 2008

Table 14:1 - World's Top 10 Oil Producers 2008
(millions of barrels per day)

Saudi Arabia	10.8
Russia	9.8
United States	8.5
Iran	4.1
China	4.0
Canada	3.3
Mexico	3.2
United Arab Emirates	3.0
Kuwait	2.7
Venezuela	2.6

Table 14:2 - World's Top Oil Consumers 2008
(millions of barrels per day)

>10	United States (19.5)
7-10	China (7.8)
4-7	Japan (4.8)
2-4	India (3.8), Russia (2.9),Germany (2.8), Brazil (2.5), Saudi Arabia (2.4), Canada (2.3), South Korea (2.2), Mexico (2.1)
1-2	France (2.0), Iran (1.7), UK (1.7), Italy (1.6), Indonesia (1.3)
<1	199 countries

Source: EIA (2010) US Energy Administration

and this remains the case today especially in the context of the rapid rise of China as a world power. Essential metals such as aluminium, bauxite, coltan, alumina, copper, iron ore, lead, nickel, zinc, and minerals such as phosphate rock, coal, and uranium are all present in Africa in large quantities. Of particular importance are the strategic minerals of chromium, cobalt, platinum group metals, and manganese as they are critical in various weapons systems and economic activity, especially as there is

no known substitute. Reserves of these minerals are concentrated geographically in South Africa, the Democratic Republic of Congo (DRC), Zimbabwe, and Zambia. Examples include:

- 33% world reserves of chromium are in South Africa and, along with Kazakhstan, accounts for 95% of world chromium reserves

- Zambia and the DRC have between them 52% of world cobalt reserves

South Africa has 77% of world manganese reserves and 88% of reserves of the platinum group metals.

Crucially, such strategic minerals are not present in sufficient quantities in the US or China to meet anticipated demand in the years immediately ahead. This geographic concentration, the lack of alternative supplies or known substitutes has contributed significantly to making Africa strategically important to China, especially in recent years.

China has been very proactive in negotiating mineral concessions in Africa; in 2008 China agreed a long-term infrastructure development programme with the DRC worth over US$9 billion in return for ending previous mining concessions and awarding new agreements to a Chinese company. South African reserves are also critical for China along with those of Angola, Equatorial Guinea and Sudan. Other countries with 'mineral-focused' agreements with China include Gabon, Egypt, Morocco, Ethiopia and Zambia. Minerals are but one key element in China's strategic interest in Africa which has witnessed a massive increase in both trade and investment from the 1990s onwards – in 2010, China was expected to replace the US and France as Africa's biggest trading partner.

China's interest in African markets and minerals has been the subject of intense debate and criticism from those concerned with human rights and labour issues (e.g. China's effective support for dictatorial and/or corrupt regimes, its policy of political 'non-interference' and significant African deaths linked to 'health and safety') or with the environmental consequences of minerals-focused *development*' (for example, concerns over the links between pollution from mines and health problems such as cholera in Zambia or the displacement of tens of thousands of local people for dam development in Sudan). Defenders of China's role in Africa argue that the debate is essentially about whether Africa's resources go East or West and that given the history of western exploitation of Africa's resources (e.g. oil in Nigeria is a prime example), criticism of China is disingenuous at best. They also point out that Chinese investment in Africa goes much further than resource extraction, building up trade and markets in both directions.

Extensive Chinese investment and engagement in Africa is now a reality, one that is attracting considerable comment and debate in the West and concern about the future of strategic minerals and potential tension and conflict between East and West.

READING

Ashok Chapagain and Stuart Orr (2004) UK Water Footprint: the impact of the UK's food and fibre consumption on global water resources: Volume Two: appendices, World Wildlife Fund, London

Ian Bannon and Paul Collier (eds., 2003) Natural Resources and Violent Conflict, Washington, World Bank (see chapters 1 and 2)

Vandana Shiva (2002) Water Wars: Privatisation, Pollution and Profit, New York, South End Press

Michael Ross (2001) 'Does Oil Hinder Democracy?' World Politics 53:3, 325-361

UNDP (2006) Human Development Report: beyond scarcity: power, poverty and the global water crisis, New York, UNDP

Worldwatch Institute (2005) State of the World Report: Redefining Global Security, Washington

Toby Shelley (2005) Oil: Politics, Poverty and the Planet, London, Zed Books

Maude Barlow and Tony Clarke (2005) Blue Gold: the fight to stop the corporate theft of the world's water, New York, New Press

MORE INFORMATION & DEBATE

http://www.worldwatch.org - Worldwatch Institute in Washington, information, analysis, briefing papers and blogs

http://www.enoughproject.org - US based site of the Enough Project; campaigns on conflict minerals and other issues

http://www.globalwitness.org - campaigns organisation on natural resource-related conflict and corruption and associated environmental and human rights abuses

http://www.who.int/water_sanitation_health/database/en/index.html - World Health Organisation database on water and sanitation issues including annual reports etc.

http://www.vandanashiva.org - site exploring the views of Vandana Shiva, earth democracy and Navdanya International

http://www.worldwatercouncil.org - World Water Council site; views of governments, business and the water sector

Photo: Gareth Bentley

[*'Education is the most powerful weapon which you can use to change the world.'*]

Nelson Mandela

Chapter 15

'... UNDERSTANDING AND INVOKING RIGHTS...'
EDUCATION AND DEVELOPMENT

Antonella (Toni) Pyke

Education is recognised as one of the most powerful drivers to individual and social development and for achieving much of the Millennium Development Goals agenda. It is promoted as a tool for reducing poverty and improving health, gender equality, peace and stability. It is a basic human right that is entrenched in international law and is seen as foundational for the enjoyment of many other rights.

This chapter argues that education directed towards women and the girl-child reaps benefits for the individual, the society and for development agendas at local, national and international levels. Focusing on the challenge of achieving Education for All by 2015, it argues that MDG 2 is crucial to realisation of the other seven goals and in challenging inequality and poverty. The chapter argues that despite the progress made in achieving the targets set for primary school enrolment there remain unacceptably high levels of children out of school and issues of poor quality education impacting negatively on both literacy and numeracy.

KEYWORDS

Education, human development, MDGs, girl-child, right to education, human rights, Democratic Republic of Congo, Kenya, Afghanistan

INTRODUCTION

In 2000 the majority of the world's political leaders agreed an agenda of eight ambitious promises to the poor of the world – the Millennium Development Goals (or MDGs) – and set a target of 2015 to realise this agenda. Sadly, it is now predicted that the majority of these goals will not be achieved, once again undermining the credibility of promises from the 'international community'. MDG 2, which aims to 'achieve universal primary education,' is seen as crucial to the achievement of many other MDGs: improving child and maternal health; reducing poverty; nutrition; tackling HIV and AIDS and other diseases, etc. It is commonly agreed that education - a good quality education – is a key element for overall development and, most importantly, for human development. It is pivotal in supporting the economic, social and political growth that is required to challenge poverty. It is highlighted as providing women and girls with knowledge and empowerment to help reduce maternal and child mortality, challenge domestic and gender-based violence and help prevent the transmission of diseases such as HIV and AIDS.

At an additional meeting in Senegal in 2000, governments agreed the Dakar Framework for Action that set out six broad goals and targets for achieving Education for All by 2015. Its overall aim was to ensure that all children begin school at an appropriate age and complete a full cycle of primary schooling. Important progress has been made towards achieving these targets, with the number of children out of school reducing, gender gaps appearing to narrow in a number of cases, and more children moving from primary to secondary education and beyond.

However, if we compare the promises made at Dakar with the reality of delivery on the ground, the indications are that the education gap is widening and, according to UNESCO, 'there could be more children out of school in 2015 than there are today' - if current trends continue, it is unlikely that this basic goal will be reached, as currently only 87 out of every 100 children in the developing world complete their primary education (UNESCO 2011).

EDUCATION FOR ALL – WHERE ARE WE?

'The world is not on track to achieve the Education for All targets set for 2015' begins the most recent UNESCO Education For All Global Monitoring Report (2011). With only four years left to reach the MDG targets, it is evident that at the current pace, the world will fail to meet these targets 'by a wide margin,' leaving over 72 million children still not attending primary school and 796 million people lacking basic literacy skills - two-thirds of whom are women. Additional to the slow pace of progress in net enrolment is the issue of the poor quality of much of the education being offered that results in many children leaving school lacking sufficient literacy and numeracy skills.

THE PROGRESS SO FAR...

- There are over 30 million more children in school today than at the beginning of the decade

- The total number of children out of school fell from 106 million to 67 million between 1999 and 2008

- Universal primary enrolment in sub-Saharan Africa has experienced an 18% rise in the past 10 years

- Between 1999 and 2008 an additional 52 million children enrolled in primary school while the number of children out of school was halved in South and West Asia

- In sub-Saharan Africa, enrolment rates rose by one third despite a large increase in the primary school-age population

- Gender parity in primary enrolment has improved significantly in the regions that started the decade with the greatest gender gaps

- The most notable progress has been in some of the poorest countries: Burundi, Rwanda, Samoa, Sao Tome and Principe, Togo and the United Republic of Tanzania which have achieved or are nearing the goal of universal primary education

- Notable improvements in key areas such as net primary enrolment rates, budgetary allocations to education and girls' education have occurred in a wide variety of countries, such as Ethiopia, Mali, Mozambique, Niger, Afghanistan, Vietnam, Yemen and India

- Crucially, two-thirds of all developing countries reached gender parity in primary schools as of 2005

- More broadly, early childhood welfare is improving, with mortality among children under-5 falling from 12.5 million in 1990 to 8.8 million in 2008

(based on UNESCO Education For All Global Monitoring Report 2011; UN MDG Report 2011; UNICEF, 2011 State of the World's Children Report, Oxfam UK MDG Briefing)

BUT...

- In 2009, more than 20% of primary-age children in least developed countries were excluded from education

- In at least 29 countries, the net enrolment rate (the proportion of children of primary school age who are enrolled in primary school) was below 80% with limited improvements since 1999 and in some cases (20 countries) it actually decreased

- The pupil/teacher ratio in many countries is dramatically in excess of the recommended 40:1 ratio, with severe teacher shortages throughout the world

- Almost half of the children out of school (29 million) live in sub-Saharan Africa and a further quarter are in South and West Asia

- One in four adults in the developing world – 872 million people – is illiterate

- 46% of girls in the world's poorest countries have no access to primary education

- Hunger is holding back progress. In developing countries, 195 million children under-5 years (1 in 3) experience malnutrition leading to irreparable damage to their cognitive development and their long-term educational prospects

- The number of children out of school is falling too slowly; in 2008, 67 million children were out of school – 36 million were girls; progress towards universal enrolment has slowed and if current trends continue, there could be more children out of school in 2015 than there are today

- Wider economic and social inequalities are restricting opportunity for schooling, e.g. in Pakistan, almost half of children aged 7 to 16 from the poorest households are out of school, compared with just 5% from the richest households

- The quality of education remains very low in many countries with millions leaving primary school with poor reading, writing and numeracy skills

- Another 1.9 million teachers will be needed by 2015 to achieve universal primary education, more than half of them in sub-Saharan Africa.

A UNIVERSAL HUMAN RIGHT FOR ALL

'The right to education is a fundamental human right and is essential and indispensable to the understanding of all other human rights.'

Irish Government White Paper on Irish Aid 2006

Education is widely understood to be the primary driver in empowering people to claim and enjoy their rights. For Brazilian educationalist Paulo Freire, illiteracy is one of the most tangible expressions of social injustice. He devoted his life to promoting education and to emphasising and developing *'consciousness'* amongst the poor to empower them to claim their most basic human rights.

A defining moment occurred on the 10th December 1948. In response to the atrocities and human rights abuses of World War II, the Universal Declaration of Human Rights (UNDHR) was developed and adopted by the United Nations. The Declaration, which took two years to develop, defines the rights and freedoms afforded to every individual by virtue of their humanity (see Grech 2007)

The UNDHR contains 30 articles in total with Article 26 referring to the right to education, stating: *'Everyone* has the right to education' and that it should be 'free' and 'compulsory':

- *Everyone has the right to education. Education shall be free, at least in the elementary and fundamental stages. Elementary education shall be compulsory. Technical and professional education shall be made generally available and higher education shall be equally accessible to all on the basis of merit.*
- *Education shall be directed to the full development of the human personality and to the strengthening of respect for human rights and fundamental freedoms. It shall promote understanding, tolerance and friendship among all nations, racial or religious groups, and shall further the activities of the United Nations for the maintenance of peace.*
- *Parents have a prior right to choose the kind of education that shall be given to their children.*

The right to education has been enshrined in various international conventions, including: the International Covenant on Economic, Social and Cultural Rights (CESCR,1966); the International Convention on the Elimination of All Forms of Racial Discrimination (CERD, 1969); the Convention on the Elimination of All Forms of Discrimination Against Women (CEDAW, 1979); the Convention on the Rights of the Child (CRC 1989) and the Convention on the Rights of Persons with Disabilities (CRPD, 2008). It is included in various regional treaties, at the African, Inter-American and European levels, and many countries have made provisions for the right to education in their national laws and constitutions.

For example, the Committee for the International Covenant on Economic, Social and Cultural Rights in its General Comment 13 states that:

'...education is the primary vehicle by which economically and socially marginalised adults and children can lift themselves out of poverty and obtain the means to participate fully in their communities...' regardless of age, gender (Art.13(1), (2)(d)(24)) or race.

Importantly, education and literacy are considered a fundamental ingredient for the full enjoyment of other human rights, for active participation in society, for dignity and freedom.

Importantly, education and literacy are considered a fundamental ingredient for the full enjoyment of other human rights, for active participation in society, for dignity and freedom.

EDUCATION FOR DEVELOPMENT

'Basic education is not just an arrangement for training to develop skills (important as that is), it is also a recognition of the nature of the world, with its diversity and richness, and an appreciation of the importance of freedom and reasoning as well as friendship.'

Amartya Sen 2003

There is a clear link between education and development. Fundamentally, education is about the most basic investment in development and directly and measurably benefits the individual, society, and the world as a whole. According to the World Bank, a good quality education is central to development as it acts as a *'powerful equaliser'* in empowering individuals to *'lift themselves out of poverty'* and to challenge inequality. It is necessary for the creation of opportunities for sustainable and viable economic growth, for reducing mother and child mortality, improving health generally and in fighting diseases such as HIV and AIDS, for promoting gender equality, encouraging accountability, good governance and stability, helping in the fight against corruption, mitigating climate change and reducing human insecurity. It means investing in a country's human resources and in its future. The specific links between education and development can be realised at the individual level, throughout society and in ultimately benefiting development at many levels:

The Individual

- *Health and nutrition* – most notably with reference to women, education has demonstrated benefits on reproductive health, in improving maternal and child mortality and welfare through improved nutrition, supporting higher immunisation rates, impacting on the challenges of epidemics, such as HIV and AIDS

- *Productivity and earnings* – each year of schooling is estimated to increase individual wages for both men and women by a worldwide average of about 10% (this can be higher in poor countries)

- *Reduced inequality* - primary education plays a crucial role in poverty reduction among those seen to be the poorest in society: girls, ethnic minorities, orphans, disabled people, and rural families. Education enables individuals to know and understand their rights and empowers them to claim those rights.

Society

- *Improved Income* - an individual's potential income can increase as much as 10% with each additional year of schooling

- *Increased productivity* - Just four years of primary schooling is said to increase a farmer's productivity by nearly 9%

- *Increased GDP* - annual GDP increases by 1% with each year of additional schooling

- *Increased growth* - an increase in one standard deviation in student scores on international assessments of literacy and mathematics is associated with a 2% increase in annual GDP per capita growth

- *Reduced poverty* - if all students in low income countries left school with basic reading skills, 171 million people could be lifted out of poverty, reducing poverty by 12%.

Development

- *Economic competitiveness* – education promotes a knowledge-based economy, technical innovation and economic growth through improved productivity. This is particularly relevant in a rapidly changing global environment.

- *Poverty reduction* - education is crucial to the realisation of poverty reduction agendas such as the Millennium Development Goals

- *Democracy* - countries with higher primary

schooling rates and smaller gender gaps in education are said to experience greater democracy. Democratic political institutions (such as power-sharing and fair elections) are more likely to exist in countries with higher literacy rates and education levels. A literate population is empowered to actively participate in their country's political process (Sen 2003)

- **Peace and stability** – education is *'central to peace in the world'* (Sen 2003). It positively impacts on issues of human security, equity, justice, and intercultural understanding which are crucial to peace-building throughout the world. At the local level, education has been linked to crime reduction - poor school environments are linked to deficient academic performance, absenteeism, and school drop-out which are viewed as precursors of delinquent and violent behaviour. The 2011 UNESCO EFA Global Monitoring Report focuses on the issue of armed conflict and education, stating that more than 40% of the world's out of school children and the highest rates of illiteracy are in countries experiencing armed conflict

- **Environmental sustainability** - education about environmental issues can encourage the sustainable management of natural resources, advance disaster prevention and support the implementation of improved environmentally friendly technologies.

Adapted from World Bank 2011
The World Bank and Education: Facts

THE WAY FORWARD - INVESTING IN THE EDUCATION OF GIRLS FOR DEVELOPMENT

'There is no justification – be it cultural, economic or social – for denying girls and women an education. It is a basic right and an absolute condition for reaching all the internationally agreed development goals. It is through education that girls and women can gain the freedom to make choices, to shape their future and to build more inclusive and just societies.'

Irina Bokova, Director-General of UNESCO

The fundamental role of education in promoting development, in particular human development and well-being at the individual, societal and global levels has been clearly demonstrated. Also, universally established is the pivotal role of the promotion of girls' education in supporting development at all levels.

Some of the reported benefits of prioritising girls' education and achieving gender parity (MDG5) include:

- **Reduced fertility rates** - women who have experienced a few years of formal education are reported to be much more likely to use reliable family planning methods, delay marriage and childbearing, and have fewer and healthier babies as compared to those with no formal education. Just one year of female schooling is estimated to reduce fertility by 10%

- **Reduced infant and child mortality rates** - women who are exposed to some formal education are more likely to seek medical care, ensure that their children are immunised, be more knowledgeable regarding their children's nutritional needs, and adopt improved sanitation practices

- **Reduced maternal mortality rates** - women with formal education tend to have better

Continued on page 274 →

SUMMARY - ADVANTAGES OF EDUCATING THE 'GIRL-CHILD':

- A 10% increase in girls' secondary enrolment in low income countries will save approximately 350,000 children's lives and reduce maternal mortality by 15,000 every year

- Girls' education is often the single most powerful factor affecting health outcomes: infant and maternal mortality, seeking safer birth options, averting risky teenage births, likelihood of having smaller and healthier families,

- HIV and AIDS infection rates drop by 50% among children who complete their primary education

- If all children received a primary education, 7 million HIV cases worldwide could be prevented each year

- An estimated 1.8 million children's lives in sub-Saharan Africa could be saved this year if their mothers had at least a secondary education

- Every additional year of schooling reduces the number of children a woman will have by 10%

- Investing in girls education could boost sub-Saharan Africa's agricultural output by 25%

- Girls' income potential increases by up to 15% with each additional year of primary education

- Increasing the number of women with secondary education by 1% can increase annual per capita economic growth by 0.3%

- There has been some progress in the goal of achieving gender parity in education (MDG 5):

- The total number of out-of-school girls in South Asia has dropped from 23 million girls to 9.5 million since 1999

- In Sub-Saharan Africa, the number of out-of-school girls has decreased from 25 million in 1999 to 17 million in 2008

- More than six million girls are now enrolled in secondary school in Bangladesh, up from 1.1 million in 1991

- Globally, the gender gap has been gradually shrinking since 1999

- In Latin America and East Asia, girls have reached parity with boys, however more boys complete primary school than girls in all other regions

Yet:

- Although the gap in gender parity has decreased substantially, there are still many more girls out of primary school than boys. In 1999, around 106 million children were out of primary school. Almost 61 million (58%) were girls compared to 45 million (42%) boys. In 2009, around 35 million girls were still out of school compared to 31 million boys.

Source: World Bank, 2011 The State of Girls' Education (see http://go.worldbank.org)

NUMBER OF OUT-OF-SCHOOL CHILDREN BY REGION AND SEX, 1990-2009

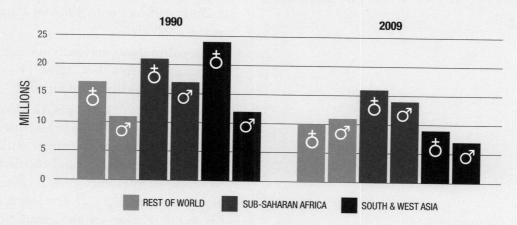

Source: UIS database, 2011 - http://www.uis.unesco.org/Education/Pages/out-of-school-children-data-release.aspx

knowledge about health-care practices, are less likely to become pregnant at a very young age, have fewer, better-spaced pregnancies and seek pre- and post-natal care. It is estimated that an additional year of schooling for 1,000 women can help prevent two maternal deaths

- *Increased protection against HIV and AIDS* reduces a woman's, and potentially an unborn baby's, vulnerability to HIV infection. It can contribute to a reduction in the spread of HIV and AIDS through greater information about the disease and awareness of prevention methods.

- *Increased participation of women in the labour force* - supports female economic independence, increased income and productivity, increased access to credit

- *Intergenerational education benefits* - an educated mother impacts on her children's educational attainment and opportunities. Even with just a few years schooling, she is more likely to send her children to school. In many countries each additional year of formal education completed by a mother translates into her children remaining in school for an additional one-third to one-half year.

However, despite this reality, two-thirds of all children who remain out of school throughout the world are girls.

BARRIERS IN ACCESS TO EDUCATION

Despite the progress being made towards achieving education for all, there remain many reasons why children still do not attend school. Although education is cited as *'free'* basic education, some schools continue to charge high school fees, there are additional *'hidden'* fees, there are safety concerns due to conflict or natural disasters, the prioritising of boys education over girls, etc. Some key factors why children remain out of school include:

- **Poverty** – the demand for children's labour, especially the girl-child, to support in the home, work in the fields, caring for the sick and providing childcare. Despite being *'free'*, there are other costs associated with going to school such as the need to buy school uniforms, PTA (parent/teacher) fees, books and other school materials, transport to school, long distances to schools, etc.

- **Poor health and nutrition** – linked to poverty, children may be too sick or hungry to attend school or to stay in school. They may find it difficult to retain information as a result of hunger or experience poor performance, they may be impacted by HIV and AIDS

- **Poor school infrastructure** – lack of proper facilities such as dilapidated or inadequate school buildings, poor sanitation (which directly impacts on girls attending schools), inadequate teaching staff, lack of facilities for children with special needs etc.

- **Socio-cultural reasons** – early marriage, teenage pregnancy, parents low value on education (especially for girls), fear of rape etc.

- **Conflict** – in 2011 UNESCO identified 48 armed conflicts in 35 countries between 1999 and 2008 which directly impact on education systems, children, teachers and schools. Spending on education in conflict zones is being diverted to military spending – 21 developing countries are spending more on arms than on primary schools, a cut of 10% in military spending would put 9.5million more children in school (UNESCO 2011).

- **Poor quality** - despite the noted achievements, the quality of education remains poor all too often with inadequately trained and supported teachers, high drop-out rates and poor infrastructure

- **Pupil-teacher ratios** globally have declined from 30 in 2000 to 24 in 2008. However, these appear to be on the increase again in some regions

- **Lack of relevant curricula, inappropriate language of instruction and poor quality learning materials** are also impacting negatively on the quality of education.

TABLE 15:1 - 10 COUNTRIES WITH THE HIGHEST PRIMARY PUPIL TEACHER RATIOS (2008-2010)

1	Central African Republic	84.3
2	Rwanda	68.3
3	Congo, Rep.	64.4
4	Chad	60.9
5	Zambia	60.5
6	Mozambique	58.5
7	Ethiopia	57.9
8	Tanzania	53.7
9	Burundi	51.4
10	Mali	50.1

Source: UNESCO Institute for Statistics in EdStats, Aug 2011
Note: Data was not available for 47 countries.

IMPACT OF THE GLOBAL FINANCIAL 'CRISIS'

Since 2002, aid to support basic education has almost doubled. However, the pledges made at Dakar and subsequent international summits indicate that agreed targets have not been met by donors. Pledges made in 2005 by the Group of Eight and the European Union show a shortfall of USD$20bn by 2010. The global financial crisis is predicted to substantially impact on the achievement of the goal of Education for All with governments scrambling to find ways of cutting national budgets, including aid budgets. While impacting directly on developed world economies, the financial 'crisis' has also forced millions more of the world's poorest deeper into poverty. With slower global economic growth and increasing food prices, a further 64 million people have been plunged into extreme poverty and 41 million more malnourished in 2009 when compared to pre-crisis levels. According to the Education for All Global Monitoring Report there is evidence to show that the additional stress on poor households is resulting in children being unable to go to school, or being too hungry or undernourished to stay in school.

The global financial crisis is predicted to substantially impact on the achievement of the goal of Education for All with governments scrambling to find ways of cutting national budgets, including aid budgets.

The EFA Global Monitoring report surveyed 18 low-income countries and 10 middle-income countries to assess changes in education sector spending in 2009 and compared this with planned spending for 2010. The results showed that:

- In seven low-income countries, spending was cut in 2009, leaving 3.7 million children out of school

- planned spending in five of the seven low-income countries showed that levels in education sector spending would revert to 2008 levels

- seven lower-middle income countries maintained or increased their spending in 2009, but planned to cut spending in 2010

- Referring to the target date of 2015 for achieving the education goal, it seems that projected spending increases will average 6% for low income countries, falling dramatically short of the required 12% to meet the goal of universal primary education.

CASE STUDIES:

DRC: MILLIONS MISS OUT ON BASIC EDUCATION

Access to basic education in the Democratic Republic of Congo (DRC) remains poor, with up to seven million children across the vast country out of school - despite a 2010 government decision to make primary education free. The seven million figure was contained in the preliminary findings of a study conducted by the DRC government with the UK Department for International Development and the UN Children's Fund, UNICEF. The study argued that 25% of primary school-aged children and 60% of adolescents were not enrolled in classes.

'Even with the announcement of free primary education, parents, many of whom are unemployed and have little means of sustaining themselves, are bearing most of the costs involved in educating their children because of delays in releasing the funds for free education.'

Representatives of teachers' unions and officials of NGOs dealing with education issues argue that the quality of education offered in public schools remains low because teachers are poorly paid. Almost 74% of primary teachers are qualified but only 33% at secondary level.

Education officials have expressed concern over the severe shortage of teachers in public schools. In primary school, the national average is one teacher for 37 pupils, according to the national statistics, but in marginalised or rural areas, there can be more than 100 pupils per class. Generally, teachers, like other Congolese workers, survive on very little, some on even less than $1 a day, yet the cost of education is borne by parents, sometimes even up to 65% of the total cost. In rural areas, some teachers supplement their earnings by working as casual labourers on farms while those in urban areas end up begging for money from their pupils' parents just to survive.

A 2007 survey by UNESCO and UNICEF suggested teachers' conditions contributed to the poor quality of tuition and found that up to 43% of sixth-grade pupils lacked basic knowledge of French, mathematics and general knowledge.

Source: http://www.irinnews.org/report.aspx?reportID=94196

THE SANITARY TOWELS CAMPAIGN PROGRAMME - KENYA

During menstruation, girls often do not go to school as a result of inappropriate or inadequate sanitation facilities; menstruation for girls without access to sanitary pads and underpants is a major contributor to girls dropping out of school. Kenyan primary and secondary schools have at least 1.5 million menstruating girls, at least three-fifths, or 872,000 of whom miss four to five days of school per month due to a lack of funds to purchase sanitary pads and underwear combined with inadequate sanitary facilities at their schools. These 3.5 million lost learning days each month impede their ability to compete in the classroom, and leads to low self-esteem, higher drop-out rates resulting in lower future wages and diminished economic productivity, and makes them vulnerable to early marriage, pregnancy, and HIV.

As a result, the Girl Child Network (GCN) launched a campaign on the issue of sanitary pads with the aim of improving the *'participation, retention and performance of girls in school'* through the provision of sanitary pads and improved sanitation facilities. Since then, the Ministry of Finance in Kenya has revoked the 16% VAT on sanitary pads, Proctor and Gamble distributed free sanitary pads and the campaign succeeded in de-stigmatising the issues surrounding sexual and reproductive health issues. The campaign also prompted the Ministry of

Education (MoE) to formulate a gender policy in education to address the issues raised and address factors that prevent girls retention in school.

http://www.zanaa.org/managing-menstruation/policy-advocacy/national-committee/
http://www.aidlink.ie/index.php/kenya/girl-child-network
http://www.commonwealtheducationfund.org

A SCHOOL BUS FOR SHAMSIA - AFGHANISTAN

Even before the men with acid came, the Mirwais Mena School for Girls was surrounded by enemies. It stood on the outskirts of Kandahar, barely 20 miles from the hometown of Mullah Muhammad Omar, the Taliban's founder. The area around the school is in the Taliban heartland and teaching girls to read was not something that would escape their notice. Across much of Afghanistan the Taliban have made the destruction of schools, particularly schools for girls, a hallmark of their war.

The attackers appeared in the morning on November 12th, 2008 as the girls were walking to school. The men came on three motorcycles, each one carrying a driver and a man on back. They wore masks. Each of the men riding on the back carried a small container filled with battery acid. The masked men circled for several minutes as the girls streamed to school. Shamsia Husseini and her sister, Atifa, were walking along the highway when they spotted the men on the motorbikes. Shamsia, then 17, was old enough to be married; she was wearing a black scarf that covered most of her face. Shamsia had seen Taliban gunmen before and figured the men on the motorcycles would pass. Then one of the bikes pulled alongside her, and the man on back jumped off. Through the mask, he asked Shamsia what seemed like a strange question *Are you going to school?'*

The masked man pulled the scarf away from Shamsia's face and, with his other hand, pumped the trigger on his spray gun. Shamsia felt as if her face and eyes were on fire. As she screamed, the masked man reached for Atifa, who was already running. He pulled at her and tore her scarf away and pumped the spray into her back. The men sped off toward another group of girls. Shamsia lay in the street holding her burning face.

In 2001, only a million Afghan children were enrolled in school, all of them boys. The education of girls was banned. Today, approximately seven million Afghan children attend school, of which 2.6 million, or roughly a third, are girls.

'My parents told me to keep coming to school even if I am killed', Shamsia said.
She exhibited a perfect grasp of the situation, both hers and her country's: 'The people who did this to me don't want women to be educated. They want us to be stupid things'.

By early summer 2008, at least 478 Afghan schools had been destroyed, damaged or threatened out of existence, the overwhelming majority of them for girls, according to the Afghan Education Ministry. The means employed to terrorise girls were inventive. In May, 61 teachers and pupils in Parwan Province, most of them girls, seemed to have been poisoned by a cloud of gas let loose into a school courtyard. It was the third suspected gas attack on a school this year. In Kandahar, Nasaji Nakhi High School and Miyan Abdul Hakim High School were set afire.

Source: Dexter Filkins, New York Times, August 17, 2009

ECOWAS COURT INSISTS EDUCATION IS A LEGAL RIGHT

In December 2010 The ECOWAS Community Court of Justice (the principal legal structure of ECOWAS - the Economic Community of West African States, a regional grouping of 15 member states including Nigeria) affirmed that Nigeria was legally accountable for ensuring the right to education. The Court also confirmed that the right to education should not be undermined by corruption and held that Nigeria's Universal Basic Education Commission has the responsibility to ensure that funds disbursed for basic education are properly used for this purpose.

While the Court stated that proof of high level corruption had not been presented to it, it noted that there was 'prima facie' evidence of embezzlement of funds on the basis of the reports of Nigeria's Independent Corrupt Practices Commission (ICPC). The Court stated that while steps were being taken to recover funds and/or prosecute the suspects, the Nigerian government was required to provide the funds necessary to cover the shortfall in order to avoid denying any of its people the right to education.

The Court stated that it would hold ECOWAS States accountable if they denied the right to education for their people. It reaffirmed an earlier ruling in October 2009 which rejected the government's claim that the right to education was not enforceable and which stated that the Court has jurisdiction to hear cases involving allegations of breaches of the African Charter on Human and Peoples' Rights (see Chapter 4). The judgment represented an important restatement of the principles of universality and indivisibility of all human rights established in both the African Charter on Human and Peoples' Rights and the Universal Declaration of Human Rights.

In its judgement, the Court ruled that the right to education could be enforced before the Court and the objections brought by the Federal Government of Nigeria which argued through the Basic Education Commission that education was '...a mere directive policy of the government and not a legal entitlement of the citizens.'

The Court also commented on broader legal and human rights issues which have significant implications for the education of women and girls:

'Public international law in general is in favour of promoting human rights and limiting the impediments against such a promotion, lends credence to the view that in public interest litigation, the plaintiff need not show that he has suffered any personal injury or has a special interest that needs to be protected to have standing. Plaintiff must establish that there is a public right which is worthy of protection which has been allegedly breached and that the matter in question is justiciable. This is a healthy development in the promotion of human rights and this court must lend its weight to it, in order to satisfy the aspirations of citizens of the sub-region in their quest for a pervasive human rights regime.'

The case (ECW/CCJ/APP/0808) was brought by the Socio-Economic Rights and Accountability Project (SERAP), supported by Amnesty International against the Federal Government of Nigeria and the Universal Basic Education Commission (UBEC). Commenting on the judgement, SERAP solicitor Femi Falana noted:

'This is the first time an international court has recognized citizens' legal right to education, and sends a clear message to ECOWAS member states, including Nigeria and indeed all African governments, that the denial of this human right to millions of African citizens will not be tolerated."

Sources: http://www.right-to-education.org and Amnesty International

READING

Amartya Sen (2003) 'The Importance of Basic Education', speech to the Commonwealth education conference, Edinburgh. Oct. 28, 2003

Action Aid (Sept 2005) Contradicting Commitments How the Achievement of Education for All is Being Undermined by the International Monetary Fund." http://www.actionaid.org.uk/doc_lib/132_1_contradicting_commitments.pdf.   Also, http://www.actionaid.org.uk/index.asp?section_id=11

Omar Grech (2007) A Human Rights Perspective on Development, Bray, 80:20 Educating and Acting for a Better World
Oxfam UK – Education Now Campaign: http://www.oxfam.org.uk/oxfam_in_action/issues/education.html;

UNESCO Education for All Global Monitoring Report 2011, UNESCO, Paris

UN Millennium Project (2005) Taking Action: Achieving Gender Equality and Empowering Women. Task Force on Education and Gender Equality, UN, New York.

United Nations Millennium Development Goals Report (2011) UN Millennium Development Goals Report 2011, New York

UNICEF, 2011 State of the World's Children Report, Oxfam UK MDG Briefing)

MORE INFORMATION & DEBATE

http://www.un.org/millenniumgoals/11_MDG%20Report_EN.pdf

http://www.unesco.org/new/en/education/themes/leading-the-international-agenda/efareport/reports/2011-conflict/

http://www.oxfam.org.uk/get_involved/campaign/index.html

http://www.unicef.org/sowc

[*'...relentless acceptance of the status quo is not acceptable in the face of the challenges we confront...'*]

Chapter 16

DEBATING DEVELOPMENT AND CHANGE

Colm Regan

This chapter outlines key debates about the nature and scale of change required for greater equity in human development. It initially briefly presents the core arguments of five divergent views of change ranging from those who argue for limited, incremental change to those who argue for radical and fundamental structural change. The chapter then explores four case studies of how change has come about through different types of actions by a broad range of actors internationally as well as nationally in South Africa and India. Finally, the chapter introduces some twenty organisations and campaigns focused on development and human rights issues and themes addressed in earlier chapters.

KEYWORDS

Views of change, **debating change**, ethical consumption, **baby milk campaign**, 1997 Mine Ban Treaty, **Treatment Action Campaign**, Self Employed Women's Association, **campaigning for change**, South Africa, **India**

INTRODUCTION

F rench novelist, philosopher and feminist, Simone de Beauvoir, expended a considerable amount of her time and writing in exploring the nature of individual responsibility in the world as she understood and experienced it, especially in the context of France, the Nazi period and the Algerian War (1954-62). In a 1948 work, *The Ethics of Ambiguity*, she explores a central difficult ethical human question – when faced with the scale and nature of good and evil in the world, what is an individual to do? How should we define our existence in such a context and what values, demands and contradictions does it present us with? She concludes:

'Regardless of the staggering dimensions of the world about us, the density of our ignorance, the risks of catastrophes to come, and our individual weakness within the immense collectivity, the fact remains that we are absolutely free today if we choose to will our existence in its finiteness...'

Simone de Beauvoir (1976 ed.) The Ethics of Ambiguity, New York, Citadel Press, 159)

For very many contemporary philosophers, activists and commentators, this question remains not simply relevant but urgent given the staggering dimensions of today's world. *80:20 Development in an Unequal World* is not simply about the details and dimensions of inequality and injustice in the world; it is not simply about arguments for and against particular theories or understandings of that world; it is ultimately about the nature of that world and our place within in it. It is about how we, as individuals and collectivities *'fit'* into that world and how we define ourselves in relation to it. It is about the fundamentally unjust character of our world and our relation to that absence of justice. It is about what the UNDP Human Development Report for 2010 described as the *'extreme poverty'* of many and the *'gracious luxury'* of some.

This chapter explores some of these thorny issues and seeks to stimulate reflection (and, ultimately action) in considered and appropriate ways. It examines the arguments and debates about our roles, responsibilities and duties in an unequal world at the personal, professional and political levels. It does not seek to prescribe a set of *'correct'* actions from whatever source they emanate – it seeks to stimulate reflection and debate on what *'action'* is, how it relates to each of us and what options or *'duties'* it places before each of us. Before proceeding further, it may be useful to review a number of stark current realities as explored elsewhere in 80:20 (of necessity, this is a limited selection):

- Of a total world population of 6.4 billion in 2005, some 1.4 billion people continued to live below a poverty line of US$1.25 per day, 1.7 billion below US$1.45 and 2.6 billion on less than US$2.*

- 925 million people were undernourished in 2010 (UN Food and Agriculture Organisation data)

- Today, nearly one billion people lack access

* For a detailed discussion on how such poverty lines are set and periodically reviewed, see S. Chen and M. Ravaillon (2008) The Developing World is Poorer Than We Thought But No Less Successful in the Fight Against Poverty, Policy Working Paper 4703, New York, World Bank

to safe water and 2.5 billion do not have improved sanitation (UNICEF data)

- 793 million adults were illiterate in 2009 (UNESCO data)
- 195 children in the Developing World suffered from severe to moderate stunting in 2009 (UNICEF data)
- One third of all deaths -18 million people a year or 50,000 per day - are due to poverty-related causes (WHO data)
- A pregnant woman in the Developing World is, on average, 50 times more likely to die during childbirth than a woman in the Developed World (WHO data)

As against this set of facts, another:

- Average life expectancy has increased by 21% to 68 years in developing countries overall since 1970 (UNDP data)
- Literacy has increased by 61% between 1970 and 2010 for all developing countries (UNESCO data)
- The share of democratic countries worldwide increased from less than a third in the early 1970s to more than half in 1996 and to three-fifths in 2008 (Freedom House data)
- Infant mortality rates (per 1,000 live births) fell from 89 to 45 between 1970 and 2006 while adult female mortality rates fell from 257 to 164 (per 1,000 adults) and that for male adults from 308 to 237 (UNICEF data)
- The share of undernourished people in developing countries fell from 25% in 1980 to 16% in 2005 (FAO data)
- In 1960, 20 million children died before their 5th birthday because of poverty. In 2007, UNICEF announced that, for the first time, the number of deaths of young children has fallen below 10 million a year

So, how do we read and understand the above selection of realities and what does such understanding suggest in terms of desired or needed change?

- *Are they inevitable and unavoidable?*
- *the result of factors beyond human intervention or control?*
- *the result of purely domestic decisions and problems in developing countries?*
- *not our responsibility or concern but for others to engage with?*
- *are they issues of welfare and charity or issues of justice and duty?*
- *are they the inevitable result of modern corporate-driven globalisation?*
- *a legacy of slavery and colonialism?*
- *the outcome of a profoundly unjust and unfair international system that condemns the many and favours the few?*
- *who is responsible and who should be accountable?*

How each of us understands and responds to the above questions has profound implications for our conception of change (and the scale of change that is needed) and our role within such change. One inevitable conclusion that arises from the above realities, questions and arguments is that change is inevitable; things do not remain the same but are constantly being re-shaped and re-formed. This chapter introduces differing models of change and the degree of challenge they present; a number of case studies of how change has actually come about and a brief review of organisations advocating change across a variety of issues and challenges.

The past decades have highlighted much that is wrong and much that may be possible. Australian philosopher Peter Singer summarises the situation as follows:

'We can liken our situation to an attempt to reach the summit of an immense mountain. For all the aeons of

human existence, we have been climbing up through dense cloud. We haven't known how far we have to go, nor whether it is even possible to get to the top. Now at last we have emerged from the mist and can see a route up the remaining steep slopes and on to the summit ridge. The peak still lies some distance ahead.'

Peter Singer 'Change a little to make a big difference', Independent Newspaper, London, March 28th, 2009

The remainder of this chapter explores a number of key aspects of change; the different worldviews and politics which inform arguments for change; a number of case studies which illustrate how change has come about and an introduction to a range of organisations campaigning for change on key issues.

One inevitable conclusion that arises from the above realities, questions and arguments is that change is inevitable; things do not remain the same but are constantly being re-shaped and re-formed.

The perspectives, case studies and organisations reviewed are by no means comprehensive – they are presented with a view to exploring the diversity and character of change today and in recent decades as it relates to human development and human rights.

ARGUING ABOUT CHANGE – FIVE DIVERGENT APPROACHES

The five approaches explored below include:

- that of the World Bank and of those who argue that change should and can be accommodated within the current dominant economic and political system
- those, such as Professor Jeffrey Sachs, who argue that radical and urgent reform is necessary and immediately possible at a variety of levels but largely within the current economic and political system
- those, such as Peter Singer, who support far-reaching radical action at individual and collective level – action which would radically transform the current economic system
- those, such as Thomas Pogge, who emphasise a 'rights perspective' and who emphasise the need to immediately transform structures and processes that harm the world's poor
- those, such as Indian activist and environmentalist Vandana Shiva, who reject the current economic and political model of development, describing it as a 'war' on the planet and who advocate alternatives such as earth democracy.

For many, the changes required in order to deliver a more equal and just world are essentially incremental and supportive of the current economic and political world order, albeit with a focus on degrees of urgent reform. For example, the World Bank's World Development Report 2010 outlines the case for integrating development and climate concerns without fundamentally challenging or damaging the current international trade regime (World Development Report 2010: 251-255). The Bank's 2009 World Development Report talks of countries *'lagging behind'* and of the need for greater *'integration'* of poorer countries into world markets (see chapter 9 of the Report for example).

In a similar vein, the Adam Smith Institute continues the argument insisting that *'free markets'* are the only viable engine of growth and that state regulation is anathema to development*. Many of these arguments are also echoed, albeit differently in Professor Paul Collier's *The Bottom Billion: Why the Poorest Countries are Failing and What Can Be Done About It* (2007) in which he examines four key development traps – conflict, resources, being landlocked and having bad governance and argues for a series of solutions including free trade with preferential access for the exports of 'bottom billion' countries while also arguing against what he terms the 'headless heart' approach to change. As with all of those summarised in this section, Collier's views have been the subject of debate and criticism; see, for example http://www.worldchanging.com or http://www.guardian.co.uk/books/2007/jun/10/politics

In recent years, one of the most cited arguments for what needs to change and for what can readily change has come from Professor Jeffrey Sachs (Director of the Earth Institute at Columbia University and closely associated with the UN MDG agenda) who has identified four great challenges – environment, demography, poverty and global politics and advanced a change agenda on world poverty and sustainable development. In *Commonwealth: Economics for a Crowded Planet* (2008), he insists that the *'relentless acceptance of the status quo is not acceptable in the face of the challenges we confront'* and critiques a negativism based on a *'state of mind, not a view based on facts'*. In a chapter entitled *'The Power of One'*, he outlines (following the arguments of economist Albert Hirschman) an *'unholy trinity'* of reaction to the possibility of radical change:

- **Futility:** the course of reform cannot work because the problem is insoluble
- **Perversity:** attempting to solve the problem will actually make matters worse
- **Jeopardy:** attempting to solve the problem will take attention and resources away from something even more important

In rejecting these *'resistances'*, he insists on the responsibility of individuals and offers an agenda of change focused around learning, interculturalism, social action and engaging at the personal, professional and political level. Sach's analysis and agenda have been the subject of much debate and criticism (see, for example the reviews of Sach's book *The End of Poverty; Economic Possibilities of Our Time* by William Easterley and by Professor Michael Blim, yet his agenda for change, especially in the context of the MDGs, remains persuasive and highly influential.

Taking the argument one stage further, one of the most controversial commentators on wealth and poverty in the world and the responsibilities of each individual is Australian philosopher Peter Singer, whose views have already been analysed in Chapter 7. Singer's conclusions regarding our responsibility to effect change in the world are far-reaching and startling; he concludes that we do have a 'duty' to respond to human suffering in the world for a variety of reasons – such suffering is evil; people 'must' respond to relieve it whenever they are in a position to do so (and it is morally wrong not to do so) and affluence places very considerable demands on the affluent.

The World Bank's World Development Report 2010 outlines the case for integrating development and climate concerns without fundamentally challenging or damaging the current international trade regime

In a well-known article in the New York Times in September 1999, Singer argued:

'In the world as it is now, I can see no escape from the conclusion that each one of us with wealth surplus to his or her essential needs should be giving most of it to help people suffering from poverty so dire as to be life-threatening. That's right: I'm saying that you shouldn't

* For example, see the Institute's briefing papers by Marc Sidwell (2008) Unfair Trade; Sam Bowman (April, 2010)'An international development policy that works' and Dalibor Roháč (May, 2011) 'Does inequality matter?'

buy that new car, take that cruise, redecorate the house or get that pricey new suit. After all, a $1,000 suit could save five children's lives…

'…If we don't do it, then we should at least know that we are failing to live a morally decent life – not because it is good to wallow in guilt but because knowing where we should be going is the first step toward heading in that direction.'

Singer's arguments on the moral obligations of affluence and on the options open to affluent individuals are developed more fully in his 2009 book *The Life You Can Save: Acting Now to End World Poverty* and have

'relentless acceptance of the status quo is not acceptable in the face of the challenges we confront'

been the subject of much debate and criticism on a number of grounds – overstating the responsibility and potential of individuals; overstating obligations to those we do not know or who live distantly and offering an 'extreme' scenario in terms of how far one's obligations to others go. As noted in Chapter 7, an extended critique of Singer's position is offered by Gareth Cullity in his 2004 work *The Moral Demands of Affluence.*

Arguments about the compelling need for fundamental change are taken a significant step further by philosopher Thomas Pogge of Yale University and co-founder of Academics Stand Against Poverty who asserts that people in wealthy Western democracies are currently simply not doing enough to tackle poverty but are actually harming the world's poor through support for, and participation in, the current dominant economic and political system. The world's poor are being actively and unjustly harmed by an international system of political and economic policies and practices that are disproportionately shaped by and for wealthy Western societies (see Chapter 7 for more on this). His analysis suggests that not

only is individual charitable action required, but so too is individual and collective political action to address this international system of harm.

For Pogge, eliminating poverty is easily achieved – in a 2007 essay, he observed:

'The collective annual consumption of the 2,735 million people reportedly living (on average 42%) below the World Bank's $2/day poverty line is about $440 billion and their collective shortfall from that poverty line roughly $330 billion per year. This poverty gap is less than 1% of the gross national incomes of the high-income countries, which sum to $35,142 billion in 2005 (World Bank 2006: 289). These countries contain 15.7% of the world's population with 79% of the global product (ibid.). The global poor are 42% of the world's population with 1% of the global product. At market exchange rates, the per capita income of the former is roughly 200 times greater than that of the latter… Eradicating severe poverty (relative to the $2/day poverty line) is a matter of raising the income of the poor from currently 2.3% of the average human income to 4%.'

See http://www2.ohchr.org/english/issues/poverty/expert/docs/Thomas_Pogge_Summary.pdf

When confronted with such facts, citizens of the rich countries should recognise that we, the world's affluent, should (and could) do a lot more to help the poor. For Pogge, the issue is that the majority in the West continue to see this as a *'demand of humanity or charity - not as a demand of justice'* and certainly not as a moral and political *'duty'* imposed on us by our acceptance of the principle of human rights. Pogge argues that *'…the rich countries' response to world poverty is mainly rhetorical…and even the rhetoric is appalling'* and insists that we have both negative duties (to not do *'harm'*) but also positive duties (to ensure the rights of others). He rejects western arguments of *'innocence'* and *'powerlessness'* when faced with the current international system and argues that the benefits and advantages western populations (and, by implication, developing world elites) gain from this system must be fundamentally challenged, as a matter of fulfilling our human rights obligations.

Pogge's analysis has been criticised on a number of grounds – that the current world order is *'imperfect'* rather than *'unjust'* (as asserted by Pogge); that his definition of *'harm'* in relation to the poor is overstated; that the scale of demands his analysis places on individuals as regards positive duties are too high (or low); which causal factors should receive greatest weight (and who decides this) and that his proposals are unworkable. For an extended discussion of Pogge's work (and his robust response), see Alison Jaggar (2010, ed.) Thomas Pogge and his Critics, Cambridge, Polity Press).

One of the most far-reaching analyses of the change that is required has come from creator of the concept of Earth Democracy, Indian environmental activist and eco-feminist, Vandana Shiva, who argues that the current international economic, political and social system amounts to little more than *'a war against the planet'*. For Shiva, this war is rooted in an economy that fails to respect either 'ecological or ethical limits' – limits to inequality, limits to injustice, limits to greed and economic concentration. In her speech on receipt of the Sydney Peace Prize in 2010, she argued:

'A handful of corporations and of powerful countries seeks to control the earth's resources and transform the planet into a supermarket in which everything is for sale. They want to sell our water, genes, cells, organs, knowledge, cultures and future.'

Shiva highlights three levels of violence that characterise non-sustainable development – the first is the violence against the earth, which manifests itself as the ecological crisis; the second is the violence against people, which is expressed as poverty, destitution and displacement. The third is the violence of war and conflict, as the powerful reach for the resources that lie in other communities and countries for their limitless appetites. The richer the world becomes, the poorer we are ecologically and culturally; the growth of affluence, measured in money, is leading to a growth in poverty at *'the material, cultural, ecological and spiritual levels'*.

Respecting the earth and essentially *'making peace'* with the earth was, for Shiva, always an ethical and ecological imperative but today it has become *'a survival imperative for our species'*, as she chillingly comments *'...dead soils and dead rivers cannot give food and water'*. In her book Earth Democracy: Justice, Sustainability and Peace (2005), she argues against what she describes as our increasing *'monoculture of the mind'* where both our society and our science and technology force us to think and act in one frame of reference only and where we have a growing *'inability to appreciate diversity'* to our own disadvantage but also that of millions of the poor and of the earth itself. She argues that we *'must change the paradigm of who we are'* because we are effectively *'shrinking citizenship'* by disenfranchising vast groups of people and perpetuating injustice.

> The world's poor are being actively and unjustly harmed by an international system of political and economic policies and practices that are disproportionately shaped by and for wealthy Western societies

> For Shiva, this war is rooted in an economy that fails to respect either 'ecological or ethical limits' – limits to inequality, limits to injustice, limits to greed and economic concentration.

As is clear from the brief reviews above, the debates about change worldwide are not simply about deciding which particular economic, social or political (or technical) response will address the problem. The debates are clearly about the nature, scale, and the *fundamentals* of the issue. As Amartya Sen argues: the *'atrocity of poverty'* will not correct itself:

'Quiet acceptance – by the victims and by others – of the inability of a great many people to achieve minimally effective capabilities and to have basic substantive freedoms acts as huge barrier to social change. And so does the absence of public outrage at the terrible helplessness of millions of people…We have to see how the actions and inactions of a great many persons together lead to this social evil, and how a change of our priorities – our policies, our institutions, our individual and joint actions – can help to eliminate the atrocity of poverty.'

Amartya Sen (2008) Preface in D. Green From Poverty to Power, Oxfam International

The remainder of this chapter examines the nature of change; how change comes about, the real and/or potential role of individuals, civil society groups, governments and international structures in such change. The choice of case studies is inevitably very limited and selective.

EXPLORING HOW CHANGE HAPPENS: FOUR CASE STUDIES

CASE STUDY 1 - ETHICAL CONSUMPTION AND CHANGE; USING CONSUMER POWER AND BOYCOTTS

'Ethical consumption is best understood as a political phenomenon rather than simply a market response to changes in consumer demand. It reflects strategies and organisational forms amongst a diverse range of governmental and non-governmental actors. It is indicative of distinctive forms of political mobilisation and representation. And it provides ordinary people with pathways into wider networks of collective action, ones which seek to link the mundane spaces of everyday life into campaigns for global justice.'

Clive Barnett, Paul Cloke, Nick Clarke and Alice Malpass (2007)

Sparked by the famous bus protest and arrest of Rosa Parks in December 1955, the Montgomery bus boycott was a 13-month mass protest that ended with the U.S. Supreme Court ruling that segregation on public buses is unconstitutional. The Montgomery Improvement Association organised the boycott led by its President, Martin Luther King, Jr., who claimed that the real value of the boycott was its power to demonstrate the potential for nonviolent mass protest and that served as an example for other civil rights campaigns that followed. Although the word boycott was not coined until 1880 (as a result of a Land League protest in Ireland), the practice itself dates back to at least 1830, when the US National Negro Convention (1830-1864) encouraged a boycott of slave-produced goods.

Boycotts offer individuals and groups the opportunity to use economic pressure on companies to change their practices or policies and have routinely been effective when governments are unwilling or unable to introduce policy change or reform (for example against the former Apartheid regime in South Africa). They have become a powerful international tool of protest and an important part of modern democracy – a direct way of rejecting the products or behaviours of companies and even countries.

The Ethical Consumer organisation in the UK has identified four specific types of boycott:

- Positive Buying – the favouring of particular ethical products, for example, energy saving light bulbs
- Negative Purchasing - avoiding products of which you disapprove of, such as battery eggs or gas-guzzling cars.
- Company-Based Purchasing - targeting a business as a whole and avoiding all the products made by one company, a famous example is the Nestle boycott over the way it markets baby food internationally, especially in the developing world.
- Fully-Screened Approach - looking both at companies and at products and evaluating which product is the most ethical overall.

See http://www.ethicalconsumer.org/Home.aspx

Companies, particular products, and even countries have become targets for boycotts for environmental, political, gender, racial and human rights reasons. Companies have become increasingly sensitive to being boycotted because of the potentially serious market and financial implications: should an individual boycott a product or company, a return to that brand is unlikely, so companies can lose a customer for life. Ethical Consumer argues that:

'The economic clout of boycotts shouldn't be underestimated. In 2011 a boycott of the state of Arizona in the US cost its economy a massive $141 million dollars in just seven months after it passed controversial immigration laws. And there are more people boycotting products than you might think. Research carried out by the Co-operative bank in 2007 put the value of UK boycotts at around £2,500million - food and drink boycotts were valued at £1,144m, travel boycotts £817m, and clothing boycotts £338m.'

For an extensive list of current international boycotts of companies and products, see http://www. ethicalconsumer.org/Boycotts/currentboycottslist. aspx where you can also download (in PDF) a list of over 50 successful boycotts in the period 1986-2009.

Boycotts are one element of a broader ethical consumerist culture which is growing internationally; for a fuller discussion of the nature and impact of ethical consumption, see Clive Barnett, Paul Cloke, Nick Clarke and Alice Malpass (2007) *The Subjects and Spaces of Ethical Consumption: doing politics in an ethical register* and (2010) *Globalising Responsibility: The Political Rationalities of Ethical Consumption*, London, Wiley Blackwell.

On May 24th, 2011, 18 NGOs working in Laos issued an open letter to Nestle claiming that the company's *'marketing of formula milk still jeopardises the health of infants and children in Laos.'* They asserted that:

'Nestle continues to make millions of dollars of profit, at the expense of infants and children in Asia, through violations of the International Code of Marketing of Breast-milk Substitutes.

'Unethical marketing by food companies, including Nestle, contributes to the situation of high infant and child mortality in Laos. Babies and children are dying in Laos because food companies such as Nestle are weakening national regulatory frameworks and aggressively flooding the market with information that dilutes public health campaigns that promote breastfeeding.'

Source: http://info.babymilkaction.org/sites/info.babymilkaction.org

The open letter listed a number of breaches of the agreed international code including advertising, using hospital-based promotion, labelling, misinformation, and incentivising both doctors and nurses. In response, Nestle denied the allegations and said it was investigating them and specifically denied providing incentives. The letter is but the latest development in an ongoing battle between Nestlé and child health groups that began in 1977 in the US and which has spread out worldwide since then and is waged in the hospitals, villages, health centres and markets of developing countries. It is also being fought out in the research, offices, institutions and conferences of health and development organisations internationally.

The boycott, one of the longest continuing campaigns is driven by fundamental disagreement about Nestle's promotion techniques for breast milk substitutes (infant formula) especially in the developing world which campaigners claim adds to the unnecessary suffering and even deaths of babies, largely among the poor. The specific concerns include the dangers of mixing formula with contaminated water; the reality of illiteracy making appropriate use of label information problematic; the availability of fuel and light and the strength of the mix used. Campaigners accuse Nestlé of using 'unethical methods' in the promotion of infant formula over breast milk to poor mothers in developing countries. The International Baby Food Action Network has accused the company of distributing free samples to hospitals and maternity wards; formula is no longer free after hospital and because the formula has interfered with lactation, mothers must continue to buy the formula. IBFAN also claims that Nestlé uses humanitarian aid to create markets, does not label its products appropriately in addition to offering incentives. Nestlé denies these claims and insists it complies with internationally agreed codes.

As a result of ongoing concerns and concerted campaigning, the World Health Assembly adopted the International Code of Marketing of Breast-milk Substitutes in 1981; the code addresses the marketing of infant formula and other milk products, foods and beverages, when presented as a partial or total replacement for breast milk. It specifically bans or limits the advertising and promotion of breast milk substitutes, the use of incentives and samples; insists information is provided in local languages and simply and that information on the fundamental importance of breast-feeding is included. In 1984, Nestlé agreed to implement the code and the boycott was suspended but in 1988 IBFAN alleged that formula companies were flooding health centres in poorer countries with free and low-cost supplies and re-launched the campaign in 1989. The campaign has continued since then and has involved the World Health Organisation, the European Parliament, the UK Advertising Standards Authority, TV and print media as well as politicians, church leaders and celebrities.

IBFAN now consists of 200 groups in over 100 countries and claims that, as a result of its work 60 countries have now introduced laws implementing most or all of the provisions of the agreed 1981 code. To date, campaigners claim that many European universities, colleges, and schools have banned the sale of Nestlé products from their shops and vending machines. In the UK alone, 73 student unions, 102 businesses, 30 faith groups, 20 health groups, 33 consumer groups, 18 local authorities, 12 trade unions, 31 MPs, and many celebrities now support the Nestle boycott. The company continues to insist that it is in compliance with the Code and investigates and takes action on all breaches of the rules. In 2010, IBFAN published its 2010 report *Breaking the Rules, Stretching the Rules* and in May 2010, the World Health Assembly called on '...*infant food manufacturers and distributors to comply fully with their responsibilities under the International Code of Marketing of Breast-milk Substitutes and subsequent relevant World Health Assembly resolutions...*'

✎ MORE INFO

- http://www.ibfan.org/index.html
- http://apps.who.int/gb/ebwha/pdf_files/WHA63/A63_R23-en.pdf
- http://www.babymilkaction.org/

CASE STUDY 2 - MAKING CHANGE HAPPEN – NGOS AND GOVERNMENTS WORKING TOGETHER, THE 1997 MINE BAN TREATY

The 1997 Mine Ban Treaty and the process that led to its agreement demonstrated the importance and the possibilities of civil society driven campaigns. Historically, armed force personnel, 'security experts' and even governments argued that the use of landmines was justified despite the collateral civilian casualties. The International Campaign to Ban Landmines, led initially by NGOs who had direct experience of the effects of such mines on civilians and who were frustrated at the lack of progress by governments in controlling the use of these weapons, mobilised public opinion over two decades and eventually built up a network of NGOs, international organisations, United Nations agencies and governments committed to a total ban. This objective was achieved in 1997 and its consequences are being realised to this day.

Antipersonnel mines were first used on a wide scale in World War II and since then have been used in many conflicts, including the Vietnam War, the Korean War and the first Gulf War. Until the 1990s, antipersonnel landmines had been used by almost all armed forces in the world, in one form or another. The mines were first developed as a defence against the removal of anti-tank mines and a key characteristic of the weapon is that it maims rather than kills enemy soldiers thus using up resources in caring for the injured. Antipersonnel landmines then began to be used on a wider scale, often in internal conflicts and became aimed at civilians, terrorising communities, denying them access to farmland and restricting population movement.

Mines killed or injured as many as 26,000 people each year, some 50 countries were manufacturing them and over 125 countries had stockpiles. As a result of the ban 39 countries have stopped production and the international trade in mines has almost halted. None of the 156 states which are party to the ban produces landmines but 12 states not party to the treaty continue to produce them or have not confirmed their non-production - China, Cuba, India, Iran, Myanmar, North Korea, South Korea, Pakistan, Russia, Singapore, the United States, and Vietnam. Of those, only three were actively manufacturing antipersonnel mines in 2009 - India, Myanmar and Pakistan.

A total of 3,956 new casualties to landmines and explosive remnants of war (ERW) were recorded in 2009, the lowest annual total since monitoring began in 1999 and 28% lower than in 2008. However, non-state armed groups or rebel groups in various countries (Afghanistan, Colombia, India, Myanmar, Pakistan, and Yemen) produce antipersonnel mines, mostly of the improvised variety. A total of 66 states and seven other areas were confirmed or suspected to be mine-affected.

The Campaign was awarded the Nobel Peace Prize in recognition of its efforts to bring about the 1997 Mine Ban Treaty. Since then, it has been campaigning for the words of the treaty to become a reality, *demonstrating on a daily basis that civil society has the power to change the world.'*

GOOD NEWS	BAD NEWS
156 countries have joined the Mine Ban Treaty—80% of the world's nations.	No state has joined the treaty since Palau acceded in November 2007.
The Cartagena Action Plan adopted at the Second Review Conference provides an ambitious and concrete five-year roadmap to implement and universalise the Mine Ban Treaty.	
There has been no need for States Parties to invoke the treaty's formal compliance provisions to clarify any compliance matters.	There are highly disturbing allegations that members of the armed forces in Turkey used antipersonnel mines in 2009; these are currently the subject of a legal investigation by Turkey.
86 states have completed the destruction of their stockpiles, collectively destroying over 45 million stockpiled antipersonnel mines.	Ukraine missed its stockpile destruction deadline in June 2010 and is in violation of the treaty; as are Belarus, Greece, and Turkey, which missed their deadlines in March 2008.
A rigorous process is in place for extending the 10-year mine clearance deadlines. As of September 2010, 22 States Parties have received or were formally seeking additional time.	Too many States Parties granted extensions in 2008 and 2009 have since made disappointing progress. Of greatest concern is Venezuela, which has not started clearance operations more than 10 years after ratifying the treaty.
In June 2010, Nicaragua formally declared that it had completed its clearance obligations. It was the 16th state to do so; Albania, Greece, Rwanda, Tunisia, and Zambia declared they fulfilled their clearance obligations in 2009.	The rate of compliance with submitting annual transparency reports is at an all-time low (56%); Equatorial Guinea is 11 years late with its initial report. Less than 40% of states have passed domestic laws to implement the treaty

In commenting on the success of the campaign, Duncan Green noted:

'In the face of opposition from the great powers, this core group created a new form of international diplomacy, built upon a willingness to operate outside the UN system, extensive NGO participation, leadership from small and medium-sized countries, rejection of consensus rules, and avoidance of regional blocs.'

Duncan Green (2008) From Poverty to Power, Oxfam, p.444

 MORE INFO

- http://www.icbl.org/intro.php

CASE STUDY 3 - 'BEING POSITIVE' – THE TREATMENT ACTION CAMPAIGN IN SOUTH AFRICA

'This is a real triumph of David over Goliath, not only for us here in South Africa, but for people in many other developing countries who are struggling for access to healthcare.'

This was the comment of Zachie Achmat (one of the founders of the South African NGO Treatment Action Campaign) in April 2001 when some 39 international pharmaceutical companies agreed to drop their legal challenge to the Government of South Africa. The companies had taken the government to court in an attempt to block proposed legislation (initiated in 1997) which would give it the power to import or manufacture cheap versions of brand-name drugs which the companies claimed would breach patent protection agreements. The struggle to challenge the power and, more specifically, the pricing policies of the pharmaceutical industry in South Africa was just one dimension of a major campaign relating to HIV and AIDS developed by TAC.

TAC was launched in South Africa on International Human Rights Day 1998, by a small group of political activists who viewed access to health care, and in particular medicines for HIV, as a human rights issue rather than simply a welfare or health issue and who saw HIV as symptomatic of deeper social and political issues facing poor people and, therefore, ultimately a question of human rights. From the beginning, TAC adopted a different approach, it emphasised building capacity directly among the poor, stressing the need to build a popular political movement 'on the ground' in contrast to what they saw as a professionalised 'AIDS and human rights movements - articulate but ineffective. TAC adopted an approach from the United States, where AIDS activists, led by people with HIV, had pioneered the idea of *'treatment literacy'* and became the first AIDS activist organisation to pioneer the concept and practice of HIV 'treatment literacy' in a developing country.

Treatment literacy involves a programme of health education and communication that aims to educate HIV-vulnerable and poor people on all aspects of the virus - information about the medical and health dimensions is linked to political science, human rights, equality, and the positive duties of the state. This programme was backed up by an extensive training agenda, popular educational materials and a primary focus on mobilising people. As a result, a new generation of human rights activists emerged and people with AIDS ceased to be silent victims and became political agitators for their human right to treatment.

TAC's campaign began in 1998 with people's immediate needs - through demanding that the South African Government introduce a national programme to prevent mother-to-child HIV transmission (PMTCT). This campaign caught the attention of young women with HIV and – for the first time in Africa – began to galvanise a social movement that was made up of people who were predominantly poor, black, and living with HIV. TAC's starting point was to insist that the excessive pricing of essential medicines by multi-national pharmaceutical companies violated a range of the human rights that had, since 1996, been entrenched in the South African Constitution. It argued that intellectual property and patents, whose protection in law had been strengthened under the World Trade Organisation's 1995 TRIPS agreement, was not an inherent human right, but a device granted by the state for a public purpose. TAC sued the South African Government and won its case on the basis of the Constitutional guarantee of the right to health-care, and the government was ordered to provide MTCT programmes in public clinics.

TAC also helped to establish the Joint Civil Society Monitoring Forum (JCSMF), a network that closely monitored and reported publicly on the expansion of access to ARV treatment and the primarily political obstacles it was encountering. Close monitoring of the programme required the government to constantly account for its omissions and weaknesses and maintained pressure for ongoing investment and expansion. As a result of the ARV programme, at least 400,000 people are alive who would have died.

Given its overtly rights-based approach and its emphasis on the political dimensions of HIV and AIDS, TAC also engaged vigorously with the 'official' position of the South African Government as expressed publicly by President Thabo Mbeki and Health Minister Diamini Zuma over AIDS 'denialism'. This view (widespread in the history of the virus worldwide) challenged the link between HIV and AIDS; consistently denied the scale and impact of the virus and delayed the effective development of policy and strategy. TAC directly challenged the government on its views and the debate became vigorous and protracted as activists organised to defeat government policy.

TAC has become an internationally acclaimed organisation for its approach to change and for its overall impact. In 2004, TAC was nominated for a Nobel Peace Prize and in August 2006, the organisation was named 'the world's most effective AIDS group' by the New York Times.

Sources:

Mark Heywood (2009) South Africa's Treatment Action Campaign: Combining Law and Social Mobilization to Realize the Right to Health, Journal of Human Rights Practice 1:1, 14–36

Steinberg, J. 2009. Three Letter Plague, A Young Man's Journey Through a Great Epidemic. London, Vintage Johannesburg and Cape Town: Jonathan Ball.

http://www.avert.org/history-aids-south-africa.htm

CASE STUDY 4 - WOMEN ORGANISING FOR CHANGE – THE SELF EMPLOYED WOMEN'S ASSOCIATION, INDIA

'There are risks in every action. Every success has the seed of some failure. But it doesn't matter. It is how you go about it. That is the real challenge.'

Ela Bhatt, founder, Self-Employed Women's Association of India

The Self-Employed Women's Association (SEWA) is a trade union registered in 1972; it has a membership today of over 1.2 million women and represents the interests and agendas of poor, self-employed women workers. These women earn a living through their own labour or through small businesses and represent the unprotected and unorganised labour force of India (making up over 90% of the labour force). Of the female labour force in India, more than 94% are unorganised and their work is not counted and remains invisible.

SEWA has its roots in the trade union movement and was formed in 1972; it grew out of the Textile Labour Association (TLA) - India's oldest and largest union of textile workers founded in 1920 by a woman, Anasuya Sarabhai, inspired by Mahatma Gandhi who led a successful strike of textile workers in 1917. Against this background of active involvement in labour issues and politics, the TLA created their Women's Wing in 1954 with the original purpose of supporting the wives and daughters of mill workers through both training and welfare activities and by the late 1960's classes in sewing, knitting embroidery, spinning, press composition, typing and stenography had been established.

The style of its work changed considerably in the 1970's following complaints by women workers (e.g. tailors) of exploitation by contractors and evidence that very large numbers of women were unprotected in work and were not unionised. In 1971, women workers in Ahmedabad (cart pullers and 'head loaders' or head carriers) approached the Women's Wing organiser Ela Bhatt about housing and work conditions and following a campaign highlighting their conditions (and those of women garment workers) and the need for improved conditions and for greater organisation to realise them, SEWA was formed. Immediately, it faced a struggle to be recognised as a trade union, given the absence of a formal employer, but SEWA stressed its role in unifying workers. Its growth through

the 70s and its links to the growing assertiveness of the women's movement plus its stand on caste issues in the early 1980s, led to SEWA being expelled from the TLA.

Following this event, SEWA grew steadily and expanded its remit to include necessary supportive services such as savings and credit, health-care, child-care, insurance, legal aid, capacity building and communications – if women were to achieve full employment and self-reliance, such services were seen to be essential.

SEWA provides these services (and their associated employment) in a decentralised and affordable manner within the workers' own communities and many services have now become self-financing. Rejecting much of what is defined as development today, Ela Bhatt captures much of the philosophy and practice of SEWA as follows:

'But if our goal is to build a society where everyone's basic needs for food, clothing and housing are fully met, and where the full potential of every human being is realised, we will need a radically different approach. We will need to get in partnership with our conscience; we will need to get in partnership our fellow human beings and we will need a long-term partnership with Mother Nature.'

Ela Bhatt UNDP Speech, June 27, 2011 Tarrytown, NY

Sources:
http://www.sewa.org/
http://www.inclusivecities.org

CAMPAIGNING FOR CHANGE

Below we list a very limited selection of organisations and campaigns internationally which engage at a variety of different levels with key aspects of the core development and human rights issues explored in 80:20; of necessity, the list is selective and by no means indicative of importance or prioritisation – it is simply representative of engaged groups and campaigns.

ON FINANCE, TAX AND CORRUPTION

http://www.taxjustice.net

The Tax Justice Network works to promote transparency and oppose secrecy in international finance; it promotes a level playing field on tax and opposes loopholes and distortions in tax and regulation, and the abuses that flow from them. TJN promotes tax compliance, opposes tax evasion, tax avoidance and all the mechanisms that enable owners and controllers of wealth to escape their responsibilities to those societies on which they and their wealth depend. Tax havens, or secrecy jurisdictions as TJN prefers to call them, lie at the centre of its concerns.

http://www.gfip.org

Global Financial Integrity (GFI) is a Washington-based not-for-profit organisation that promotes national and multilateral policies, safeguards, and agreements aimed at curtailing the cross-border flow of illegal money. In suggesting solutions, building strategic partnerships, and conducting research, GFI works to curtail illicit financial flows and promote global development and security.

http://www.transparency.org/

Transparency International is the leading global civil society organisation fighting against corruption; it brings people and organisations together in a worldwide coalition to challenge the devastating impact of corruption on men, women and children around the world. TI's mission is to create change towards a world free of corruption.

ON TRADE AND DEBT

http://www.tjm.org.uk/home.html

The Trade Justice Movement is one of many organisations worldwide focused on the ever increasing gap between many of the stated aims and objectives of the world trading regime and the global reality of growing inequalities and environmental degradation. Although directly focused on the World Trade Organisation, the Movement's activities are also relevant to other trade and finance institutions that impact on local, national and international trade policy.

For a representative NGO campaign on trade issues, see Oxfam's trade campaign at http://www.oxfam.org/en/campaigns/trade; on Fairtrade and its growth and work, see http://www.fairtrade.org.uk and for a different perspective on trade see also http://www.valueaddedinafrica.org

http://www.jubileedebtcampaign.org.uk

The Jubilee Debt Campaign is a UK based charity campaigning on debt issues – currently, the world's poorest countries pay almost $23 million every day to the rich world; read some of the basics about the issue and how debt actually works and why it should be challenged.

In this context, also see a recent study on Ireland's tax regime and how it can hurt the world's poor at http://www.debtireland.org

ON RESEARCH AND INFORMATION

http://www.odi.org.uk

ODI is a leading independent think tank on international development and humanitarian issues; it seeks to stimulate and inform policy and practice which will lead to the reduction of poverty, the alleviation of suffering and the achievement of sustainable livelihoods in developing countries.
http://www.worldwatch.org

Environment focused research think-tank;

Worldwatch focuses on the challenges of climate change, resource degradation, population growth, and poverty by developing and disseminating solid data and strategies for achieving a sustainable society. Worldwatch was one of the first research institutes devoted to the analysis of global environmental concerns.

http://www.socialwatch.org

Social Watch is an international network of citizens' organisations struggling to eradicate poverty and the causes of poverty, to ensure an equitable distribution of wealth and the realisation of human rights. Social Watch is committed to social, economic and gender justice, and emphasises the right of all people not to be poor. It publishes a wide range of research and highlights the developing world perspective on issues.

ON HUMAN RIGHTS

http://www.hrw.org/home

Human Rights Watch is one of the world's leading independent organisations dedicated to defending and protecting human rights; it focuses international attention on human rights abuses and seeks to establish accountability for such abuses. HRW carries out high quality research and supports a wide variety of campaigns on key issues such as women's rights, torture etc. It is an independent human rights organisation.

http://www.amnesty.org

Amnesty International is a household name and a global movement of more than 3 million supporters and activists in over 150 countries who educate and campaign to end grave abuses of human rights and to promote a culture and practice of human rights. Amnesty was inspired by the values enshrined in the Universal Declaration of Human Rights and other international human rights standards. It is independent of any government, political ideology, economic interest or religion and is funded mainly by its membership and public donations.

http://www.euromedrights.org/en/

A network of more than 80 human rights organisations, institutions and individuals based in 30 countries in the Euro-Mediterranean region; the Network is increasingly important for human rights issues in the Arab World. The Network focuses on a range of issues including freedom of association, gender, justice, migration and refugees, human rights education and Palestine, Israel and the Palestinians.

ON WOMEN'S AND MEN'S RIGHTS

http://www.genderjustice.org.za

The Sonke Gender Justice Network works across Africa to strengthen government, civil society and citizen capacity to support men and boys to take action to promote gender equality; to prevent domestic and sexual violence and to reduce the spread and impact of HIV and AIDS. Sonke works with government on policy issues; with communities on education and mobilisation; with the media on communication and with other networks on building impact and effectiveness. Sonke places primary emphasis on transforming attitudes and practices within individuals, communities, organisations and public institutions.

http://www.ungift.org/knowledgehub/

A UN based inter-organisation campaign the United Nations Global Initiative to Fight Human Trafficking (UN.GIFT) was established in 2007 to promote the fight on human trafficking, on the basis of international agreements reached at the UN. It includes the International Labour Organisation, the Office of the United Nations High Commissioner for Human Rights, the United Nations Children's Fund, the United Nations Office on Drugs and Crime, the International Organisation for Migration and the Organisation for Security and Cooperation in Europe. The organisation works to increase knowledge and awareness; to provide technical assistance; promote effective rights-based responses; build the capacity of state and non-state structures and to promote joint action.

http://www.madre.org

MADRE is an independent New York based international women's human rights organisation that works in partnership with community-based women's groups worldwide focused on issues of health and reproductive rights, economic development, education and other human rights. Madre provides resources and training to enable organisations to address immediate local needs and in developing long-term solutions to the challenges they face. It focuses on gender-based violence, economic and environmental justice and peace building.

http://www.girlchildnetwork.org/home.html

The Girl Child Network is an umbrella organisation based in Kenya that brings together over 300 international and national NGO and community groups interested in the welfare and rights of girls in Kenya. The Network works on policy issues; on awareness raising; on a range of practical programmes (e.g. a sanitary towels campaign, education intervention) as well as on human rights legislation.

ON ENVIRONMENT AND SUSTAINABLE DEVELOPMENT

http://www.worldwildlife.org/home-full.html

For over 50 years, the World Wildlife Fund has worked to protect the environment and the future of nature; it is a well-respected and independent conservation NGO working in 100 countries with a membership of over 5 million globally. WWF's mission is to use scientific knowledge to preserve the diversity and abundance of life on Earth and the health of ecological systems through protecting natural areas and wild populations of plants and animals, including endangered species; promoting sustainability and promoting the more efficient use of resources and energy and the maximum reduction of pollution.

http://www.greenpeace.org/international/en/

Beginning with a voyage of the ship 'Greenpeace' to Amchitka island to try to stop a US nuclear weapons test, Greenpeace grew to become one of the best-known environmental campaigning organisations, now with offices in more than 40 countries and over 2.9 million members worldwide. Greenpeace focuses on action on a range of key issues including climate change, forests, oceans, agriculture, toxic pollution and nuclear power.

http://www.awaaz.org

The Awaaz Foundation is an independent environmental NGO based in Mumbai, India, which focuses on building awareness of environmental issues, carrying out advocacy and education projects to protect the environment and prevent environmental pollution. It has engaged in litigation on issues such as noise pollution, banning tobacco sales to minors and on politically supported sand mining projects. Awaaz is just one of a growing range of environmental organisations in India.

http://www.fsc.org

The Forest Stewardship Council is an independent NGO, based in Germany, established following the UN Conference on Sustainable Development in 1992 to promote the responsible management of the world's forests. The Council was established by concerned business representatives, social groups and environmental organisations to promote responsible production and consumption of forest products via an approved and certified system of forest product accreditation and, through it, to improve forest management worldwide. FSC has developed a series of chain of custody standards for timber, a series of accreditation assurances and a trademark to facilitate consumers in ensuring use of sustainable forest products.

FSC principles and criteria for sustainable forest products can be accessed at http://www.fsc.org.

'THE INTERNET DOES NOTHING WITHOUT PEOPLE DOING SOMETHING WITH IT.'

In his report to the Human Rights Council of the UN in May 2011, the UN Special Rapporteur on the promotion and protection of the right to freedom of opinion and expression, Frank La Rue argued:

'...unlike any other medium, the Internet enables individuals to seek, receive and impart information and ideas of all kinds instantaneously and inexpensively across national borders. By vastly expanding the capacity of individuals to enjoy their right to freedom of opinion and expression, which is an 'enabler' of other human rights, the Internet boosts economic, social and political development, and contributes to the progress of humankind as a whole'.

He also argued that the Internet has become a means for exercising individual freedom of opinion and expression as guaranteed by article 19 of the Universal Declaration of Human Rights and the International Covenant on Civil and Political Rights. His focus in reporting to the Council was twofold: to highlight the ways in which governments seek to control access to (and use of) internet content and the issue of access to the necessary infrastructure arguing that the internet has created 'fear amongst governments and the powerful'. This has led to increased restrictions through sophisticated technologies to block content, monitor and identify activists, criminalise legitimate expression and then adoption of restrictive legislation.

There are many ongoing debates about the internet and its potential to transform economic, social, cultural and political development and one of the most vigorous is about whether access to the internet is a human right or not. Ever since the 1994 indigenous Zapatista rebellion in the southern state of Chiapas, Mexico made extensive use of the internet and became dubbed as the world's first 'postmodern' movement led by 'information-age activists', the internet has continued to grow as a platform for different types of social, economic and political activism and as a focal point for debate and creativity.

THE UNEVEN GEOGRAPHY OF INFORMATION

Traditionally, information and knowledge about the world have been geographically limited as the transmission of information required the movement of people or the availability of other means of communication. Until the late 20th century, almost all forms of information – books, newspapers, journals, patents etc. were characterised by similar constraints. Europe and the Western World consumed and controlled the vast majority of the world's codified knowledge with other parts of the globe largely left out. It is a truism to say that the internet era has changed all this utterly, but is this actually the case?

Every hour, enough information is consumed by internet traffic to fill 7 million DVDs and it is estimated that in four years from now, it will be four times larger again with over 2 billion people now online (most of them in the Developing World) and with few parts of the world disconnected, even if there are very large groups of people in such locations for whom access to the internet remains a pipe dream. While claims by IBM that the *digital divide will cease to exist* shortly are clearly exaggerated, the internet is now within the grasp of many. But, how does this reality sit alongside other realities where profound 'digital divisions' reflect and reinforce global inequalities?

The UK Guardian newspaper recently reported that, for example, Google's databases contain more indexed user-generated content about the Tokyo metropolitan region than the entire continent of Africa and on Wikipedia, there is more written about Germany than South America and Africa combined – '...*in other words, there are massive inequalities that cannot simply be explained by uneven internet penetration rates. A range of other physical, social, political and economic barriers reinforce the digital divide, amplifying the informational power of the already powerful and visible...' and concluded '... As we rely increasingly on user-generated platforms, there is a real possibility that we will see the widening of divides between digital cores and peripheries. It is crucial to keep asking where visibility, voice and power reside in an increasingly networked world'.*

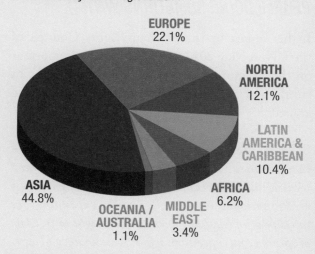

Internet users
Distribution by world regions 2011

Based on 2,267, 233, 742 users in December 2011
Source: www.internetworldstats.com

TABLE 16:1 - WORLD INTERNET USE AND POPULATION, DECEMBER 2011

WORLD REGIONS	Est. Population 2011	Internet users 2000	Internet users 2011	% penetration (of population	Growth 2000-2011	% of users worldwide
Africa	1,037,524,058	4,514,400	139,875,242	13.5 %	2,988.4 %	6.2 %
Asia	3,879,740,877	114,304,000	1,016,799,076	26.2 %	789.6 %	44.8 %
Europe	816,426,346	105,096,093	500,723,686	61.3 %	376.4 %	22.1 %
Middle East	216,258,843	3,284,800	77,020,995	35.6 %	2,244.8 %	3.4 %
North America	347,394,870	108,096,800	273,067,546	78.6 %	152.6 %	12.0 %
Latin America and the Caribbean	347,394,870	108,096,800	273,067,546	78.6 %	152.6 %	12.0 %
Oceania and Australia	35,426,995	7,620,480	23,927,457	67.5 %	214.0 %	1.1 %
World Total	6,930,055,154	360,985,492	2,267,233,742	32.7 %	528.1 %	100.0 %

Sources: International Telecommunications Union (2011) Measuring the Information Society, Geneva

DEBATING THE INTERNET AS A 'FORCE FOR CHANGE'

FOR

The Internet is not just an enabler of human rights: it is a human right - for many people the Internet has become speech itself: the Internet has become a key means by which individuals can exercise their right to freedom of opinion and expression and this is a catalyst for other rights (e.g. the right to education, the right to take part in cultural life etc.) as well as facilitating civic and political rights (e.g. the rights to freedom of association and assembly).

A tactical toolbox for everyone - Facebook and other social networks have become invaluable tools for many info-activists in gathering support (witness the role of the internet in the 'Arab Spring') and as an accessible platform for self-publishing. People can witness, record, interact and share events as they are happening using photographs, audio and text stories.

A platform for global civic communities and public space - individuals are no longer passive recipients of information - they can be active creators and distributors of information thus amplifying people's voices. In countries without independent media, people can now share critical views and find multiple sources of alternate information. Internet democracy allows greater active participation.

Geography matters - the Internet opens up the world and its regions to scrutiny; the dramatic events in Tunisia and Egypt in recent years encouraged 'the street' in Libya and Bahrain. Smaller, sympathetic protests took place throughout the region and in social spaces online. Part of what is powerful about the network is its potential to connect and as such, the Internet is inherently a force for democracy - it is not the technology itself, but the way people use it and build it that matters.

AGAINST

It is not a priority - the cost of providing access and the necessary infrastructure to make it universally available would divert resources from other important, higher priority development and human rights projects. It would be irresponsible to allocate aid and funding to it when roads, hospitals, schools and sanitation remain severely underfunded.

The geography of knowledge is hugely uneven – for many, many people access to the internet is simply an impossibility; literacy remains poor, costs remain high, access is restricted in very practical ways. The dominant content of the internet is Developed World oriented and the dominant language remains English. These realities effectively exclude.

The Internet is an unregulated anarchy where anything goes - uncritical trust in the Internet's democratic power could in fact be making it harder for people to escape 'authoritarian' control; Evgeny Morozov's book 'The Net Delusion' examines how the Internet can prop up even the most unpalatable regimes and can be used as effectively by despots as democrats. The tension between Internet freedom and civic freedom is evident even in established democracies, where criminals and extremists routinely use the web to circumvent democratically created laws and the very tools of web repression are often supplied by businesses in democratic and Internet-soaked states.

(Adapted, in part, from 'This house believes that the Internet is not inherently a force for democracy' debate on The Economist website, 23rd February 2011 http://www.economist.com/debate/overview/196)

See also http://www.tacticaltech.org/ - an international NGO helping human rights advocates use ICT in their work and 2011 Report by the UN Special Rapporteur on the Promotion and Protection of the Right to Freedom of Opinion and Expression http://www2.ohchr.org/english/bodies/hrcouncil/docs/17session/A.HRC.17.27_en.pdf

CASE STUDY ONE - MOBILE PHONES, 'APPROPRIATE' TECHNOLOGY, BANKING AND FISHING

Mobile phones, 'appropriate' technology, banking and fishing

Mobile phone use and development in Africa provides an excellent example of appropriate technology which goes far beyond simply maintaining communications between people at the level of business or social life. In the last decade, there has been an incredible boom in mobile phone use in Africa. In 1998 there were less than four million phones in total but now, it is estimated that there are around 500 million. For growing numbers of Africans, this technology is fast becoming a way of life in a continent where the majority do not have access to a computer, never mind the internet and where mobile phones present huge possibilities. One example is mobile phone banking which has emerged recently; networks such as Safaricom and Vodafone have introduced a service - 'M-Pesa' - which allows people to store money in their phones and allows customers to pay bills, send, and receive money via text which can then be cashed in local offices.

Ken Banks is a strong advocate of local innovation and argues that *Africans are not the passive recipients of technology many people seem to think they are*' and points to internet site Afrigadget which highlights African ingenuity including the example of Pascal Katana, an engineering student at the University of Nairobi. Katana developed an electronic device that 'automates' fishing using a trap that uses amplification of the sounds made by fish while feeding, this attracts fish and a 'good catch' is detected by a net weighing mechanism which, in turn, triggers a GPRS device which informs the fisherman. Katana is now in the process of developing a control system to ensure this doesn't cause over fishing.

See http://www.afrigadget.com and http://www. kiwanja.net/

CASE STUDY TWO - MAPPING DARFUR

Google Earth allows those who have downloaded its software to focus on satellite images and maps of most of the world. In 2007, Google Earth and the US Holocaust Memorial Museum joined forces to create high resolution satellite images of Sudan to document destroyed villages, displaced people and refugee camps in Darfur where tens of thousands of people are estimated to have been killed since the conflict erupted there in 2003.

When users view the Darfur region of West Sudan on Google Earth, they will see a number of icons of flames, tents or cameras. Each of these icons represents destroyed villages or refugee camps. Clicking on one of the icons opens up separate windows of information on specific villages, such as the village's name and statistics on the extent of the destruction. Images have been enhanced so users can even see remnants of burnt houses. The map of the region also includes icons with links to a presentation by the Holocaust Museum on the conflict with photos, video, historical background and also testimonies on the atrocities.

Sudanese officials, including their president, Omar Al-Bashir, denied that these atrocities had occurred in Darfur; however the evidence is in plain view on Google maps. The technology allows anyone to challenge the views and arguments of the Government.

See http://www.ushmm.org/maps/projects/darfur/ and http://earth.google.com/outreach/cs_darfur. html

READING

Paul Collier (2007) The Bottom Billion: Why the Poorest Countries are Failing and What Can Be Done About It, Oxford University Press

Garrett Cullity (2004) The Moral Demands of Affluence, Oxford, Oxford University Press

Duncan Green (2008) From Poverty to Power: How Active Citizens and Effective States Can Change the World, Oxfam International

Thomas Pogge (2008 2nd ed.) World Poverty and Human Rights, New York, Polity Press

Jeffrey Sachs (2008) Common Wealth: Economics for a Crowded Planet, London, Penguin

Vandana Shiva (2005) Earth Democracy: Justice, Sustainability and Peace, London, Zed Books

Peter Singer (2009) The Life You Can Save: Acting Now to End World Poverty, New York, Random House

World Bank (2010) World Development Report 2010: Development and Climate Change, New York, World Bank

MORE INFORMATION & DEBATE

http://www.guardian.co.uk/global-development - articles, blogs, reviews and campaigns for change; see in particular the Poverty Matters blog on http://www.guardian.co.uk/global-development/poverty-matters/2011

http://www.ted.com - rich site for presentations including by Sachs, Collier, Pogge and especially Hans Rosling (see, for example http://www.ted.com/talks/hans_rosling_shows_the_best_stats_you_ve_ever_seen.html

http://www.vandanashiva.org - site exploring the views of Vandana Shiva, earth democracy and Navdanya International

'...extreme poverty, gracious luxury'

THE WORLD IN NUMBERS

95:1 the ratio of the average per capita wealth of the world's richest and poorest people

The richest **1%** of the world's people own **40%** of the entire world's wealth

Almost **75%** of the world's wealthiest individuals live in just 4 countries – the US, Japan, the UK and France

82% the share of world wealth owned in the high income OCED countries (2005)

50% the reduction in the % of world population living in poverty between 1981 and 2005

1 IN 2 AFRICANS LIVE IN POVERTY

1.4 BILLION LIVE IN POVERTY ON LESS THAN **$1.25 PER DAY**

The wealthiest **1%** of Americans control **40%** of all US wealth

3 countries have a lower human development in 2010 than in 1970 – the Democratic Republic of Congo, Zimbabwe and Zambia

60% of people now live in democracies, up from 33% in 1970

1% – the average increase in life expectancy in developing countries overall since 1970

1/2 for every 1 country in which inequality has reduced it has increased in more than 2 since the 1980s

32:1 the rate at which an individual in the 'developed world' consumes resources and generates waste when compared with an individual in the 'developing world'

9 the number of countries with life expectancy in 2010 **below** 1970 levels – 6 in Sub-Saharan Africa and 3 in former Soviet Union countries

61% the increase in literacy rates 1970- 2010 for all developing countries. 183% in Sub-Saharan Africa and 149% in the Arab States

81 YEARS is the average life expectancy of a Norwegian – in Afghanistan it is **44.6 YEARS**

27% the share of the world's poor living in Sub-Saharan Africa, up from 11% in 1980

25% the decline in the number of undernourished people in developing countries between 1980 and 2005

95% of the world's people in 155 countries experienced increases in real per capita income since 1970 to an average of $10,760 today – almost 15 times that of 20 years ago and twice that of 40 years ago

500 MILLION+ THE NUMBER OF PEOPLE IN AFRICA WHO ARE EXPECTED TO BE AFFECTED BY SERIOUS WATER SHORTAGES BY 2050

1 the number of toilets for every 1,440 inhabitants in one Dharavi, a slum in Mumbai, India

200-300 LITRES LITRES PER PERSON IS THE AVERAGE DAILY WATER USE RANGE FOR MOST EUROPEAN COUNTRIES, IT IS **LESS THAN 10 LITRES** PER DAY IN COUNTRIES SUCH AS **MOZAMBIQUE**, **RWANDA** AND **HAITI**

915 the average maternal mortality rate (per 100,000 live births) in the 'bottom 20 countries' – the global average is 273